C0-GHU-314
William F. Maag Library
Youngstown State University

Selected Titles in This Subseries

48 **L. Lerman, G. Polotovskiĭ, and L. Shilnikov, Editors,** Methods of qualitative theory of differential equations and related topics (TRANS2/200)

47 **R. A. Minlos, Senya Shlosman, and Yu. M. Suhov, Editors,** On Dobrushin's way. From probability theory to statistical physics (TRANS2/198)

46 **Vladimir Arnold, Maxim Kontsevich, and Anton Zorich, Editors,** Pseudoperiodic topology (TRANS2/197)

45 **Ya. Eliashberg, D. Fuchs, T. Ratiu, and A. Weinstein, Editors,** Northern California symplectic geometry seminar (TRANS2/196)

44 **Alexander Astashkevich and Serge Tabachnikov, Editors,** Differential Topology, Infinite-Dimensional Lie Algebras, and Applications (D. B. Fuchs' 60th Anniversary Collection) (TRANS2/194)

43 **A. Yu. Morozov and M. A. Olshanetsky, Editors,** Moscow Seminar in Mathematical Physics (TRANS2/191)

42 **S. Tabachnikov, Editor,** Differential and Symplectic Topology of Knots and Curves (TRANS2/190)

41 **V. Buslaev, M. Solomyak, and D. Yafaev, Editors,** Differential Operators and Spectral Theory (M. Sh. Birman's 70th anniversary collection) (TRANS2/189)

40 **M. V. Karasev, Editor,** Coherent Transform, Quantization, and Poisson Geometry (TRANS2/187)

39 **A. Khovanskiĭ, A. Varchenko, and V. Vassiliev, Editors,** Geometry of Differential Equations (TRANS2/186)

38 **B. Feigin and V. Vassiliev, Editors,** Topics in Quantum Groups and Finite-Type Invariants (Mathematics at the Independent University of Moscow) (TRANS2/185)

37 **Peter Kuchment and Vladimir Lin, Editors,** Voronezh Winter Mathematical Schools (Dedicated to Selim Krein) (TRANS2/184)

36 **V. E. Zakharov, Editor,** Nonlinear Waves and Weak Turbulence (TRANS2/182)

35 **G. I. Olshanski, Editor,** Kirillov's Seminar on Representation Theory (TRANS2/181)

34 **A. Khovanskiĭ, A. Varchenko, and V. Vassiliev, Editors,** Topics in Singularity Theory (TRANS2/180)

33 **V. M. Buchstaber and S. P. Novikov, Editors,** Solitons, Geometry, and Topology: On the Crossroad (TRANS2/179)

32 **R. L. Dobrushin, R. A. Minlos, M. A. Shubin, and A. M. Vershik, Editors,** Topics in Statistical and Theoretical Physics (F. A. Berezin Memorial Volume) (TRANS2/177)

31 **R. L. Dobrushin, R. A. Minlos, M. A. Shubin, and A. M. Vershik, Editors,** Contemporary Mathematical Physics (F. A. Berezin Memorial Volume) (TRANS2/175)

30 **A. A. Bolibruch, A. S. Merkur′ev, and N. Yu. Netsvetaev, Editors,** Mathematics in St. Petersburg (TRANS2/174)

29 **V. Kharlamov, A. Korchagin, G. Polotovskiĭ, and O. Viro, Editors,** Topology of Real Algebraic Varieties and Related Topics (TRANS2/173)

28 **L. A. Bunimovich, B. M. Gurevich, and Ya. B. Pesin, Editors,** Sinai's Moscow Seminar on Dynamical Systems (TRANS2/171)

27 **S. P. Novikov, Editor,** Topics in Topology and Mathematical Physics (TRANS2/170)

26 **S. G. Gindikin and E. B. Vinberg, Editors,** Lie Groups and Lie Algebras: E. B. Dynkin's Seminar (TRANS2/169)

25 **V. V. Kozlov, Editor,** Dynamical Systems in Classical Mechanics (TRANS2/168)

24 **V. V. Lychagin, Editor,** The Interplay between Differential Geometry and Differential Equations (TRANS2/167)

23 **Yu. Ilyashenko and S. Yakovenko, Editors,** Concerning the Hilbert 16th Problem (TRANS2/165)

(*Continued in the back of this publication*)

Methods of Qualitative Theory of Differential Equations and Related Topics

William F. Maag Library
Youngstown State University

Dedicated to the memory of
Evgeniya Aleksandrovna Leontovich-Andronova
(1905–1996)

American Mathematical Society
TRANSLATIONS
Series 2 • Volume 200

Advances in the Mathematical Sciences — 48

(*Formerly Advances in Soviet Mathematics*)

Methods of Qualitative Theory of Differential Equations and Related Topics

L. Lerman
G. Polotovskiĭ
L. Shilnikov
Editors

American Mathematical Society
Providence, Rhode Island

William F. Maag Library
Youngstown State University

ADVANCES IN THE MATHEMATICAL SCIENCES EDITORIAL COMMITTEE

V. I. ARNOLD
S. G. GINDIKIN
V. P. MASLOV

Translation edited by A. B. Sossinski

2000 *Mathematics Subject Classification.* Primary 34Cxx, 37Cxx; Secondary 14H99.

ABSTRACT. This book contains papers written by collaborators and students of E. A. Leontovich-Andronova. The papers are mainly devoted to new results in qualitative theory of ordinary differential equations and in the theory of smooth dynamical systems.

This volume is of interest to researchers and graduate students working in differential equations, dynamical systems, and related areas.

Library of Congress Card Number 91-640741
ISBN 0-8218-2663-8
ISSN 0065-9290

Copying and reprinting. Material in this book may be reproduced by any means for educational and scientific purposes without fee or permission with the exception of reproduction by services that collect fees for delivery of documents and provided that the customary acknowledgment of the source is given. This consent does not extend to other kinds of copying for general distribution, for advertising or promotional purposes, or for resale. Requests for permission for commercial use of material should be addressed to the Assistant to the Publisher, American Mathematical Society, P. O. Box 6248, Providence, Rhode Island 02940-6248. Requests can also be made by e-mail to reprint-permission@ams.org.

Excluded from these provisions is material in articles for which the author holds copyright. In such cases, requests for permission to use or reprint should be addressed directly to the author(s). (Copyright ownership is indicated in the notice in the lower right-hand corner of the first page of each article.)

© 2000 by the American Mathematical Society. All rights reserved.
The American Mathematical Society retains all rights
except those granted to the United States Government.
Printed in the United States of America.

⊗ The paper used in this book is acid-free and falls within the guidelines
established to ensure permanence and durability.
Visit the AMS home page at URL: http://www.ams.org/

10 9 8 7 6 5 4 3 2 1 05 04 03 02 01 00

QA
3
A572
Ser. 2
v. 200

William F. Maag Library
Youngstown State University

Contents

Evgeniya Aleksandrovna Leontovich-Andronova (1905–1996)
J. P. Shilnikov 1

On the abundance of traveling waves in coupled expanding circle maps
V. Afraimovich and M. Courbage 15

On cusp-bifurcations of periodic orbits in systems with a saddle-focus homoclinic curve
S. A. Alekseeva and L. P. Shilnikov 23

Collapse and continuity of geodesic frameworks of surface foliations
S. Aranson, V. Medvedev, and E. Zhuzhoma 35

Homoclinic and heteroclinic linkages in concrete systems: Nonlocal analysis and model maps
Vladimir N. Belykh 51

On the mutual arrangement of a conic and a quintic in the real projective plane
Anna A. Binstein and Grigoriĭ M. Polotovskiĭ 63

The structure of the boundary curve for planar controllability domains
N. N. Butenina 73

Orbit structure in a neighborhood of a separatrix cycle containing two saddle-foci
V. V. Bykov 87

Example of a blue sky catastrophe
Nikolaĭ Gavrilov and Andrey Shilnikov 99

Dynamics and moduli of Ω-conjugacy of 4D-diffeomorphisms with a structurally unstable homoclinic orbit to a saddle-focus fixed point
S. V. Gonchenko 107

Topological classfication of amply situated attractors of A-diffeomorphisms of surfaces
V. Z. Grines and R. V. Plykin 135

A proof of Shilnikov's theorem for C^1-smooth dynamical systems
Mikhail V. Shashkov and Dmitry V. Turaev 149

A new simple bifurcation of a periodic orbit of "blue sky catastrophe" type
LEONID P. SHILNIKOV AND DMITRY V. TURAEV 165

On the splitting of the complex loop of a separatrix
V. P. TAREEV 189

Amer. Math. Soc. Transl.
(2) Vol. **200**, 2000

Evgeniya Aleksandrovna Leontovich-Andronova (1905–1996)

L. P. Shilnikov

Evgeniya Aleksandrovna Leontovich-Andronova was born on September 19, 1905 in Kiev in the family of Aleksandr Vasilievich Leontovich, an instructor at the University of Kiev (later to become a full member of the Ukrainian Academy of Sciences). Her mother, Vera Victorovna Kirpicheva, was an oculist. Besides Evgeniya, there were two other children in the family: her older brother, Mikhail (he later became a physicist and was elected to full membership of the Soviet Academy of Sciences) and her younger sister, Vera.

In 1913 the Leontovich family moved to Moscow in connection with the appointment of A. V. Leontovich to a professorship at the Petrovski-Razumovski Institute (now the Timiryazev Academy). A. V. Leontovich worked there until 1936, when he returned to Kiev. Evgeniya's mother worked in a military hospital during World War I (1914–1918), then in a children's sanatorium. She died in 1919.

In 1921 Zhenya, as she was called in the Leontovich family, graduated from high school and entered the Physics and Mathematics Faculty of Moscow State University, in the mathematics section[1]. Her older brother Mikhail was already studying there (but in the physics section). At that time Mikhail was part of a very unusual group of people, forming a commune[2]. Besides Mikhail, members of this commune included A. A. Andronov, P. S. Novikov, I. E. Tamm (later all three became full members of the Soviet Academy of Sciences, and Tamm received the Nobel prize in physics), A. A. Vitt (who achieved the title of professor, but tragically perished during the Stalin repressions in 1938), N. I. Pariĭskiĭ (an astronomer, later corresponding member of the Soviet Academy of Sciences), and others. The men listed above, who determined the spirit of the group, all studied at the same Faculty of Physics and Mathematics of MSU. There were also some women in the group, in particular, L. V. Keldysh. E. A. Leontovich, thanks to her brother, also joined them.

2000 *Mathematics Subject Classification.* Primary 01A65.

[1]The biographical data about this period of the E. A. Leontovich's life are borrowed from the article by L. V. Pariĭskaya, *Sixty years together*, in: Reminiscences on Academician M. A. Leontovich, "Nauka", Moscow, 1996, pp. 13–34.

[2]Life in Russia at the time, after two very destructive wars, was very difficult, and for that reason students (and not only students) would group together into communes to jointly take care of household chores and thus help overcome the hardships of the times.

©2000 American Mathematical Society

All these people were strong personalities, and of course very different from each other. However, many things held them together, and the main backbone was, of course, science. For them, the meaning and goal of their existence was learning and discovering scientific facts; everything else was secondary. It is also important to note that they did not care much about their own scientific careers, although they were well aware of their role in science.

It is only natural that several couples were formed in the commune. In 1927, E. A. Leontovich became A. A. Andronov's wife. Among the others, let us mention the mathematical couple L. V. Keldysh and P. S. Novikov.

E. A. Leontovich's first steps in scientific research began in the seminar of N. N. Luzin. Luzin came into the history of science, in particular, as the founder of the first large Russian scientific school—"Luzitania". Its origin goes back to the time of World War I, and so it incorporated several generations: the elder one, which included V. V. Stepanov, I. I. Privalov, P. S. Aleksandrov, M. Ya. Suslin, D. E. Men′shov, A. Ya. Khinchin, and others; the middle one, with P. S. Urysohn, A. N. Kolmogorov, V. V. Nemytskiĭ, N. K. Bari, V. I. Glivenko, L. A. Lyusternik, L. G. Shnirelman and others; the younger generation included P. S. Novikov, L. V. Keldysh, and E. A. Leontovich.

The years 1922–1926 were the period when Luzitania flourished. By that time, many of the accessible problems from the theory of functions of a real variable and descriptive set theory were solved. At the same time, starting in 1925–1928, many participants of Luzin's seminar made their way into other branches of mathematics. It is during this period that E. A. Leontovich's series of reports at the seminar took place. The titles and the contents of those reports are not known to the present author, because E. A. Leontovich did not publish those results. It is only known from her own words that the topics of the talks were of her own choice (a fact of which she was especially proud).

In 1928 E. A. Leontovich graduated from Moscow University, and until 1932 worked as an assistant professor at the Mendeleev Chemical-Technological Institute.

In 1931 the university at Gor′kiĭ was reopened, and, in its framework, the Gor′kiĭ Physical-Technical Research Institute (GPTRI) was organized. A. A. Andronov applied for a position in Gor′kiĭ. On November 1, 1931, he was enlisted in the university faculty, and in 1932 his entire family moved to Gor′kiĭ.

In 1932 E. A. Leontovich became a research collaborator in the department of the theory of oscillations at GPTRI, and her main research interest was the qualitative theory of dynamical systems. It is clear that the choice of this new research topic was made under the influence of A. A. Andronov, who by then had become a prominent expert in the theory of nonlinear oscillations. Although A. A. Andronov had graduated from the Physics Department of Moscow University, he possessed a very solid mathematical background. This was due to the fact that the mathematics courses were the same for future mathematicians and physicists, and A. A. Andronov, seriously studying mathematics, acquired a mathematical culture far surpassing what was normally required of a theoretical physicist. This enabled him to maintain contacts with mathematicians, among whom L. S. Pontryagin, A. A. Markov, P. S. Novikov, I. G. Petrovskiĭ, and V. V. Stepanov should be mentioned.

In Gor′kiĭ, A. A. Andronov secured the collaboration of A. G. Maier, also a graduate of the mathematics division of the Department of Mathematics and

Physics at at Moscow university, who had been sent to the Gor′kiĭ Pedagogical Institute after completing his graduate work in analytic number theory under A. Ya. Khinchin. N. N. Bautin, an alumnus of the Gor′kiĭ Pedagogical Institute, became A. A. Andronov's PhD student and later his collaborator.

This group, which constituted the section of oscillation theory at GPTRI, became the nucleus of the Gor′kiĭ school in oscillation theory.

At that time the following question was crucial in the theory of oscillations of two-dimensional systems: what is necessary to know about the system in order to completely determine the qualitative structure of the behavior of trajectories. Or, in contemporary language, what is a complete invariant of such a system with respect to topological conjugation?

E. A. Leontovich-Andronova (1934)

The main aspects of this question were, in fact, presented in A. A. Andronov's talk at the All-Union Physics Congress in 1931. In 1937, A. A. Andronov and L. S. Pontryagin solved this problem for the class of rough systems, which they had introduced. The complete invariant of a rough system, which they called a scheme, is a set of trajectories: the special ones (equilibrium states, periodic motions, and separatrices of saddles) together with one trajectory from each cell of wandering trajectories. The finiteness of the invariant followed from the roughness of the system for compact domains in the plane.

The notion of rough system gave a solid mathematical foundation to the theory of nonlinear systems. However, the strong necessity of extending the notion of scheme to nonrough systems remained. This was important because physical models are explicitly given finite parameter families of vector fields, whose phase portrait may change as the parameters vary. Usually, dynamical phenomena which are of great importance for the applications are related to these bifurcational reconstructions. Andronov involved E. A. Leontovich and A. G. Maier in the study of this problem.

However, the first joint work of E. A. Leontovich and A. G. Maier was their note "On an inequality related to Fourier integrals" published in Soviet Math. Doklady in 1934. It belonged to the theory of linear systems and answered a question of L. I. Mandelshtam on the relation between the duration of a radio impulse and its diffusion, which was of fundamental importance for communication theory. On the other hand, in that note E. A. Leontovich and A. G. Maier gave an exact statement of the analog of the quantum indeterminacy principle.

The next joint paper by E. A. Leontovich and A. G. Maier was "On trajectories determining the qualitative structure of the decomposition into trajectories". It appeared in the same year as the previously mentioned paper of Andronov and Pontryagin on rough systems, and was devoted to finding a complete invariant for two-dimensional systems on the sphere without the roughness assumption. In that context an extra condition was needed: the finiteness of the number of specific trajectories. Let us note that this paper turned out to be technically more involved than the Andronov–Pontryagin paper, since it required the consideration of equilibrium states of an arbitrary order of degeneracy. The construction of the appropriate homeomorphism for systems whose equilibrium states possess elliptic, parabolic, and hyperbolic sectors was a very painstaking and laborious matter. The final formulation appeared in the note "On the scheme determining the topological structure of the decomposition into trajectories", published in 1955 (i.e., after the death of the second coauthor).

Naturally, the further research work of E. A. Leontovich was related to the study of nonrough systems and their bifurcations. Thus, in 1938 the paper "On the theory of changes in the qualitative structure of the decomposition of the phase plane into trajectories", coauthored by A. A. Andronov and E. A. Leontovich, appeared. Here they study the so-called systems of the first degree of nonroughness, which was natural to regard as "relatively rough, because they are rough with respect to all nonrough" systems. It was shown that such systems are distinguished by nine conditions. In particular, systems of the first degree of nonroughness are allowed to have one of the following types of nonrough elements:

1) an equilibrium state of the type of the weak focus with nonzero first focus value;
2) a limit cycle with zero characteristic exponent and nonzero first Lyapunov value (a double cycle);
3) a separatrix loop to the simplest nonrough equilibrium state of the saddle-node type which does not lie in the boundary of the node region;
4) a separatrix loop to a saddle for which the saddle value, equal to the sum of characteristic roots, is nonzero.

Generally, in modern language, the systems of the first degree of nonroughness locally constitute, in the space of C^3-smooth vector fields, a bifurcation submanifold of codimension 1, outside a small neighborhood of which there are no other nonrough systems. Therefore such systems cannot have

1) a saddle separatrix which winds (unwinds) on the separatrix loop of another saddle;
2) a saddle separatrix which winds around a double cycle if there is a separatrix of some saddle that unwinds from this same cycle.

Systems possessing such a nonrough element also constitute a bifurcational set of codimension 1, but in any of its neighborhoods there are nonrough systems with a scheme different from that of the given system. Therefore in any neighborhood of any point of this bifurcational set there are countable many bifurcational surfaces, also of codimension 1, which accumulate to the given one.

From this 1938 paper, the following problem arises: to find all main bifurcations of limit cycles of two-dimensional systems. This problem was solved by A. A. Andronov and E. A. Leontovich in a series of four papers. The first one, "Some cases of the dependence of limit cycles on the parameters", was published in 1939 in the poorly accessible journal *Scientific Notes of Gor'kiĭ University*. There are grounds to believe that this paper was written before the 1938 paper mentioned above. A special feature of the 1939 paper is the consideration of bifurcations in a one-parameter family of systems. For the first two local bifurcations (see the list of nonrough elements above), this question was solved exhaustively. Namely, in the case of a weak focus with nonzero first Lyapunov value, the birth of a unique limit cycle was established. This allowed the authors to give a mathematical foundation to the generation of oscillations and to explain the phenomena of soft and rigid regimes of oscillations. Bifurcations of the double cycle were also studied completely and the bifurcation diagram was given.The basis of this study was the investigation of the one-dimensional Poincaré map. In the modern literature, this bifurcation is known as the tangent bifurcation for the related Poincaré map.

The two remaining cases are global, i.e., they require the consideration not only of the neighborhood of the equilibrium state, but also of a certain trajectory leaving this neighborhood and returning to it. This is the birth of a limit cycle from the separatrix loop of a saddle-node or from the separatrix loop of a saddle. The former case first appeared in a paper by A. A. Andronov and A. A. Vitt published in 1930 and was the basis for explaining the passage from the synchronization regime to the beat regime (or synchronization) in the framework of the averaged van der Pol equation. In studying this bifurcation, A. A. Andronov and E. A. Leontovich established that when the saddle-node disappears, at least one limit cycle is born from the separatrix loop.

The three cases listed above describe transitions from systems of first degree of nonroughness. As to the fourth case, the condition of nonroughness of the first degree was not explicitly formulated there, so that the main statement, dealing with the existence of the limit cycle when the loop splits, was, in fact, a sufficient condition (in the language of the Poincaré map). Thus the two last cases required additional study. This was done in the same years and rather rapidly. In the case of the separatrix loop of a saddle-node, the uniqueness of the newborn limit cycle was proved, while in the case of the separatrix loop of a saddle (under the condition that the saddle value is nonzero) conditions for the birth of a limit cycle and its uniqueness were given. However, for a series of reasons (the war, the hardships of the postwar period, the death of Andronov (1952) and of Maier (1951)), publications on this topic were delayed until 1955. Thus the study of the main cases of bifurcation of limit cycles on the plane was completed.

During these same prewar years, E. A. Leontovich turned her attention to the study of the bifurcation of limit cycles from a separatrix loop in the general mathematical context.

The behavior of trajectories in a small neighborhood of the separatrix loop of a saddle had been the object of study of Dulac in the analytic case. He established that this behavior may be one of the following:

1) all the trajectories wind towards the loop;

2) all the trajectories are closed.

The first case occurs if at least one of the quantities in the sequence

$$\sigma_0, A_1, \sigma_1, A_2, \dots, \sigma_k, A_{k+1}, \dots$$

is nonzero. Here, in particular, the quantity σ_0 is the sum of the eigenvalues of the saddle, while $\sigma_1, \dots, \sigma_k, \dots$ are none other than the coefficients of the normal form of the saddle if $\sigma_0 = 0$ (i.e., if the saddle index equals 1). For obvious reasons, the quantities σ_i were called saddle values. In their turn, the quantities A_i are called global, or separatrix, values. Essentially, they are the derivatives of the Poincaré map along the separatrix arc lying outside a neighborhood of the saddle.

Starting the study of the bifurcation of the limit cycles from the separatrix loop, E. A. Leontovich was immediately faced with an important consideration: in Dulac's paper the normal forms in the neighborhood of the saddle were variable, depending on whether the saddle index (the absolute value of the ratio of the characteristic roots of the saddle) is rational or irrational. Hence she had to construct the necessary normal form in a special way, involving the killing of nonresonance functions. As a result, the normal form (for the case where the saddle index equals 1) can be written as

$$\frac{dy}{dx} = -\frac{y}{x}\,(1 + \nu + c_1 xy + c_2 x^2 y^2 + \cdots + c_n x^n y^n + x^n y^n F_1(x, y)),$$

where $F_1(0,0) = 0$. In order to obtain this form, one must require the C^r-smoothness of the right-hand sides, where $r \geqslant 4n + 2$.

In the general case, where $\sigma_0 \neq 0$, the separatrix loop can give birth to no more than one cycle. This result, as we mentioned above, was obtained by A. A. Andronov and E. A. Leontovich, and it is important to note that this was done for C^1-smooth families of vector fields without using normal forms.

The case where $\sigma_0 = 0$, while $A_1 \neq 0$, corresponds, in contemporary language, to codimension 2. In that case A_1 is nothing but the integral of the divergence of the vector field along the separatrix loop. Here the separatrix loop can give birth to no more than two limit cycles. As to the general case, as E. A. Leontovich established, if

$$\sigma_0 = A_1 = \cdots = \sigma_{n-2} = A_{n-1} = 0, \qquad \sigma_{n-1} \neq 0,$$

then the separatrix loop can give birth to no more than $2n - 1$ limit cycles, while if

$$\sigma_0 = A_1 = \cdots = A_{n-1} = \sigma_{n-1} = 0, \qquad A_n \neq 0,$$

to no more than $2n$. At the same time she showed that there exist perturbations such that the separatrix loop gives birth to exactly $2n-1$ cycles in the first case and $2n$ in the second, so that these estimates are sharp.

We have presented this result in more detail here because, unlike the previous ones, it appears only in a short note, "On the birth of limit cycles from a separatrix", published in 1951 in Doklady.

These results constituted E. A. Leontovich's kandidat dissertation (PhD thesis), defended in 1946.[3] At about the same time a very nontrivial event occurred: a Russian translation of Poincaré's papers in the qualitative theory of differential equations was published under the title "On curves defined by differential equations" in the series *The Classics of Natural Sciences* (1947, GITTL Publishers). The translation was done by E. A. Leontovich and A. G. Maier under the editorship of A. A. Andronov. The book contained the following supplements: "General qualitative theory", by E. A. Leontovich and A. G. Maier, "Centers", by A. G. Maier, "Integral curves on the surface of a torus", by V. V. Stepanov, "On the behavior of integral curves of systems of ordinary differential equations in the neighborhood of a singular point", by I. G. Petrovskiĭ, and "Singular points of vector fields" by Yu. Rozhanskaya. The book also contains numerous comments with additional references, thus forming a very unified presentation of many problems of the theory of dynamical systems as it was explored after Poincaré.

At this point, the problem of writing a detailed exposition of the results obtained in the thirties arose for the three leading members of the Gor′kiĭ school. At the end of the forties, they decided to write a book. But fate would not let this decision to be implemented: in 1951 A. G. Maier died at the age of 46, and the next year, at the age of 51, so did A. A. Andronov. The task of writing the book was entirely left in the hands of E. A. Leontovich.

* * *

During the life of A. A. Andronov and A. G. Maier, it seemed that E. A. Leontovich played a subordinate role. But after their death, everything changed, E. A. Leontovich became its informal leader. Naturally, in this role she was helped by her high moral standards, typical of the Russian *intelligentsiya* of the older generation, her professionalism, based on a first-rate mathematical culture, her kindness and sense of duty, which made contacts with her so agreeable.[4]

First of all, E. A. Leontovich took over the scientific guidance of all the last pupils and graduate students of A. A. Andronov, most of whom were radiophysicists, since Andronov had headed the Chair of Oscillation Theory and Automatic Control of the Radiophysics Department at Gor′kiĭ University. Here her task was mainly in giving a mathematical expertise of the work. But her pedagogical activity did not stop there. In 1954 she initiated a course in the theory of dynamical systems for students in mathematics. This two-semester course included Poincaré–Bendixon theory, the theory of rough systems, and the main bifurcations of limit

[3]Later she published her thesis in VINITI in two preprints. But in this form the paper was not accessible to western readers.

[4]The first of these traits was clearly manifested in 1952, when E. A. Leontovich, together with another Gor′kiĭ mathematician, Prof. A. G. Sigalov, openly defended G. S. Gorelik, accused of the allegation that his book "Oscillation and Waves" was written from the positions of "idealism". At that time, when Stalin was still alive, such a stand was quite dangerous.

William F. Maag Library
Youngstown State University

cycles. First it was given to fourth year students, and in the sixties to third year students. This course was part of the curriculum of several generations of Gor′kiĭ (Nizhny Novgorod) university students. Moreover, this course turned out to be the basis of the scientific debut of several dozen future researchers. E. A. Leontovich remained the carrier and keeper of many scientific ideas of the Gor′kiĭ school, which she had in a sense inherited. In this connection her report in 1956 at the Third All-Union mathematics congress must especially be noted. Already the title of the report "Some mathematical works of the Gor′kiĭ school of A. A. Andronov" speaks for itself. In this report the results described above were presented, as well as some work of A. G. Maier, who had described the possible types of behavior of trajectories on surfaces and had solved the Birkhoff problem on the transfinite number of steps to reach the center of a dynamical system (in the sense of G. Birkhoff) in the nonanalytic case. The report also set several problems, first stated by A. A. Andronov, as well as some formulated by E. A. Leontovich herself. We shall dwell in more detail on one of these problems, that of generalizing the notion of roughness to multidimensional systems, or, to be more precise, to two-dimensional diffeomorphisms and to three-dimensional flows; to quote the commentary of E. A. Leontovich herself:

"One must not think that the notion of roughness can be carried over to these cases trivially. Not to mention the fact that in the case of a second order nonautonomous system (periodically depending on time) this question is intimately connected to the question of specific and ordinary trajectories, which is rather unclear, and there are also several difficulties of principle. Similar difficulties also arise in the case of an autonomous dynamical system in three-dimensional space. I do not have the possibility of dwelling on these questions. I shall only mention that the root of these difficulties is related to homoclinic points for transformations of the plane into itself."

The quoted commentary essentially has to do with what is now known as the Andronov problem: can rough diffeomorphisms of the two-dimensional sphere have a countable set of periodic points?

By that time E. A. Leontovich had published (of course, as joint work with A. A. Andronov) the paper "On the birth of limit cycles from a nonrough focus or center or from a nonrough limit cycle" in *Mat. Sbornik*. It is curious that a footnote indicates that this paper was actually written in 1937. The same remark appears in the next paper, "On the birth of limit cycles from the separatrix loop or from the separatrix of an equilibrium state of the saddle-node type" published in *Mat. Sbornik* in 1959. This publication completes the 20-year long epic story of the appearance in print of the main cases of the birth of limit cycles for two-dimensional dynamical systems. And if until then these results were well known and used mostly in Gor′kiĭ, they then became available to a wide scientific audience. This was also supported by the appearance of the second edition of *The Theory of Oscillations* under the editorship of N. A. Zheleztsov in 1949, which contained several addenda, written by E. A. Leontovich, including a detailed exposition of bifurcations of the birth of limit cycles from a weak focus and from a nonrough limit cycle, and, in more concise form, of global bifurcations.

Besides the activities described above, in the fifties E. A. Leontovich was intensively engaged in the education of future scientists. She guided term papers and diploma papers of undergraduates, as well as kandidat (PhD) dissertations of graduate students. Under her supervision two dissertations were successfully defended during this period. Their topics are entirely related to problems from the theory of two-dimensional systems. But to her next PhD student, R. M. Mints, E. A. Leontovich proposes to study problems arising in multidimensional systems.

In her thesis, R. M. Mints investigated the structure of nonrough equilibrium states with one zero eigenvalue. Thanks to her, the notions of multidimensional saddle-node and multidimensional saddle-saddle naturally become standard notions. The latter can be regarded as the one obtained by merging two rough saddles, but of a different topological type (with stable manifolds of different dimensions). Later, on the basis of this work, I studied the main global bifurcation related to the disappearance of such equilibrium states. And I also generalized to the multidimensional case global bifurcations related to the birth of a stable periodic solution from the separatrix loop of a saddle-node. Basically, at that time the study of multidimensional systems and, especially, of their bifurcation theory, became one of the leading topics of study at the department chaired by E. A. Leontovich. The main role of this shift of emphasis was played by E. A. Leontovich. As we can clearly see today, this shift was very timely, since the intensive development of the theory of rough multidimensional systems started in the sixties.

The multidimensional subjectmatter of the department deserves special mention. It is clear from the above that by 1937 the theory of rough two-dimensional systems and, especially, the corresponding bifurcation theory had been developed and a complete topological invariant (the scheme) had been found. As E. A. Leontovich told the author of the present article, "Our intention then was to study multidimensional systems". At that time the department had carried out a "mobilization of information" (the expression is due to A. A. Andronov), i.e., had collected the main publications in this area, including those by Poincaré, Birkhoff, Morse, Hedlund, and others, and not only in qualitative theory, but also in ergodic theory. However, after 1937 A. A. Andronov's interests had shifted to automatic control theory. A basic feature of automatic control systems is the presence, as a rule, of discontinuous nonlinearities, due to dry friction, relay switches, etc. To solve the corresponding problems, A. A. Andronov used the continuation method, which allows us to obtain the appropriate Poincaré map, usually in parametric form. In the case of three-dimensional problems, the Poincaré map turns out to be two-dimensional. Therefore, in a sense, this is also a multidimensional problem, requiring the use of the machinery of two-dimensional maps in nonstandard situations. To this subjectmatter A. A. Andronov also attracted A. G. Maier and N. N. Bautin. As to E. A. Leontovich, she was interested, in particular, in the problems of bifurcation theory. Here we must note the diploma paper of 1939 of A. F. Leont′ev, written under her supervision, in which the main bifurcation of the birth of a limit cycle from a weak focus was studied for three-dimensional analytic systems. The methods of that paper are easily generalized to the multidimensional case, provided that the weak focus has no positive Lyapunov exponents. The results of that study were substantially used in N. N. Bautin's dissertation on the theory of dangerous and safe boundaries of stable domains of equilibrium states. Bautin's dissertation was published after the war, but Leont′ev's paper remained unpublished. After his demobilization, A. F. Leont′ev changed his research interests and became a prominent expert in the theory of complex functions (and corresponding member of the Soviet Academy of Sciences). Among the postwar work on the theory of dynamical systems, let us also note A. G. Maier's paper on the Birkhoff problem on the ordinal number of central trajectories.

* * *

At the beginning of the sixties, several important events occurred. In 1961, at the conference on nonlinear oscillations in Kiev, Steve Smale gave a positive solution to the Andronov problem: he presented an example of a rough diffeomorphism of the plane with a countable set of periodic points. E. A. Leontovich also attended the conference, where she met S. Smale and his teacher, S. Lefshetz. After returning to Gor′kiĭ, E. A. Leontovich spoke of Smale's discovery with great enthusiasm, adding that Smale reminded her of Huckleberry Finn. It is precisely after these discussions in our department the term "Smale horseshoe" was coined, and was later used among the specialists in the Soviet Union. After the end of the conference, Smale came to Moscow, where he put forward the assertion that geodesic flows on compact manifolds of negative curvature and algebraic automorphisms of the torus are rough. The validity of this statement was established by D. V. Anosov, who also introduced a wide class of rough systems with hyperbolic behavior, now known as Anosov systems. This disproved the widely held opinion that conservative systems cannot be rough.

In the department of oscillation theory headed by E. A. Leontovich, an important event also occurred. In December 1962, I defended my thesis, in which the multidimensional analog of the theorem of Andronov and Leontovich on the birth of a stable limit cycle from a separatrix loop was proved. The main condition for birth of the cycle consisted in the negativity of the saddle value

$$\sigma_0 = \lambda_n + \max_{1 \leqslant i \leqslant n-1} \operatorname{Re} \lambda_i < 0, \quad \text{where } i = 1, \dots, n-1, \text{ while } \lambda_n > 0.$$

The following question then came to mind: What will happen if $\sigma_0 > 0$? At first, it seemed that nothing essential. But already at the end of January 1963 I established that for three-dimensional systems with $\sigma_0 > 0$ for which the equilibrium state is a saddle-focus, in any neighborhood of the saddle-focus loop there is a countable number of periodic trajectories of saddle type. Moreover, the Poincaré map has a countable set of Smale horseshoes. Naturally, the first person I told this to was E. A. Leontovich. She described to me her first reaction to this much later: "I immediately wanted to say that this simply cannot be". For understandable reasons this result also surprised L. S. Pontryagin, when E. A. Leontovich described it to him.

From that moment the topics related to the study of multidimensional systems with complicated dynamics became one of the most important ones for the department. This was also furthered by another important event. In 1964 our institute (GIFTI) moved to a new building. The following idea arose: to organize a new institute, the Institute of Applied Mathematics and Cybernetics, in the old vacated building. Naturally, E. A. Leontovich was one of the founders of the institute. Her petition to the president of the Academy of sciences M. V. Keldysh, to the rector of Moscow State University I. G. Petrovskiĭ, and to the head of the Siberian Branch of the Academy M. M. Lavrent′ev received full support—the high reputation of the Gor′kiĭ school of oscillation theory was not doubted by anyone.

In the new institute E. A. Leontovich also headed a department, but its name was changed. It was called the Department of Differential Equations. The reason is easy to understand: the department consisted of university graduates who had majored in mathematics and had taken E. A. Leontovich's course in differential equations. It was notable that the oldest collaborator of the department headed by E. A. Leontovich, the author of the present article, was only 30 at the time. Naturally, our research seminar met once a week. The atmosphere there was extremely democratic. No authorities were unquestionable, anyone could ask any question in order to make things clear, independently of the rank of the speaker. And of course this was achieved thanks to E. A. Leontovich. In scientific questions she was extremely exacting, to herself and to her collaborators. She required that scientific articles and reports be highly polished, that each word be weighted, and was especially demanding in questions related to quoting other authors working on the same problems. This made talks at the seminar a great school for its younger participants and helped them mature as scientists. For many years this seminar, whose traditions were established by E. A. Leontovich, became an influential center of attraction for many experts in qualitative theory. Of course, the main topics of the seminar were both the traditional ones related to the qualitative theory of two-dimensional dynamical systems and bifurcation theory, and the newer ones, related to multidimensional systems with complicated dynamics, the qualitative theory of nonautonomous systems, the Hamiltonian dynamics of integrable and nonintegrable systems, and several other topics.

Systems with complicated dynamics and their bifurcations especially interested E. A. Leontovich. And although she did not participate in their investigation herself, when the question of a report at the conference on nonlinear oscillations in Kiev arose in 1969, she suggested that the two of us should present a joint talk. At the end of this report, written in E. A. Leontovich's own hand, she expressed the hope that systems with complicated dynamics would find their inherent place in the natural sciences. I am stressing this in particular because researchers of the older generation did not find it easy to comprehend this new complicated nonlinear world. And this was when the theory of strange attractors as the mathematical model of chaos was soon to appear.

Of course, one must understand this was easier for E. A. Leontovich than for others, because of the influence of A. A. Andronov, with his understanding of science.

At the same time work on the monograph was proceeding. In the early sixties, E. A. Leontovich finished writing a detailed proof of the theorem asserting that a scheme is a complete invariant of topological conjugacy and defended her doctoral dissertation at Moscow University. In his report, one of the reviewers, M. A. Krasnosel′skiĭ, noted that the proof of the topological invariance (which takes several hundred pages) is one of the longest proofs that he had ever come across.

It became clear that the size of the monograph, which was to include the results on the scheme of dynamical systems, would be so large that the book must be divided into two volumes. The first was called *The Qualitative Theory of Dynamical Systems*, the second, *The Theory of Bifurcations of Dynamical Systems on the Plane.* To speed things up, E. A. Leontovich invited I. I. Gordon to be a coauthor. In 1966 the first volume appeared, in 1967, the second one. Both were soon translated into English. It is difficult to overestimate the importance of these monographs. It suffices to note that they have become the standard reference for numerous experts in nonlinear dynamics and are often quoted in numerous papers even at the present day. However, E. A. Leontovich did not stop there. Jointly with N. N. Bautin, she wrote the monograph-handbook "Methods and tools of the qualitative study of dynamical systems in the plane". The book appeared in the

Soviet Union in 1976, its second edition came out in 1990. For E. A. Leontovich this was a final summing up of the work that had begun with such ardor in the thirties.

Then there remained one unfinished business, namely, the results on estimates of the number of limiting cycles born from a separatrix loop, published as a short note in Doklady in 1951. Of course, there was the text of E. A. Leontovich's dissertation, but that was only available to the narrow circle of collaborators of her department. Now E. A. Leontovich began preparing it for publication. First she prepared the first part, dealing with the construction of normal forms, and then the second, devoted to the proof of the main statement. All together this was a 164-page text. It is probably because of its length that E. A. Leontovich did not include it into the monograph. But its size made it too long for a journal article, so it was deposited at VINITI, a kind of publication that makes the text available to anyone needing it for personal use. The drawback of this form of publication was that the text was not actually available to potential foreign readers.

By then one could think that the main goals of E. A. Leontovich's life had been achieved. In 1982 she resigned her position of department head, but remained in the department in a senior research position, and several years later retired for reasons of health. Nevertheless, until the last years she retained a very lively interest in the work of her former collaborators, who were her pupils in one way or another.

In 1996 Nizhny Novgorod hosted the international conference "Contemporary problems in the theory of dynamical systems". The participants of the conference by common consent wrote a letter of greetings to the oldest representative of the Gor′kiĭ school of oscillation theory (N. N. Bautin had died in 1993). Then a very representative delegation from the conference visited her at her home. E. A. Leontovich was very pleased. She joked a good deal, recalled various stories from the scientific life of the twenties and thirties. As it turned out, this was her last encounter with the scientific community. Evgeniya Aleksandrovna Leontovich-Andronova died half a year later.

The list of scientific works of E. A. Leontovich-Andronova

[1] *Laplace* (Series "Life of Famous People"), Gosizdat: Moskovskiĭ Rabochiĭ, 1930 (A. A. Andronov, coauthor). (Russian)

[2] *On an inequality related with the Fourier integral*, Dokl. Akad. Nauk SSSR **4** (1934), no. 7, 353–360 (A. G. Maier, coauthor). (Russian)

[3] *On trajectories determining the qualitative structure of orbit foliation on a sphere*, Dokl. Akad. Nauk SSSR **14** (1937), no. 5, 251–254 (A. G. Maier, coauthor). (Russian)

[4] *To the theory of changes of the orbit foliation on the plane*, Dokl. Akad. Nauk SSSR **21** (1938), no. 9, 427–430 (A. A. Andronov, coauthor). (Russian)

[5] *Several cases of dependence of limit cycles of parameters*, Uch. Zap. Gor′kov. Univ. **6** (1939), 3–24 (A. A. Andronov, coauthor). (Russian)

[6] *General qualitative theory*, in the book H. Poincaré, *On the curves defined by differential equations*, Transl. to Russian, Moscow–Leningrad, Gostekhizdat, 1947, 267–300 (A. G. Maier, coauthor).

[7] *On the birth of limit cycles from a separatrix*, Dokl. Akad. Nauk SSSR **78** (1951), no. 4, 641–644. (Russian)

[8] *The birth of limit cycles from a weak focus or a center and from a weak limit cycle*, Dokl. Akad. Nauk SSSR **99** (1954), no. 6, 885–888 (A. A. Andronov, coauthor). (Russian)

[9] *On a scheme determining the topological structure of the orbit foliation*, Dokl. Akad. Nauk SSSR **103** (1955), no. 4, 557–560 (A. G. Maier, coauthor). (Russian)

[10] *The ideal topographic system in a neighborhood of a rough limit cycle and focus*, Uch. Zap. Gor′kov. Univ. **28** (1955), 3–19. (Russian)

[11] *Some directions of research in the qualitative theory of differential equation in the Gor′kiĭ School of oscillation theory and prospects of development of these directions*, Proc. of 3d All-Union mathematical Congress, The edition of USSR Academy of Science, vol. 2, 1956, 17–18. (Russian)

[12] *The birth of limit cycles from a weak focus or a center and from a weak limit cycle*, Mat. Sb. **40 (82)** (1956), no. 2, 179–224 (A. A. Andronov, coauthor). (Russian)

[13] *Some mathematical works of the Gor′kiĭ School by A. A. Andronov*, Proc. of 3d All-Union Mathematical Congress, The edition of USSR Academy of Science, vol. 3, 1958, 116–125. (Russian)

[14] *On the birth of limit cycles from a separatrix loop and a separatrix of an equilibrium point of saddle-node type*, Mat. Sb. **48 (90)** (1959), no. 3, 335–376 (A. A. Andronov, coauthor). (Russian)

[15] *On some analogies between of plane algebraic curves and algebraic dynamical systems on the plane*, Dokl. Akad. Nauk SSSR **129** (1959), no. 3, 503–506. (Russian)

[16] Sections 3, 4 in Chapter VI of the second edition of the book A. Andronov, A. Witt, and S. Khaikin, *The oscillation theory* (N. A. Zheleztsov, ed.), Fizmatgiz, Moscow, 1959. (Russian)

[17] *Bifurcation theory of dynamical systems of the second order and its application to the study of nonlinear problems of the oscillation theory*, in: Qualitative Methods in the Theory of Nonlinear Vibrations (Proc. Internat. Sympos. Nonlinear Oscillations, Vol. 2, Kiev, 1963), pp. 7–28 (L. N. Belyustina, coauthor).

[18] *On the definition of a rough dynamical system*, in: Nonlinear Vibration Problems, vol. 5, Warszawa, 1964, 291–295.

[19] *Dynamical systems of the first degree of nonroughness on the plane*, Symposium on qualitative theory of differential equations and its applications, Abstracts, Samarkand, 1964, 5–7 (A. A. Andronov, coauthor). (Russian)

[20] *On a singular bifurcation of a dynamical system in the 3-dimensional space related to the disappearance of a periodic solution*, Proc. Politekhn. Institute, vol. 6, Tashkent, 1964, 132–139 (Yu. I. Neimark, coauthor). (Russian)

[21] *Principal directions and new results of the Gor′kiĭ School by A. A. Andronov*, Proc. of 4th All-Union Mathematical Congress, Leningrad, vol. 2, 1964, 415–424 (N. N. Bautin, L. N. Belyustina, and Yu. I. Neimark, coauthors). (Russian)

[22] *Letter to the editor on the article by N. F. Otrokov "Multiple limit cycles"*, Mat. Sb. **64 (106)** (1964), no. 1, 140–144. (Russian)

[23] *Dynamical systems of the first degree of nonroughness on a plane*, Mat. Sb. **68 (110)** (1965), no. 3, 328–372 (A. A. Andronov, coauthor). (Russian)

[24] *Qualitative theory of dynamical systems of the second order*. "Nauka", Moscow, 1966 (A. A. Andronov, I. I. Gordon, and A. G. Maier, coauthors); English transl. Israel Program of Scientific Translations, Jerusalem–London, 1973.

[25] *Theory of bifurcations of dynamical systems on the plane*, "Nauka", Moscow, 1967 (A. A. Andronov, I. I. Gordon, and A. G. Maier, coauthors); English transl. Israel Program of Scientific Translations, Jerusalem–London, 1973.

[26] *The state-of-the-art in the bifurcation theory of dynamical systems*, 5th Intern. Conf. on Nonlinear Oscillations, Abstracts, Kiev, vol. 2, 1969, 130–131 (L. P. Shilnikov, coauthor). (Russian)

[27] *Sufficient conditions for "the first degree of nonroughness" for a dynamical system on the plane*, Differentsial′nye Uravneniya **6** (1970), no. 12, 2121–2134. (A. A. Andronov, coauthor). (Russian)

[28] *To the problem on the birth of integral curves from a separatrix contour of a two-dimensional complex dynamical system*, Dokl. Akad. Nauk SSSR **207** (1972), no. 6, 1277–1280 (V. P. Tareev, coauthor). (Russian)

[29] *Studying bifurcations of a double equilibrium with two zero eigenvalues*, 3d All-Union Conf. on the Qualitative Theory of Differ. Equat., Abstracts, Samarkand, 1973, 126 (I. P. Fedosenko, coauthor). (Russian)

[30] *Bifurcations of a double equilibrium of a dynamical system with two zero eigenvalues*, Methods of Qualitative Theory of Differ. Equat., Gor′kiĭ State Univ., Gor′kiĭ, vol. 1, 1975, 88–100 (I. P. Fedosenko, coauthor). (Russian)

[31] *The ideal topographic system in a neighborhood of a weak limit cycle and a limit cycle*, All-Union Conf. on the Qualitative Theory of Differ. Equat., Abstracts, Ryazan′, 1976, 194 (I. P. Fedosenko, coauthor). (Russian)

[32] *Methods and modes of the qualitative study of dynamical systems on the plane*, "Nauka", Moscow, 1976 (N. N. Bautin, coauthor). (Russian)

[33] *On the canonical form for a differential equation depending on a parameter in a neighborhood of a singular point from the Ziegel domain*, Methods of Qualitative Theory of Differ. Equat., Gor′kiĭ State Univ., Gor′kiĭ, 1980, 108–122. (V. P. Tareev, coauthor). (Russian)

[34] *Comments to the Hopf theorem*, Supplement to the book J. Marsden and M. McCraken, *Hopf bifurcation and its applications*, Russian transl., Mir, Moscow, 1980, 336–344.

[35] *Remarks on the problem of the splitting of complex separatrices*, Methods of Qualitative Theory of Differ. Equat., Gor′kiĭ State Univ., Gor′kiĭ, 1983, pp. 59–70. (V. P. Tareev, coauthor). (Russian)

[36] *Reduction to canonical form of an autonomous dynamical system in a neighborhood of a saddle with a ratio of eigenvalues close to one*, Preprint VINITI No. 3777-B87, 1987. (Russian)

[37] *The birth of limit cycles from a separatrix loop of a saddle on the plane in the case of a zero saddle value*, Preprint VINITI No. 161-B88, 1988. (Russian)

[38] *To the problem of the splitting of the complex separatrix by means of "the rotation of the field"*, Methods of Qualitative Theory and Bifurcation Theory, Gor′kiĭ State Univ., Gor′kiĭ, 1989, 75–83 (V. P. Tareev, coauthor). (Russian)

[39] *To the problem of the splitting of the complex separatrix*, 7th All-Union Conf. on the Qualitative Theory of Differ. Equat., Abstracts, Riga, 1989, 144. (V. P. Tareev, coauthor).

Translated by A. B. SOSSINSKI

Amer. Math. Soc. Transl.
(2) Vol. **200**, 2000

On the Abundance of Traveling Waves in Coupled Expanding Circle Maps

V. Afraimovich and M. Courbage

To the memory of Professor E. A. Leontovich-Andronova

ABSTRACT. We show that in some dynamical extended systems possessing space–time chaos, traveling wave solutions are everywhere dense

1. Introduction

Coupled map lattices (CML) are special models of spatially extended nonequilibrium media (see, for instance [**1**]). They can be treated as discrete versions of partial differential equations (PDE), being systems with discrete spatial and temporal variables. The discrete nature of the variables allows one to avoid many mathematical PDE difficulties in studies of such qualitative features of the media as space-time chaos. A rigorous definition of space–time chaos was introduced in [**2**], based on the notion of Sinai–Ruelle–Bowen (SRB)-measure, which is invariant under both space and time translations and has strong mixing properties. The authors of [**2**] studied special classes of coupled expanding circle maps. In [**3, 4**] the results of [**2**] were generalized to general expanding circle maps, and in [**5, 6, 7**], to maps with multidimensional phase space (see, for instance, the survey [**8**]). All these results are related to ergodic properties of CML with space-time chaos.

Topological properties of such CML are not so well known. There are some particular results about traveling wave solutions in specific CML (for example, [**9**]), and some kind of theory of behavior of traveling wave solutions in CML with small coupling [**10**], but there is no general theory of topological behavior in CML admitting space-time chaos.

One of the authors of the present paper studied cellular automata with chaotic behavior [**11**]. It was found that traveling wave solutions play an important role—in the sense that they are as abundant as periodic orbits in hyperbolic sets of dynamical systems with chaotic behavior. In the present work we generalize some results of [**11**] to coupled extended circle maps.

2000 *Mathematics Subject Classification.* Primary 37D45; Secondary 37B15.

V. Afraimovich was supported by the National Science Council, Republic of China, Grant NSC86-2115-M-007-018, and by a French Grant from the Université Denis Diderot.

©2000 American Mathematical Society

We prove that traveling wave solutions with different rational velocities are dense in the space of all configurations of CML. We believe that this result is in the spirit of a qualitative theory of coupled map lattices.

2. Statement of the problem

Let S be the unit circle, $S = \{x \pmod 1\}$. Set $S^{\mathbb{Z}} = \bigotimes_{i\in\mathbb{Z}} S_i$, where S_i are copies of S and $\mathbb{R}^{\mathbb{Z}} = \bigotimes_{i\in\mathbb{Z}} \mathbb{R}_i$, where $\mathbb{R}_i$ are copies of $\mathbb{R}$, i.e., $\mathbb{R}^{\mathbb{Z}} = \{\underline{x} = (\dots, x_{i-1}, x_i, x_{i+1}, \dots)\}$ are two-sided infinite sequences of real numbers. Fix a number $q > 1$ and introduce the Hilbert space $B = \{\underline{x} \in \mathbb{R}^{\mathbb{Z}} \mid \|\underline{x}\| < \infty\}$, where $\|\underline{x}\| = \sqrt{\langle \underline{x}, \underline{y}\rangle}$, and the scalar product is

$$\langle \underline{x}, \underline{y}\rangle = \sum_{i=-\infty}^{\infty} \frac{x_i y_i}{q^{|i|}}\,.$$

This type of metric with weights is used (see [**2, 5, 10, 12, 13**]) in studies of CML, since it induces the Tikhonov topology in $\mathbb{R}^{\mathbb{Z}}$. Denote by $\pi\colon \mathbb{R}^{\mathbb{Z}} \to S^{\mathbb{Z}}$ the projection $\pi((\dots x_i, x_{i+1}, \dots)) = (\dots x_i \pmod 1, x_{i+1} \pmod 1, \dots)$. Endow $S^{\mathbb{Z}}$ with the metric

$$d(\pi\underline{x}, \pi\underline{y}) = \inf \|\underline{\tilde{x}} - \underline{\tilde{y}}\|, \tag{2.1}$$

where the infimum is taken over all elements $\underline{\tilde{x}}, \underline{\tilde{y}} \subset B$ with $\pi\underline{\tilde{x}} = \pi\underline{x}$, $\pi\underline{\tilde{y}} = \pi\underline{y}$. We denote $S^{\mathbb{Z}}$ endowed with the metric (2.1) by $\mathbb{T}$.

Let $f\colon S \to S$ be an expanding C^r-map, $r > 1$, and $F\colon \mathbb{R} \to \mathbb{R}$ be a lift of f. We assume that

$$1 < \lambda_0 \leqslant |F'(x)| \leqslant \lambda_1 < \infty. \tag{2.2}$$

Without loss of generality we may assume that $F(0) = 0$.

We consider a specific type of coupling, the so-called short-ranged coupling ([**2, 5, 7**]). We say that a map $G\colon B \to B$ is *short-ranged* if for any fixed k and any $\underline{x} = (x_j)$, $\underline{y} = (y_j) \in B$ with $x_j = y_j$ for all $j \in \mathbb{Z}$, $j \neq k$, we have $|(G\underline{x} - G\underline{y})_i| \leqslant C\alpha^{|i-k|}|x_k - y_k|$, where $C > 0$ and α, $0 < \alpha < 1$, are constants. In particular, G is short-ranged if there is an $r \in \mathbb{Z}_+$ such that

$$(G\underline{x} - G\underline{y})_i = 0, \qquad |i - k| > r, \tag{2.3}$$

(the number r is called sometimes the "*radius of rule*"). We assume that the map $G\colon B \to B$ satisfies the following conditions:

(i) it satisfies (2.3) with some $r \in \mathbb{Z}_+$;
(ii) it is translation-invariant, i.e., $G\sigma\underline{x} = \sigma G\underline{x}$ for any $\underline{x} \in B$, where σ is the translation map: $(\sigma\underline{x})_i = x_{i+1}$ if $(\underline{x})_i = x_i$;
(iii) G is a lift of the map $g\colon \mathbb{T} \to \mathbb{T}$, i.e., for any $\underline{x} \in B$ and for any $\underline{y}$ such that $\pi\underline{y} = \pi\underline{x}$ we have $\pi G\underline{x} = \pi G\underline{y}$.

By properties (i)–(iii), the map G can be expressed as follows:

$$(G\underline{x})_i = \beta \sum_{j-i=-r}^{r} a_{j-i}(x_j),$$

where $a_k(x)$ is a 1-periodic function of $x \in \mathbb{R}$, $k = [-r, r]$, and β is the coupling parameter. We assume that the functions $a_k(x)$ are C^r-smooth, $r \geqslant 1$. Therefore, they are bounded:

$$|a_k(x)| \leqslant A, \tag{2.4}$$

where A is some constant. It is easy to see that for a map G satisfying (2.3), (2.4) we have $G(B) \subset B$. Indeed,

$$\|G\underline{x}\| = \sqrt{\sum_{i=-\infty}^{\infty} \frac{1}{q^{|i|}} \Big(\sum_{j-i=-r}^{r} a_{j-i}(x_j) \Big)^2} \leqslant (2r+1)A \sqrt{\sum_{i=-\infty}^{\infty} \frac{1}{q^{|i|}}} < \infty.$$

Fix some G satisfying (2.3), (2.4) and introduce a CML generated by a map $\hat{\theta}\colon B \to \mathbb{R}^{\mathbb{Z}}$ in the form

$$(\hat{\theta}\underline{x})_i = F(x_i) + (G(\underline{x}))_i = F(x_i) + \beta \sum_{j-i=-r}^{r} a_{j-i}(x_j). \tag{2.5}$$

First of all, it is simple to see that $\hat{\theta} B \subset B$, and we have $\|\hat{\theta}\underline{x}\| \leqslant \lambda_1 \|\underline{x}\| + \|G\underline{x}\|$ from (2.5) and (2.2). Second, the map $\hat{\theta}$ is a lift of the map $\theta\colon \mathbb{T} \to \mathbb{T}$ given by

$$(\theta x)_i = F(x_i) + \beta \sum_{j-i=-r}^{r} a_{j-i}(x_j) \pmod 1. \tag{2.6}$$

The following proposition can be proved in the same way as in [**10, 12, 13**].

PROPOSITION 1. *Both the map $\hat{\theta}$ and the map θ are Lipschitz-continuous.*

Thus, the map $\hat{\theta}|B$ (as well as the map $\theta|\mathbb{T}$) generates a classical dynamical system.

Consider the shift map $\hat{\sigma}\colon B \to B$, $(\sigma\underline{x})_i = x_{i+1}$.

PROPOSITION 2. *The linear operators $\hat{\sigma}$ and $\hat{\sigma}^{-1}$ are bounded and*

$$\|\hat{\sigma}\| = \|\hat{\sigma}^{-1}\| = \sqrt{q}. \tag{2.7}$$

PROOF. We know that

$$\|\underline{x}\|^2 = \sum_{i=-\infty}^{\infty} \frac{x_i^2}{q^{|i|}}.$$

Then

$$\begin{aligned}
\|\hat{\sigma}\underline{x}\|^2 &= \cdots + \frac{x_{-1}^2}{q^2} + \frac{x_0^2}{q} + x_1^2 + \frac{x_2^2}{q} + \cdots \\
&= \cdots + \frac{1}{q}\frac{x_{-1}^2}{q} + \frac{1}{q} x_0^2 + q\frac{x_1^2}{q} + q\frac{x_2^2}{q^2} + \cdots \\
&= \cdots + \frac{(x_{-1}/\sqrt{q})^2}{q} + \Big(\frac{x_0}{\sqrt{q}}\Big)^2 + \frac{(\sqrt{q}x_1)^2}{q} + \frac{(\sqrt{q}x_2)^2}{q^2} + \cdots = \|D_1\underline{x}\|^2,
\end{aligned}$$

where $D_1 = \operatorname{diag}(\dots, 1/\sqrt{q}, 1/\sqrt{q}, \sqrt{q}, \sqrt{q}, \dots)$. Thus, $\|D_1\underline{x}\| = \|\hat{\sigma}\underline{x}\|$, $\|\hat{\sigma}\| = \|D_1\| = \sqrt{q}$.

In the same way we can show that

$$\|\hat{\sigma}^{-1}\underline{x}\|^2 = \cdots + \frac{(\sqrt{q}x_{-2})^2}{q^2} + \frac{(\sqrt{q}x_{-1})^2}{q} + \Big(\frac{x_0}{\sqrt{q}}\Big)^2 + \frac{(x_1/\sqrt{q})^2}{q} + \cdots = \|D_2\underline{x}\|^2,$$

where $D_2 = \operatorname{diag}(\dots, \sqrt{q}, \sqrt{q}, 1/\sqrt{q}, 1/\sqrt{q}, \dots)$, and $\|D_2\| = \sqrt{q}$. □

Hence, $\hat{\sigma}|B$ generates a dynamical system, the so-called *translational* dynamical system. The shift map $\hat{\sigma}$ is a lift of the continuous shift map $\sigma\colon \mathbb{T} \to \mathbb{T}$, $(\sigma \underline{x})_i = x_{i+1} \pmod 1$, $x \in \mathbb{T}$, which generates the translational dynamical system on $\mathbb{T}$. Let us remark that since $\hat{\sigma}\hat{\theta} = \hat{\theta}\hat{\sigma}$ (and $\sigma\theta = \theta\sigma$), i.e., the pair $(\hat{\sigma}, \hat{\theta})$ (as well as (σ, θ)) generates the $\mathbb{Z} \times \mathbb{Z}_+$ action on B (on $\mathbb{T}$).

In the present work we consider traveling wave solutions. In [**9, 10**] and other articles the authors studied traveling wave solutions of the form

$$x_j(n) = \psi(kj + mn), \tag{2.8}$$

where $x_j(n) = (\mathcal{F}^n \underline{x})_j$, $\mathcal{F}$ is an evolution operator of a coupled map lattice (in our work $\mathcal{F}$ is either θ or $\hat{\theta}$), ψ is a function, and $k \in \mathbb{Z}_+$, $m \in \mathbb{Z}$. The fact that $x_j(n)$ depends only on a linear combination of temporal and spatial variables allowed us to construct a dynamical system with finite-dimensional phase space and give a description of solutions of type (2.8) in terms of this dynamical system ([**9, 10**]).

Let us emphasize that the solution $\underline{x}(n)$ of the form (2.8) satisfies the equations $\mathcal{F}^k \underline{x}(n) = \sigma^m \underline{x}(n)$, and $\sigma^{-m}\mathcal{F}^k \underline{x}(n) = \underline{x}(n)$ for each $n \in \mathbb{Z}_+$, where σ is the translation operator. In other words, $\underline{x}(0)$ (and $\underline{x}(n)$) is a fixed point of the operator $\sigma^{-m}\mathcal{F}^k$. Let us introduce the following definition.

DEFINITION. A configuration $\underline{x}$ is said to be a traveling wave solution of the (k, m)-type if $\underline{x}$ is a fixed point of the operator $\sigma^{-m}\mathcal{F}^k$.

The solution $\underline{x}(n)$ of the form (2.8) satisfies the definition. We believe that, perhaps under additional conditions, the fixed points of the operator $\sigma^{-m}\mathcal{F}^k$ can be represented in the form (2.8). But we are not going to prove this in the present work, and we will use the above definition in the traveling wave solutions. This is natural because of the following observation. Let $\underline{x} = \sigma^{-m}(\mathcal{F}^k \underline{x})$. Then

$$\mathcal{F}^k(\mathcal{F}^s \underline{x}) = \mathcal{F}^s(\mathcal{F}^k \underline{x}) = \mathcal{F}^s(\sigma^m \underline{x}) = \sigma^m(\mathcal{F}^s x)$$

for any $s \in \mathbb{Z}_+$, i.e., $\mathcal{F}^s \underline{x} = \sigma^{-m}(\mathcal{F}^k(\mathcal{F}^s \underline{x}))$, i.e., $\mathcal{F}^s \underline{x}$ is a traveling wave solution of the (k, m)-type, as well as $\underline{x}$.

The main result of our work can be stated in the following form.

THEOREM. *There exists a $\beta_0 > 0$ such that for each $\beta \in [0, \beta_0]$ and for every $m \in \mathbb{Z}$, the traveling wave solutions of the (n, m)-type of* (2.6) *are dense in $\mathbb{T}$.*

3. Proofs of the results

The proof of Theorem can be presented as a collection of lemmas.

LEMMA 1. (i) *The space $\mathbb{T}$ is complete and compact.*

(ii) *Any open set $V \subset \mathbb{T}$ contains an open set U which can be represented in the form*

$$U = \bigotimes_{i=-\infty}^{-k_1-1} S_i \otimes \bigotimes_{i=-k_1}^{k_2} I_i \otimes \bigotimes_{i=k_2+1}^{\infty} S_i, \tag{3.1}$$

where S_i are copies of S, $I_i = \pi(\widehat{I}_i)$, $\widehat{I}_i \subset \mathbb{R}$ is an open interval of length $(\widehat{I}_i) < 1$.

PROOF. Statement (i) follows directly from the definition of the space $\mathbb{T}$ and the metric. Let us prove (ii). Given an open set $V \subset \mathbb{T}$, choose an arbitrary point $\underline{x} = (\dots, x_{-1}, x_0, \dots, x_k, \dots) \in V$. Then, there is an $n \in \mathbb{Z}_+$ such that any point

$$\underline{y} = (\dots, y_{-n-1} \pmod 1, x_{-n}, \dots, x_{-1}, x_0, \dots, x_n, y_{n+1} \pmod 1, y_{n+2} \pmod 1, \dots)$$

belongs to V independently of the choice of the values of y_k, $|k| > n$, i.e., the set

$$\bigotimes_{i=-\infty}^{-k_1-1} S_i \otimes \bigotimes_{i=-n}^{n} \{x_i\} \otimes \bigotimes_{i=n+1}^{\infty} S_i$$

belongs to V. This implies the desired result because V is open. □

LEMMA 2. *For arbitrary $\underline{x}, \underline{y} \in B$, we have $\|\hat\theta\underline{x} - \hat\theta\underline{y}\| \geqslant \lambda\|\underline{x} - \underline{y}\|$, where*

(3.2)

$$\lambda \geqslant \left\{(\lambda_0 - \beta A)^2 - 4(\lambda_1 + \beta A)\beta A \sum_{m=1}^{r} (\sqrt{q})^m - \beta^2 A^2 \sum_{\substack{0\neq m_{1,2}=-r \\ m_1\neq m_2}}^{r} (\sqrt{q})^{m_1+m_2}\right\}^{1/2}$$

provided that β is so small that the expression in the braces is positive.

PROOF. It is clear that

$$(\hat\theta\underline{x} - \hat\theta\underline{y})_i = \sum_{|j-i|\leqslant r} b_{ij}(x_j - y_j), \tag{3.3}$$

where $b_{ij} = \beta a_j'(x_{ij})$ for $i \neq j$, $b_{ii} = (F' + \beta a_i')(x_{ii})$, $x_{ij} = x_i + t_j(y_j - x_j)$, $0 \leqslant t_j \leqslant 1$.

Thus, by assumptions (2.2), (2.4), we have

$$\lambda_0 - \beta A \leqslant |b_{ii}| \leqslant \lambda_1 + \beta A, \tag{3.4}$$

$$|b_{ij}| \leqslant \beta A. \tag{3.5}$$

Denoting $\beta A = \gamma$, from (3.3)–(3.5) we obtain

$$\begin{aligned}(\hat\theta\underline{x} - \hat\theta\underline{y})_i^2 &\geqslant b_{ii}^2(x_i - y_i)^2 + 2b_{ii}(x_i - y_i)^2 + 2b_{ii}(x_i - y_i) \sum_{\substack{|j-i|\leqslant r \\ j\neq i}} b_{ij}(x_j - y_j) \\ &\quad + 2 \sum_{\substack{j_1\neq j_2 \\ j_{1,2}\neq i}} b_{ij_1} \cdot b_{ij_2}(x_{j_1} - y_{j_1})(x_{j_2} - y_{j_2}) \\ &\geqslant (\lambda_0 - \gamma)^2(x_i - y_i)^2 - 2(\lambda_1 + \gamma)\gamma \sum_{\substack{|j-i|\leqslant r \\ j\neq i}} |x_i - y_i||x_j - y_j| \\ &\quad - 2\gamma^2 \sum_{\substack{j_1\neq j_2 \\ j_{1,2}\neq i}} |x_{j_1} - y_{j_1}|\,|x_{j_2} - y_{j_2}|.\end{aligned} \tag{3.6}$$

Let $\underline{z} = (\dots|x_{-1} - y_{-1}|, |x_0 - y_0|, \dots, |x_n - y_n|, \dots)$, so that $\|\underline{z}\| = \|\underline{x} - \underline{y}\|$. It follows from (3.6) that

$$\begin{aligned}\|\hat\theta\underline{x} - \hat\theta\underline{y}\|^2 &\geqslant (\lambda_0 - \gamma)^2\|\underline{x} - \underline{y}\|^2 - 2(\lambda_1 + \gamma)\gamma \sum_{m=1}^{r} (\langle \underline{z}, \hat\sigma^m \underline{z}\rangle + \langle \underline{z}, \hat\sigma^{-m}\underline{z}\rangle) \\ &\quad - \gamma^2 \sum_{\substack{0\neq m_{1,2}=-r \\ m_1\neq m_2}}^{r} \langle \hat\sigma^{m_1}\underline{z}, \hat\sigma^{m_2} z\rangle,\end{aligned}$$

and because of (2.7), this implies that

$$\|\hat\theta\underline{x} - \hat\theta\underline{y}\|^2 \geqslant \|\underline{x} - \underline{y}\|^2 \left[(\lambda_0 - \gamma)^2 - 4(\lambda_1 + \gamma)\gamma \sum_{m=1}^{r} (\sqrt{q})^m - \gamma^2 \sum_{\substack{0\neq m_{1,2}=-r \\ m_1\neq m_2}}^{r} (\sqrt{q})^{m_1+m_2}\right].$$

Lemma 2 is proved. □

COROLLARY 1. *There is a $\beta_0 > 0$ such that for any $\beta \in [0, \beta_0]$ the expression in the braces in (3.2) is greater than one.*

LEMMA 3. *For any $\beta \in [0, \beta_0]$ and for any $m \in \mathbb{Z}_+$ there is a $k_0 \in \mathbb{Z}_+$, $k_0 = k_0(\beta, m)$, such that the map $\hat{\sigma}^{-m}\hat{\theta}^k$, $k \geqslant k_0$, is an expansion, i.e., there exists a $\lambda_{m,k} > 1$ such that $\|\hat{\sigma}^{-m}\hat{\theta}^k \underline{x} - \hat{\sigma}^{-m}\hat{\theta}^k \underline{y}\| \geqslant \lambda_{m,k}\|\underline{x} - \underline{y}\|$ for arbitrary $\underline{x}, \underline{y} \in B$.*

PROOF. First, let us show that

$$\|\hat{\sigma}^{-m}\underline{w} - \hat{\sigma}^{-m}\underline{z}\| \geqslant (1/\sqrt{q})^m \|\underline{w} - \underline{z}\|. \tag{3.7}$$

Indeed, we can represent $\underline{w}$, $\underline{z}$ as follows: $\underline{w} = \hat{\sigma}^m \underline{w}'$, $\underline{z} = \hat{\sigma}^m \underline{z}'$ for some $\underline{w}', \underline{z}' \in B$. Then, Proposition 2 implies that $\|\hat{\sigma}^m \underline{w}' - \hat{\sigma}^m \underline{z}'\| \leqslant (\sqrt{q})^m \|\underline{w}' - \underline{z}'\|$, which is exactly the inequality (3.7).

Second, due to (3.7),

$$\|\hat{\sigma}^{-m}\hat{\theta}^k \underline{x} - \hat{\sigma}^{-m}\hat{\theta}\underline{y}\| \geqslant (1/\sqrt{q})^m \|\hat{\theta}^k \underline{x} - \hat{\theta}^k \underline{y}\|,$$

and by Lemma 2,

$$\|\hat{\sigma}^{-m}\hat{\theta}^k \underline{x} - \hat{\sigma}^{-m}\hat{\theta}\underline{y}\| \geqslant \lambda^k (1/\sqrt{q})^m \|\underline{x} - \underline{y}\|.$$

For $\beta \in [0, \beta_0]$ we have $\lambda > 1$, therefore for

$$k \geqslant k_0 = E\left(\frac{m}{2}\frac{\ln q}{\ln \lambda}\right) \tag{3.8}$$

(where $E(a)$ is the integer part of the number a) the map $\hat{\sigma}^{-m}\hat{\theta}^k$ is an expansion and $\lambda_{m,k} \geqslant \lambda^k (1/\sqrt{q})^m$. □

COROLLARY 2. *For any k satisfying (3.8), the map $\sigma^{-m}\theta^k$ is also an expansion.*

LEMMA 4. *For any $\beta \in [0, \beta_0]$ and $m \in \mathbb{Z}_+$ and an arbitrary set U of the form (3.1) there exists a $k_1 \geqslant k_0(\beta, m)$ such that $\sigma^{-m}\theta^k U \supset \overline{U}$ for any $k \geqslant k_1$.*

PROOF. Without loss of generality one can consider a set U for which the lift $\widehat{U} \subset B$ has the form

$$\widehat{U} = \bigotimes_{i=-\infty}^{-k_1-1} J_i \otimes \bigotimes_{i=-k_1}^{k_2} \widehat{I}_i \otimes \bigotimes_{i=k_2+1}^{\infty} J_i,$$

where J_i are copies of the unit interval [0, 1], and $\widehat{I}_i = (\alpha_i, \beta_i) \subset [0, 1]$. Given $i \in \mathbb{Z}$, let

$$(a_i, b_i) = \begin{cases} (\alpha_i, \beta_i), & -k_1 \leqslant i \leqslant k_2, \\ (0, 1), & i < -k_1 \text{ or } i > k_2. \end{cases}$$

Let $\underline{x}^a = (\ldots, x_{i-1}, a_i, x_{i+1}, \ldots)$, $\underline{x}^b = (\ldots, x_{i-1}, b_i, x_{i+1}, \ldots)$, where x_j are arbitrary numbers such that $\underline{x}^a, \underline{x}^b \in B$. Then

$$|\hat{\theta}\underline{x}^a - \hat{\theta}\underline{x}^b| \geqslant (\lambda_0 - \gamma)(b_i - a_i) \geqslant \lambda(b_i - a_i),$$

where λ is the constant in Lemma 2. Therefore, the set $(\hat{\theta}\widehat{U})_i$ contains an interval of length greater than $\lambda(b_i - a_i)$. This means that

$$\hat{\theta}\widehat{U} \supset \bigotimes_{i=-\infty}^{-n_1-1} J_i^{(1)} \otimes \bigotimes_{i=-n_1}^{n_2} I_i^{(1)} \otimes \bigotimes_{i=n_2+1}^{\infty} J_i^{(1)},$$

where $\text{length}(J_i^{(1)}) \geqslant \lambda > 1$, $\text{length}(\widehat{I}_i^{(1)}) \geqslant \lambda \, \text{length}(\widehat{I}_i)$. Proceeding in this way, we find for any $k \in \mathbb{Z}_+$ that

$$\hat{\theta}^k \widehat{U} \supset \bigotimes_{i=-\infty}^{-n_1-1} J_i^{(k)} \otimes \bigotimes_{i=-n_1}^{n_2} I_i^{(k)} \otimes \bigotimes_{i=n_2+1}^{\infty} J_i^{(k)},$$

where $\text{length}(J_i^{(k)}) \geqslant \lambda^k$, $\text{length}(I_i^{(k)}) \geqslant \lambda^k \, \text{length}(\widehat{I}_i)$. Assume that $k_1 \geqslant k_0$ is so large that $\min_{-n_1 \leqslant i \leqslant n_2}(\lambda^{k_1} \, \text{length}(I_i)) > 1$. Then $\hat{\theta}^{k_0} \widehat{U} \supset \bigotimes_{i=-\infty}^{\infty} \widetilde{J}_i$ with $\text{length}(\widetilde{J}_i) > 1$. In other words, the set $\hat{\theta}^{k_0} \widehat{U}$ contains a fundamental region of the projection π, i.e., $\pi \hat{\theta}^{k_0} \widehat{U} = \mathbb{T}$. Now $\sigma^{-m} \hat{\theta}^{k_0} \widehat{U} = \bigotimes_{i=-\infty}^{\infty} \widetilde{J}_{i-m}$ for any m, i.e., $\sigma^{-m} \theta^{k_0} U \supset \bar{U}$. □

COROLLARY 3. *The set $\overline{U}$ contains a fixed point of the map $\sigma^{-m}\theta^k$ for $k \geqslant k_0$.*

PROOF. The map $\sigma^{-m}\theta^k$ is an expansion (Lemma 3) and $\sigma^{-m}\theta^k U \supset \overline{U}$ (Lemma 4). Therefore, $(\sigma^{-m}\theta^k)^{-1}$ has a branch which is well defined on $\underline{U}$ and is a contraction. □

References

[1] *Theory and application of Coupled Map Lattices* (K. Kaneko, ed.), Wiley, New York, 1993.

[2] L. Bunimovich and Ya. G. Sinai, *Space-time chaos in coupled map lattices*, Nonlinearity **1** (1988), 491–516.

[3] J. Bricmont and A. Kupiainen, *Coupled analytic maps*, Nonlinearity **8** (1995), 379–396.

[4] ———, *High temperature expansions and dynamical systems*, Comm. Math. Phys. **178** (1996), 703–732.

[5] Ya. Pesin and Ya. Sinai, *Space–time chaos in chains of weakly interacting hyperbolic mappings*, Adv. Soviet Math., vol. 3, Amer. Math. Soc., Providence, RI, 1991, pp. 165–198.

[6] M. Jiang, *Equilibrium states for lattice models of hyperbolic type*, Nonlinearity **8** (1994), 631–659.

[7] M. Jiang and Ya. Pesin, *Equilibrium measures for coupled map lattices: existence, uniqueness and finite-dimensional approximations*, Comm. Math. Phys. **193** (1998), 675–711.

[8] L. Bunimovich, *Coupled map lattices: one step forward and two steps back*, Physica D **86** (1995), 248–255.

[9] V. S. Afraimovich and V. I. Nekorkin, *Chaos of travelling waves in a discrete chain of diffusively coupled maps*, Internat. J. Bifur. Chaos Appl. Sci. Engrg. **4** (1994), 631–637.

[10] V. S. Afraimovich and Ya. B. Pesin, *Traveling waves in lattice models of multi-dimensional and multi-component media. I. General hyperbolic properties*, Nonlinearity **6** (1973), 429–455; II. (with A. Tempelman) *Ergodic properties and dimension*, Chaos **3** (1993), 233–241.

[11] M. Courbage, *On the abundance of traveling waves in ID infinite cellular automata*, Physica D **103** (1997), 133–144; M. Courbage and S. Yasmieh (to appear).

[12] V. S. Afraimovich and Ya. B. Pesin, *Hyperbolicity of infinite-dimensional drift systems*, Nonlinearity **3** (1990), 1–19.

[13] V. S. Afraimovich, L. Y. Glebsky, and V. I. Nekorkin, *Stability of stationary states and topological spatial chaos in lattice dynamical systems*, Random Comput. Dynam. **2** (1994), 287–303.

NATIONAL TSING HUA UNIVERSITY, DEPARTMENT OF MATHEMATICS, HSINCHU, TAIWAN, 30043, R.O.C.

LABORATOIRE DE PHYSIQUE THÉORIQUE, UNIVERSITÉ DENIS DIDEROT (PARIS 7) PLACE JUSSIEU, 75251 PARIS, FRANCE; LABORATOIRE DE PROBABILITÉS, CNRS, 4, PLACE JUSSIEU, 75252 PARIS, FRANCE

Translated by THE AUTHORS

Amer. Math. Soc. Transl.
(2) Vol. **200**, 2000

On Cusp-Bifurcations of Periodic Orbits in Systems with a Saddle-Focus Homoclinic Curve

S. A. Alekseeva and L. P. Shilnikov

To the memory of Professor E. A. Leontovich-Andronova

ABSTRACT. The space B^1 of dynamical systems with saddle-focus homoclinic curve is considered under conditions leading to a complicated behavior of trajectories. It is proved that the systems with triple circuit periodic orbits with one unit multiplier and zero first Lyapunov value are dense in B^1.

1. Introduction

Systems with saddle-focus homoclinic curve and positive saddle value form one of the basic objects of investigation from both the theoretical and the practical points of view. The following essential reasons explain this interest: on the one hand, these systems have complicated dynamics due to existence of a countable set of saddle periodic orbits, a continuum of Poisson-stable orbits, etc., in any neighborhood of the homoclinic curve [**1, 2**] and on the other hand, a spiral chaos observed in many concrete models and experimental investigations is associated with these systems. However, the structure under consideration is nonrough. This fact leads to essentially new phenomena connected with the existence of hyperbolic trajectories of different topological types as well as nonrough trajectories in a small neighborhood of the homoclinic orbit.

Namely, it was established in [**3**] that if the unstable manifold of a saddle-focus is one-dimensional, then systems with two-circuit saddle-node periodic orbits are dense in the bifurcation set B^1 of vector fields possessing a homoclinic orbit to a saddle-focus with positive saddle value (recall that a periodic orbit lying entirely in a small neighborhood of the homoclinic orbit is called a k-circuit with respect to the homoclinic orbit if it goes around the homoclinic orbit k times). Moreover, in B^1, systems with nonrough Poincaré homoclinic orbit are dense. But it was shown in [**4**] that in the vicinity of a system with nonrough Poincaré homoclinic orbit of the third type [**5, 6**] there always exists a system with a nonrough Poincaré homoclinic orbit along which the stable and unstable manifolds have quadratic tangency.

2000 *Mathematics Subject Classification.* Primary 37C15.

This research was supported in part by the Russian Foundation for Basic Research (grant 99-01-00231), INTAS-grant-97-804, and the program "Universities of Russia" (project No. 1905).

©2000 American Mathematical Society

For systems with a structurally unstable Poincaré homoclinic orbit of the first and second types [**5, 6**], there exists a complete description for all trajectories lying entirely in some small neighborhood of the homoclinic orbit (in the first case it consists of the homoclinic orbit and the periodic orbit, and in the second case it is equivalent to a quotient system obtained from the topological Bernoulli shift with three symbols); as to systems with homoclinic orbit of the third type, the situation is more complicated. All systems of this type possess a nontrivial hyperbolic set, but this set does not exhaust all trajectories in the neighborhood of homoclinic orbits. Moreover, both systems with a countable number of arbitrarily degenerate periodic orbits and systems with a countable number of arbitrarily degenerate homoclinic orbits are dense in the bifurcational space of systems of the third type [**4**]. As it follows from the proof, these occurrences take place only for periodic orbits of high circuitness.

However, periodic orbits of low circuitness are most interesting from the practical point of view. In this connection the following questions arise: what degeneracies may occur for the low-circuit periodic orbits and which parameters control the corresponding bifurcations?

One-circuit periodic orbits of any system from B^1 are rough [**1**]). As for the two-circuit periodic orbits, they may have only the simplest degeneracy, which means that they may possess only one multiplier equal to 1 or -1, but the first Lyapunov value cannot vanish. In the present paper we show that systems with three-circuit periodic orbits with one unit multiplier and zero first Lyapunov value are dense in B^1. This fact is a generalization of the result obtained for the three-dimensional systems in the earlier paper of the authors [**7**].

This paper consists of five sections. The first section is the introduction. In the second section we formulate the theorems and indicate the parameters which control the bifurcations of periodic orbits. In the third section we choose two local cross-sections S_0 and S_1 and construct the Poincaré map T_1T_0, where $T_0\colon S_0 \to S_1$ is given by trajectories in the neighborhood of O and $T_1\colon S_1 \to S_0$ is given by the trajectories in the neighborhood of a global piece of Γ. In the forth section we indicate the conditions of existence of the three-circuit nonrough periodic orbits and reduce the considered problem to a system of transcendental equations. In the fifth section we prove the theorems. It is worth mentioning that conceptually, the proof is close to that in [**8**] of a similar result for systems with homoclinic tangency. The solution of a system of two linear equations for three positive integers is the basis for such considerations.

2. The main results

We consider a C^{r+2}-vector field ($r \geqslant 3$) on a $(n+1)$-dimensional C^{r+2}-manifold ($n \geqslant 2$). Suppose that the following conditions hold:

A) *The system has an equilibrium point O of the saddle-focus type. Namely, characteristic exponents $\lambda_1, \dots, \lambda_{n+1}$ satisfy the following conditions*: $\lambda_{1,2} = -\rho_1 \pm i\omega_1$, $\omega_1 \neq 0$, $\rho_1 > 0$, $\operatorname{Re}\lambda_j < -\rho_1$, $j = 3, \dots, n$, $\lambda_{n+1} > 0$.

B) *The saddle value $\sigma = -\rho_1 + \lambda_{n+1}$ is positive, that is index $\rho = \rho_1/\lambda_{n+1} < 1$.*

C) *One of two trajectories, leaving O, comes back at $t \to +\infty$. Denote this trajectory as Γ. Then $O \cup \Gamma$ is called a homoclinic loop.*

D) *A certain number b, which characterizes the behavior of trajectories of the system in a neighborhood of the global piece of Γ, is nonzero.*

E) *As* $t \to +\infty$, *the trajectory* Γ *tends to* O *being tangent to the two-dimensional leading subspace.*

Systems with these properties form a codimension one submanifold B^1 in the space of vector fields. The main results of this paper are given by two following theorems.

THEOREM 2.1. *The systems possessing triple-circuit periodic orbits with one unit multiplier and zero first Lyapunov value are dense in* B^1.

In fact this statement is also valid for finite-parameter families. Investigation of such families is important from the point of view of nonlinear dynamics, which deals with models. In our case these parameters are the quantities ρ, ω and some quantity τ (that can be determined by means of the variational equation along Γ and calculated in terms of coefficients of the Poincaré map). The point is that these three essential parameters are contained in the main part of the following equations:

$$j = \rho i - \tau - \tau_1(\rho, \omega) + \dots, \qquad k = \rho j - \tau - \tau_2(\rho, \omega) + \dots, \tag{2.1}$$

where

$$\tau_1(\rho, \omega) = \frac{\omega}{2\pi} \ln \frac{\omega}{\sqrt{\omega^2 + \rho^2}} + \frac{1-\rho}{2\pi} \arctan \frac{\rho}{\omega},$$

$$\tau_2(\rho, \omega) = \frac{\omega}{2\pi} \ln \frac{\omega}{\sqrt{\omega^2 + \rho^2}} + \frac{\rho}{2\pi} \arctan \frac{\rho}{\omega} + \frac{1}{4}.$$

To obtain the results this system has to be solved with respect to i, j, k ($i > j > k$). It is also clear that we may restrict ourselves by only two parameters. That is why we fix ω and consider the two-parametric family $X_{\rho,\tau} \subset B^1$. Define the strip $L = \{(\rho, \tau) : 0 < \rho < 1\}$.

The following theorem will be proved.

THEOREM 2.2. *The set of points* $(\rho^*, \tau^*) \in L$ *corresponding to the system* X_{ρ^*,τ^*} *having a triple-circuit periodic orbit with one multiplier* 1 *and the zero first Lyapunov number is dense in* L.

Note that Theorem 2.1 immediately follows from Theorem 2.2.

3. Construction of the Poincaré map

Usually, the bifurcations problem of periodic orbits lying entirely in a small neighborhood of Γ is reduced to the study of bifurcations of the Poincaré map of some cross-section. This map is constructed as a composition of two maps T_0 and T_1, where T_0 is a local map for the orbits lying close to O, and T_1 is a global map induced by the orbits lying in a neighborhood of a global piece of Γ.

3.1. The construction of T_0. The suitable form for the system and its solution in a neighborhood of O. It is known that the systems of the family under consideration in some small neighborhood of O can be written in local coordinates in the following form [**3**]:

$$\begin{cases} \dot{x} = Ax + F_{11}(x, y, z)x + F_{12}(x, y, z)y, \\ \dot{y} = By + F_{21}(x, y, z)x + F_{22}(x, y, z)y, \\ \dot{z} = z. \end{cases} \tag{3.1}$$

Here

$$x = (x_1, x_2) \in \mathbb{R}^2, \quad y = (y_1, y_2, \dots, y_{n-2}) \in \mathbb{R}^{n-2}, \quad z \in \mathbb{R}^1,$$

$$A = \begin{pmatrix} -\gamma & -\omega \\ \omega & -\gamma \end{pmatrix}, \quad \begin{pmatrix} \rho \\ \omega \end{pmatrix} = \frac{1}{\lambda_{n+1}} \begin{pmatrix} \rho_1 \\ \omega_1 \end{pmatrix}$$

and the eigenvalues of the matrix B are $(\nu_1, \dots, \nu_{n-2}) = \lambda_{n+1}^{-1}(\lambda_3, \dots, \lambda_n)$. The class C^{r+1} matrix-valued functions F_{ij} take values in the matrices of the size 2×2 (for F_{11}), $2 \times (n-2)$ (for F_{12}), $(n-2) \times 2$ (for F_{21}), and $(n-2) \times (n-2)$ (F_{22}) and satisfy the following conditions:

$$F_{1j}(x, y, 0) = 0, \qquad \left. \frac{\partial^{k+l} F_{1j}(x, y, z)}{\partial x^k \partial y^l} \right|_{z=0} = 0,$$

$$F_{i1}(0, 0, z) = 0, \qquad \left. \frac{\partial^k F_{i1}(x, y, z)}{\partial z^k} \right|_{x=0, y=0} = 0, \qquad i, j = 1, 2.$$

In these coordinates the manifolds W^s and W^u are locally straightened: $W^s_{\text{loc}} = \{z = 0\}$, $W^u_{\text{loc}} = \{x = 0,\ y = 0\}$.

The orbit Γ intersects the cross-section $\{x_1 = 0\}$ to W^s_{loc} at an infinite number of points. Choose one of them with coordinates $x = (0, x_2^+)$, $y = (y_1^+, \dots, y_{n-2}^+)$, $z = 0$. We take a small $\delta > 0$ and define the following cross-sections to the flow: $S_0 = \{x, y, z \mid x_1 = 0, |x_2 - x_2^+| < \delta, \|y\| + |z| < \delta\}$ transverse to W^s_{loc} and $S_1 = \{x, y, z \mid z = d, 0 < d < \delta, \|x\| + \|y\| < \delta\}$ transverse to W^u_{loc}. Since S_0 is transverse to the flow, the Poincaré map $T\colon S_0 \to S_0$ is defined by the orbits of system; here $T = T_1 T_0$, where $T_0\colon S_0 \to S_1$ is defined by the orbits lying near O and $T_1\colon S_1 \to S_0$ is defined by the pieces of orbits passing along Γ.

To calculate T_0, we write down the solution of the system (3.1) (use the form of solution obtained from [**3**]):

$$\begin{cases} x(t) = x^0 e^{At} + x_\alpha(t, x^0, y^0, z^0, \rho), \\ y(t) = y_\alpha(t, x^0, y^0, z^0, \rho), \\ z(t) = z^0 e^t, \end{cases} \tag{3.2}$$

$$\|x_\alpha\| + \|y_\alpha\| \leqslant M e^{(\alpha - \epsilon_1)t},$$

$$\left\| \frac{\partial^{k+l+q}(x_\alpha, y_\alpha)}{\partial(x^0)^k \partial(y^0)^l \partial(z^0)^q} \right\| \leqslant M_1 e^{qt - (\alpha - \epsilon_1)t}, \qquad 0 \leqslant k + q + l \leqslant r + 1. \tag{3.3}$$

Here $\alpha = \min(2\rho, -\nu_i)$, $i = 1, \dots, n-2$, M, M_1, and ϵ_1 are positive values which can be chosen arbitrarily small by the choice of sufficiently small neighborhood of O.

The following estimate can be obtained similarly to the estimate (3.3) in [**3**]:

$$\left\| \frac{\partial^{s+1}(x_\alpha, y_\alpha)}{\partial(x^0)^k \partial(y^0)^l \partial(z^0)^q \partial\rho} \right\| \le M_2 t e^{qt - (\alpha - \epsilon_1)t}, \qquad s = k + q + l,\ s \leqslant r. \tag{3.4}$$

Note that

$$e^{At} = e^{-\rho t} \begin{pmatrix} \cos \omega t & -\sin \omega t \\ \sin \omega t & \cos \omega t \end{pmatrix}.$$

We use the following notations. Let $x^0 = (x_2^0, \dots, x_n^0, z^0)$ be a point on S_0, and $x^1 = (x_1^1, \dots, x_n^1)$ on S_1 (with the implicit notation $y_i = x_{i+2}$, $i = 1, \dots, n-2$).

From the last equation we obtain the following expression for the flight time from S_0 to S_1: $t_0 = -\ln z_0/d$. Using this expression we write the map T_0 as follows:

$$
(3.5)\qquad \begin{cases} x_1^1 = x_2^0\left(\dfrac{z_0}{d}\right)^\rho \sin\omega\ln\dfrac{z_0}{d} + \varphi_1(x_2^0,\dots,x_n^0,z^0,\rho),\\ x_2^1 = x_2^0\left(\dfrac{z_0}{d}\right)^\rho \cos\omega\ln\dfrac{z_0}{d} + \varphi_2(x_2^0,\dots,x_n^0,z^0,\rho),\\ x_i^1 = \varphi_i(x_2^0,\dots,x_n^0,z), \qquad i=3,\dots,n, \end{cases}
$$

where φ_i satisfy the following conditions:

$$
(3.6)\qquad \left|\frac{\partial^{k_2+\dots+k_n+q}\varphi_j(x_2^0,\dots,z^0,\rho)}{\partial(x_2^0)^{k_1}\dots\partial(x_n^0)^{k_n}\partial(z^0)^q}\right| \leqslant M_3 z_0^{-q+(\alpha-\epsilon_1)},
$$

$$
(3.7)\qquad \left|\frac{\partial^{k_2+\dots+k_n+q+1}\varphi_j(x_2^0,\dots,z^0,\rho)}{\partial(x_2^0)^{k_1}\dots\partial(x_n^0)^{k_n}\partial(z^0)^q\partial\rho}\right| \leqslant M_4 \ln(z_0) z_0^{-q+(\alpha-\epsilon_1)},
$$

$$
j=1,\dots,n, \qquad k_2+\dots+k_n+q\leqslant r.
$$

3.2. The calculation of $T_1: S_1 \to S_0$. The map T_1 can be written in the form

$$
(3.8)\qquad \begin{pmatrix} \bar x_2^0\\ \vdots\\ \bar x_n^0\\ \bar z^0 \end{pmatrix} = \begin{pmatrix} x_2^+\\ \vdots\\ x_n^+\\ 0 \end{pmatrix} + A_1\begin{pmatrix} x_1^1\\ \vdots\\ x_{n-1}^1\\ x_n^1 \end{pmatrix} + \dots,
$$

where A_1 is nondegenerate matrix. The dots mean terms of higher order with respect to the written terms. These terms smoothly depend on x_2^+ and d.

By combining (3.5) and (3.8) we obtain (omit the index "0" in the notation for coordinates on S_0):

$$
(3.9)\qquad \begin{cases} \bar x_k = x_k^+ + b_k x_2\left(\dfrac{z}{d}\right)^\rho \cos\left(\omega\ln\dfrac{z}{d}+\theta_k\right) + \psi_k(x_2,\dots,x_n,z),\\ \bar z = b x_2\left(\dfrac{z}{d}\right)^\rho \cos\left(\omega\ln\dfrac{z}{d}+\theta_{n+1}\right) + \psi_{n+1}(x_2,\dots,x_n,z), \qquad k=2,\dots,n, \end{cases}
$$

where functions ψ_i $(i=2,\dots,n+1)$ satisfy the estimates similar to (3.6), (3.7) (here and in the future we omit the argument ρ of functions),

$$
b_k = \sqrt{(A_1)^2_{k-1,1} + (A_1)^2_{k-1,2}}, \qquad b = \sqrt{(A_1)^2_{n,1} + (A_1)^2_{n,2}}.
$$

We require that $b \neq 0$ (condition D of Section 2). Rescaling the variables by the formulas

$$
z \to dz\, e^{-\theta_{n+1}/\omega}, \quad b_k \to b_k e^{\rho\theta_{n+1}/\omega}, \quad b \to db\, e^{-(1-\rho)\theta_{n+1}/\omega}.
$$

we can write the Poincaré map in the form

$$
(3.10)\qquad \begin{cases} \bar x_k = x_k^+ + b_k x_2 z^\rho \cos(\omega\ln z + \tilde\theta_k) + \phi_k(x_2,\dots,x_n,z),\\ \bar z = b x_2 z^\rho \cos(\omega \ln z) + \phi_{n+1}(x_2,\dots,x_n,z), \qquad k=2,\dots,n, \end{cases}
$$

where $\tilde\theta_k = \theta_k - \theta_{n+1}$.

REMARK. The functions ϕ_i depend on x_2^+, d, b, b_k, θ_k smoothly because the equations (3.10) have been obtained from (3.9) by a smooth changing of variables. Moreover, the derivatives of ϕ_i with respect to x_2, d, b, b_k, θ_k satisfy the estimates similar to (3.6).

4. Conditions of existence nonrough triple circuit periodic orbit

The range of definition of the return map T can be represented by a infinite number of regions σ_i:

$$\sigma_i = \{(0, y, z) \in S_0,\ -2\pi(i + 1/2) < \omega \ln z \leqslant -2\pi(i - 1/2)\}.$$

Instead of coordinate z on σ_i we consider a pair (i, ξ_i) [**1, 2**]: i and ξ_i connected with z as follows:

$$\omega \ln z = -2\pi i - \xi_i, \qquad -\pi \leqslant \xi_i < \pi.$$

Let us denote a point belonging to σ_i by (x, ξ_i).

A triple circuit periodic orbit of our system corresponds to a periodic point of the map T of a period of three. Suppose that a triple circuit orbit pierces S_0 successively in regions σ_i, σ_j, σ_k. Thus (y, ξ_i), $(\bar{y}, \xi_j)$, $(\bar{\bar{y}}, \xi_k)$ are consequent points of intersection of the periodic orbit and S_0. We are interested in following fixed point M of map $T_iT_jT_k$ (where $T_j = T|_{\sigma_j}$, and i, j, k are distinct one from another).

We have:

$$(4.1)\qquad \begin{cases} \bar{x}_m = x_m^+ + b_m x_2 e^{-\rho(2\pi i+\xi_i)/\omega} \cos(\xi_i - \tilde{\theta}_m) + \tilde{\psi}_i^m(x_2, \dots, x_n, \xi_i), \\ e^{-(2\pi j+\xi_j)/\omega} = b x_2 e^{-\rho(2\pi i+\xi_i)/\omega} \cos \xi_i + \tilde{\psi}_i(x_2, \dots, x_n, \xi_i), \\ \bar{\bar{x}}_m = x_m^+ + b_m \bar{x}_2 e^{-\rho(2\pi j+\xi_j)/\omega} \cos(\xi_j - \tilde{\theta}_m) + \tilde{\psi}_j^m(\bar{x}_2, \dots, \bar{x}_n, \xi_j), \\ e^{-(2\pi k+\xi_k)/\omega} = b \bar{x}_2 e^{-\rho(2\pi j+\xi_j)/\omega} \cos \xi_j + \tilde{\psi}_j(\bar{x}_2, \dots, \bar{x}_n, \xi_j), \\ x_m = x_m^+ + b_m \bar{\bar{x}}_2 e^{-\rho(2\pi k+\xi_k)/\omega} \cos(\xi_k - \tilde{\theta}_m) + \tilde{\psi}_k^m(\bar{\bar{x}}_2, \dots, \bar{\bar{x}}_n, \xi_k), \\ e^{-(2\pi i+\xi_i)/\omega} = b \bar{\bar{x}}_2 e^{-\rho(2\pi k+\xi_k)/\omega} \cos \xi_k + \tilde{\psi}_k(\bar{\bar{x}}_2, \dots, \bar{\bar{x}}_n, \xi_k), \end{cases}$$

where $m = 2, \dots, n$, and i, j, k are sufficiently larger.

By the implicit function theorem, the system (4.1) may be resolved with respect to: $x_2, \dots, x_n$, $\bar{x}_2, \dots, \bar{x}_n$, $\bar{\bar{x}}_2, \dots, \bar{\bar{x}}_n$.

$$\bar{x}_m = x_m^+ + b_m x_2 e^{-\rho(2\pi i+\xi_i)/\omega} \cos(\xi_i - \tilde{\theta}_m) + \widetilde{\Omega}_{ikj}^m(\xi_i, \xi_k, \xi_j),$$

$$\bar{\bar{x}}_m = x_m^+ + \Omega_{jik}^m(\xi_j, \xi_i, \xi_k), \quad x_m = x_m^+ + \Omega_{kji}^m(\xi_k, \xi_j, \xi_i), \qquad m = 2, \dots, n,$$

$$\lim_{i,j,k\to\infty} \|\widetilde{\Omega}_{ikj}^m\|_{C^{r+1}} e^{\rho 2\pi i/\omega} = 0, \qquad \lim_{i,j,k\to\infty} \|\Omega_{jik}^m\|_{C^{r+1}} = 0,$$

$$\lim_{i,j,k\to\infty} \left| \frac{\partial^s \Omega_{jik}^m(\xi_j, \xi_i, \xi_k)}{\partial \xi_i^t \partial \xi_j^q} \right| e^{\rho 2\pi i/\omega} = 0,$$

$$\lim_{i,j,k\to\infty} \left| \frac{\partial^s \Omega_{kji}^m(\xi_k, \xi_j, \xi_i)}{\partial \xi_i^t \partial \xi_j^q} \right| e^{\rho 2\pi i/\omega} = 0,$$

$$0 < t \leqslant r + 1, \quad s = q + t, \quad 0 < s \leqslant r + 1.$$

Substituting these expressions into the system (4.1) and denoting θ_2 as θ we obtain

$$(4.2)\qquad \begin{cases} e^{-(2\pi j+\xi_j)/\omega} = b x_2^+ e^{-\rho(2\pi i+\xi_i)/\omega} \cos \xi_i + h_{ikj}^1(\xi_i, \xi_k, \xi_j), \\ e^{-(2\pi k+\xi_k)/\omega} = b x_2^+ e^{-\rho(2\pi j+\xi_j)/\omega} \cos \xi_j (1 + b_2 e^{-\rho(2\pi i+\xi_i)/\omega} \cos(\xi_i - \theta)) \\ \qquad + h_{jik}^2(\xi_j, \xi_i, \xi_k), \\ e^{-(2\pi i+\xi_i)/\omega} = b x_2^+ e^{-\rho(2\pi k+\xi_k)/\omega} \cos \xi_k + h_{kji}^1(\xi_k, \xi_j, \xi_i), \end{cases}$$

where the functions h^l_{ikj} satisfy the following conditions:

$$\lim_{i,j,k\to\infty} \|h^l_{ikj}\|_{C^{r+1}} e^{\rho 2\pi i/\omega} = 0, \qquad l = 1, 2, \tag{4.3}$$

$$\lim_{i,j,k\to\infty} \left| \frac{\partial^s h^2_{jik}(\xi_j, \xi_i, \xi_k)}{\partial \xi_i^t \partial \xi_j^q} \right| e^{\rho 2\pi(i+j)/\omega} = 0, \tag{4.4}$$

$$\lim_{i,j,k\to\infty} \left| \frac{\partial^s h^1_{kji}(\xi_k, \xi_j, \xi_i)}{\partial \xi_i^t \partial \xi_j^q} \right| e^{\rho 2\pi(i+k)/\omega} = 0, \tag{4.5}$$

$$0 < t, \quad s = q + t, \quad 0 < s \leqslant r + 1.$$

Let us assume that $i > j$, $i > k$. Then from the last equation of the system (4.2) we have $\cos\xi_k = \dots$ (here and later the dots denote functions tending to zero together with their derivatives as $i, j, k \to \infty$). Therefore ξ_k will be close to $\pi/2$ or to $-\pi/2$. By applying the method of successive approximations and taking into account the estimates (4.3), (4.5), we obtain

$$\xi_k = \mp\frac{\pi}{2} \pm \frac{1}{bx_2^+} e^{-(2\pi(i-\rho k)\mp\rho\pi/2+\xi_i)/\omega}(1 + \dots) + \chi^\pm(\xi_i, \xi_j, i, j, k),$$

$$\|\chi^\pm(\xi_i, \xi_j, i, j, k)\|_{C^{r+1}} \to 0, \qquad i, j, k \to \infty,$$

$$\lim_{i,j,k\to\infty} \left| \frac{\partial^s \chi^\pm(\xi_i, \xi_j, i, j, k)}{\partial \xi_i^t \partial \xi_j^q} \right| e^{-\rho 2\pi i/\omega} = 0, \qquad 0 < t,\ s = q + t,\ 0 < s \leqslant r + 1.$$

Substituting this expression into the first two equations (4.2) we obtain

(4.6)

$$\begin{cases} e^{-(2\pi j+\xi_j)/\omega} = bx_2^+ e^{-\rho(2\pi i+\xi_i)/\omega} \cos\xi_i + h_1^\pm(\xi_i, \xi_j, i, j, k), \\ e^{-(2\pi k\pm\pi/2)/\omega}\left(1 \pm \dfrac{1}{\omega bx_2^+} e^{-(2\pi(i-\rho k)\mp\rho\pi/2+\xi_i)/\omega}(1 + \dots) + \bar\chi^\pm(\xi_i, \xi_j, i, j, k)\right) \\ \quad = bx_2^+ e^{-\rho(2\pi j+\xi_j)/\omega} \cos\xi_j (1 + b_2 e^{-\rho(2\pi i+\xi_i)/\omega} \cos(\xi_i - \theta)) + h_2^\pm(\xi_i, \xi_j, i, j, k). \end{cases}$$

Here the index '$\pm$' reflects the fact that in reality we have two systems corresponding to two branches of the solution. Since both systems are solved in the same way, we restrict ourselves by the study of one of them (choosing upper sign in the equation and omit the index '$\pm$').

We shall multiply the first and the second equations of (4.6) on $e^{(2\pi i+\xi_i)/\omega}$ and on $e^{\rho(2\pi j+\xi_j)/\omega}$ respectively. In this case the system takes the form

(4.7)

$$\begin{cases} e^{-2\pi(j-\rho i+\xi_j-\rho\xi_i)/\omega} = bx_2^+ \cos\xi_i + e^{\rho(2\pi i+\xi_i)/\omega} h_1(\xi_i, \xi_j, i, j, k), \\ e^{-(2\pi(k-\rho j)+\pi/2+\xi_j)/\omega}\left(1 + \dfrac{e^{-(2\pi(i-\rho k)-\rho\pi/2+\xi_i)/\omega}}{\omega bx_2^+}(1 + \dots) + \bar\chi(\xi_i, \xi_j, i, j, k)\right) \\ \quad = bx_2^+ \cos\xi_j (1 + b_2 e^{-\rho(2\pi i+\xi_i)/\omega} \cos(\xi_i - \theta)) + e^{\rho(2\pi j+\xi_j)/\omega} h_2(\xi_i, \xi_j, i, j, k). \end{cases}$$

We consider this system in the following asymptotics:

$$j - \rho i = O(1), \quad k - \rho j = O(1) \quad \text{as } i, j, k \to \infty. \tag{4.8}$$

Then the left-hand sides of equations of the system (4.7) are different from zero. This fact and the estimates (4.3) and (4.4) allow us to conclude that $\cos\xi_i$ and $\cos\xi_j$

are different from zero also. Therefore we can take the logarithm of the system and write it in the following form:

(4.9)

$$\begin{cases} \dfrac{2\pi(j-\rho i)+\xi_j-\rho\xi_i}{\omega}+\ln(bx_2^+\cos\xi_i)+\Phi_1(\xi_i,\xi_j,i,j,k)=0, \\ \dfrac{2\pi(k-\rho j)+\pi/2-\rho\xi_j}{\omega}-\dfrac{e^{-(2\pi(i-\rho k)-\rho\pi/2+\xi_i)/\omega}}{\omega bx_2^+}(1+\dots)+\ln(bx_2^+\cos\xi_j) \\ \quad +b_2e^{-\rho(2\pi i+\xi_i)/\omega}\cos(\xi_i-\theta)(1+\dots)+\Phi_2(\xi_i,\xi_j,i,j,k)=0, \end{cases}$$

$$\lim_{i,j,k\to\infty}\|\Phi_l(\xi_i,\xi_j,i,j,k)\|_{C^{r+1}}=0, \qquad l=1,2,$$

$$\lim_{i,j,k\to\infty}\left|\frac{\partial^s\Phi_2(\xi_i,\xi_j,i,j,k)}{\partial\xi_i^t\partial\xi_j^q}\right|e^{-\rho 2\pi i/\omega}=0, \qquad 0<t,\ s=q+t,\ 0<s\leqslant r+1.$$

Thus the problem about triple circuit periodic orbits is reduced to the question about triple solutions of the system (4.9).

Let us consider the first equation. It is trivially solved with respect to ξ_j. Thus we obtain $\xi_j=\xi_1(\xi_i)$. The graph of the function $\xi_1(\xi_i)$ on the plane (ξ_i,ξ_j) is a smooth curve and has just one extremum point: $\tilde{\xi}_i=\xi_0+\dots$, $\tilde{\xi}_j=\xi_1(\tilde{\xi}_i)$ (here ξ_0 is the point of maximum of the function $\ln(e^{-\rho\xi/\omega}\cos\xi)$ on the interval $[-\pi/2,\pi/2]$, $\xi_0=-\arctan(\rho/\omega)$).

As for the second equation, its analysis depends on the ratio of values $i-\rho k$ and ρi as $i,j,k\to\infty$. By (4.8) we have $i-\rho k=i-\rho^3 i+O(1)$. Let $\tilde{\rho}$ be the unique solution of the equation $1-\rho-\rho^3=0$ in the interval $[0,1]$. For $\rho>\tilde{\rho}$ we have $\rho i\gg(i-\rho k)$ for $\rho<\tilde{\rho}$: $(i-\rho k)\gg\rho i$. The case $\rho=\tilde{\rho}$ can be easily reduced to one of the cases considered above by a small variation of ρ. Moreover, we may consider only the case $\rho>\tilde{\rho}$, because the other case may be analyzed in an analogous way. Then $\rho i\gg(i-\rho k)$. In this case it is easy to see that the term

$$\frac{1}{\omega bx_2^+}e^{-(2\pi(i-\rho k)+\xi_i-\rho\pi/2)/\omega}$$

is the largest among the terms containing ξ_i and the equation is solved with respect to ξ_i. Thus we obtain $\xi_i=\xi_2(\xi_j)$. The graph of the function $\xi_2(\xi_i)$ on the plane (ξ_i,ξ_j) is a smooth curve and has a single extremum point: $\tilde{\tilde{\xi}}_j=\xi_0+\dots$, $\tilde{\tilde{\xi}}_i=\xi_2(\tilde{\tilde{\xi}}_j)$.

Now we demonstrate that in a neighborhood of the point (ξ_0,ξ_0) the curves representing $\xi_1(\xi_i)$ and $\xi_2(\xi_j)$ on the plane (ξ_i,ξ_j) can be approximated by parabolas. In the general case, these parabola-like curves have four intersection points. However, under certain conditions the case of cubic tangency can be realized.

Let us consider the system (4.9) for ξ_i and ξ_j close to ξ_0. Let us make the change of variables $\xi_i=\xi+\xi_0$, $\xi_j=\eta+\xi_0$. In this case we have the new form of the system

$$\begin{cases} \dfrac{2\pi(j-\rho i)+(1-\rho)\xi_0}{\omega}+\ln(bx_2^+\cos\xi_0)+\dfrac{1}{\omega}\eta \\ \quad -\dfrac{1+(\rho/\omega)^2}{2}\xi^2+O(\xi^3)+\widetilde{\Phi}_1(\xi,\eta,i,j,k)=0, \\ \dfrac{2\pi(k-\rho j)+\pi/2-\rho\xi_0}{\omega}+\ln(bx_2^+\cos\xi_0)+\varepsilon\xi \\ \quad -\dfrac{1+(\rho/\omega)^2}{2}\eta^2+O(\eta^3)+\widetilde{\Phi}_2(\xi,\eta,i,j,k)=0, \end{cases}$$

where

$$\varepsilon = \frac{1}{\omega^2 b x_2^+} e^{-(2\pi(i-\rho k)-\rho\pi/2+\xi_0)/\omega}$$

and the functions $\widetilde{\Phi}_l(\xi,\eta,i,j,k)$ satisfy the following condition:

$$\lim_{i,j,k\to\infty} \|\widetilde{\Phi}_l(\xi,\eta,i,j,k)\|_{C^{r+1}} = 0, \qquad l=1,2,$$

$$\lim_{i,j,k\to\infty} \left|\frac{\partial^s \widetilde{\Phi}_2(\xi,\eta,i,j,k)}{\partial\xi^t\partial\eta^q}\right| e^{\rho 2\pi(i-\rho k)/\omega} = 0, \qquad 0<t,\ s=q+t,\ 0<s\leqslant r+1.$$

Let us assume that the coefficient at ξ vanishes in the first equation and the coefficient at η vanishes in the second equation in the linear parts of the left-hand sides of the system. This can be achieved by an appropriate small shift of the origin.

In order to get the normalized coordinates, we change variables $\xi=\alpha u$, $\eta=\beta v$ (where α and β are small parameters to be defined). The system takes the form

$$\begin{cases} \dfrac{2\pi(j-\rho i)+(1-\rho)\xi_0}{\omega} + \ln(bx_2^+\cos\xi_0) + \dfrac{1}{\omega}\beta v \\ \qquad - \dfrac{1+(\rho/\omega)^2}{2}\alpha^2u^2 + O(\alpha^3) + \Psi_1(u,v,i,j,k) = 0, \\ \dfrac{2\pi(k-\rho j)+\pi/2-\rho\xi_0}{\omega} + \ln(bx_2^+\cos\xi_0) + \varepsilon\alpha u \\ \qquad - \dfrac{1+(\rho/\omega)^2}{2}\beta^2v^2 + O(\beta^3) + \Psi_2(u,v,i,j,k) = 0, \end{cases}$$

(4.10)
$$\begin{gathered} \Psi_1(u,v,i,j,k) = \Psi_1(0,0,i,j,k) + (\alpha^2+\beta)p_1(u,v,i,j,k), \\ \Psi_2(u,v,i,j,k) = \Psi_2(0,0,i,j,k) + (\alpha e^{-2\pi\rho(i-\rho k)/\omega}+\beta^2)p_2(u,v,i,j,k), \\ \lim_{i,j,k\to\infty}\Phi_l(0,0,i,j,k)=0, \qquad \lim_{i,j,k\to\infty}\|p_l(u,v,i,j,k)\|_{C^{r+1}}=0, \\ p_l(0,0,i,j,k)=0, \qquad l=1,2. \end{gathered}$$

Let us choose α and β so that the following relations take place:

$$\frac{1}{\omega}\beta = \frac{1+(\rho/\omega)^2}{2}\alpha^2, \qquad \varepsilon\alpha = \frac{1+(\rho/\omega)^2}{2}\beta^2.$$

It is easy to find α and β from the equations:

$$\alpha = 2\omega^{-2/3}(1+(\rho/\omega)^2)^{-1}\varepsilon^{1/3}, \qquad \beta = 2\omega^{-1/3}(1-(\rho/\omega)^2)^{-1}\varepsilon^{2/3},$$

and they are indeed small (since $i>k$).

Let us multiply the first and the second equations of the system by $2(1+(\rho/\omega)^2)^{-1}\alpha^{-2}$ and by $2(1+(\rho/\omega)^2)^{-1}\beta^{-2}$ respectively. Using the estimates (4.10) we obtain

(4.11)
$$\begin{cases} v = u^2 + A + \delta_1(u,v,i,j,k), \\ u = v^2 + B + \delta_2(u,v,i,j,k), \end{cases}$$

$$A = \left(\frac{2\pi(j-\rho i)+(1-\rho)\xi_0}{\omega} + \ln(bx_2^+\cos\xi_0) + \dots\right)2\omega^{4/3}\left(1+\frac{\rho^2}{\omega^2}\right)\varepsilon^{-2/3},$$

$$B = \left(\frac{2\pi(k-\rho j)+\pi/2-\rho\xi_0}{\omega} + \ln(bx_2^+\cos\xi_0) + \dots\right)2\omega^{2/3}\left(1+\frac{\rho^2}{\omega^2}\right)\varepsilon^{-4/3},$$

$$\lim_{i,j,k\to\infty}\|\delta_l(u,v,i,j,k)\|_{C^{r+1}}=0, \quad \delta_l(0,0,i,j,k)=0, \qquad l=1,2.$$

Thus we have reduced the problem to the study of intersections of two parabola-like curves on the plane (u, v). It is easy to see that the parabolas have a cubic tangency when $A = -3/4 + \dots$, $B = -3/4 + \dots$. Therefore the conditions of existence of the triple solution can be written as follows:

$$\left(\frac{2\pi(j-\rho i)+(1-\rho)\xi_0}{\omega}+\ln(bx_2^+\cos\xi_0)+\dots\right)2\omega^{4/3}\left(1+\frac{\rho^2}{\omega^2}\right)\varepsilon^{-2/3}=-\frac{3}{4}+\dots,$$

$$\left(\frac{2\pi(k-\rho j)+\pi/2-\rho\xi_0}{\omega}+\ln(bx_2^+\cos\xi_0)+\dots\right)2\omega^{2/3}\left(1+\frac{\rho^2}{\omega^2}\right)\varepsilon^{-4/3}=-\frac{3}{4}+\dots,$$

which is equivalent to

$$\begin{aligned}\frac{2\pi(j-\rho i)+(1-\rho)\xi_0}{\omega}+\ln(bx_2^+\cos\xi_0)+\dots&=0,\\ \frac{1}{\omega}\,\frac{2\pi(k-\rho j)+\pi/2-\rho\xi_0}{\omega}+\ln(bx_2^+\cos\xi_0)+\dots&=0.\end{aligned}\tag{4.12}$$

We recall that $\xi_0 = -\arctan(\rho/\omega)$.

5. Proof of the theorems

Let us rewrite (4.12) in the form

$$\begin{aligned}j&=\rho i-\tau-\frac{\omega}{2\pi}\ln\frac{\omega}{\sqrt{\omega^2+\rho^2}}-\frac{1-\rho}{2\pi}\arctan\frac{\rho}{\omega}+f_1(\rho,\tau,\omega,i,j,k),\\ k&=\rho j-\tau-\frac{\omega}{2\pi}\ln\frac{\omega}{\sqrt{\omega^2+\rho^2}}-\frac{\rho}{2\pi}\arctan\frac{\rho}{\omega}-\frac{1}{4}+f_2(\rho,\tau,\omega,i,j,k),\end{aligned}\tag{5.1}$$

where $\tau = (\omega/2\pi)\ln(bx_2^+)$. Due to the estimates (3.4) and the remark after (3.10), the functions $f_l(\rho,\tau,\omega,i,j,k)$ $(l = 1, 2)$ smoothly depend on τ and ρ, and $\|f_l(\rho,\tau,\omega,i,j,k)\|_{C^1} \to 0$ as $i, j, k \to \infty$, $l = 1, 2$.

In these conditions, three independent parameters ρ, τ, ω appear. To prove the theorems, it suffices to use only two parameters. Therefore we fix ω and consider the two-parametric family in B^1. Let us define the strip $L = \{(\rho,\tau) : 0 < \rho < 1\}$ and the set $L^* \subset L$ such that for any $(\rho^*, \tau^*) \in L^*$ the system (5.1) has an integer solution (i, j, k) (where $i > j > k$).

We shall prove the following auxiliary statement.

LEMMA 5.1. *The set L^* is dense in L.*

PROOF. Let us choose arbitrary ρ and τ. Choose any positive ϵ and take $\delta(\epsilon) < \epsilon(1-\rho)/4$.

There always exist rational primitive fractions p/q, r/q, m/q, l/q such that

$$\left|\rho-\frac{p}{q}\right|+\left|\tau-\frac{r}{q}\right|+\left|\tau_1(\rho)-\frac{m}{q}\right|+\left|\tau_2(\rho)-\frac{l}{q}\right|<\delta,\tag{5.2}$$

$$\tau_1(\rho)=\frac{\omega}{2\pi}\ln\frac{\omega}{\sqrt{\omega^2+\rho^2}}+\frac{1-\rho}{2\pi}\arctan\frac{\rho}{\omega},$$

$$\tau_2(\rho)=\frac{\omega}{2\pi}\ln\frac{\omega}{\sqrt{\omega^2+\rho^2}}+\frac{\rho}{2\pi}\arctan\frac{\rho}{\omega}+\frac{1}{4}.$$

We carry out the proof under the assumption that $\tau > 0$ and $\tau_i > 0$. The other cases can be analyzed similarly.

Let us consider the linear system for unknown integers i, j, k:

$$\begin{cases} j = \dfrac{p}{q} i - \dfrac{r+m}{q}, \\ k = \dfrac{p}{q} j - \dfrac{r+l}{q}. \end{cases}$$

It is equivalent to the system

$$\begin{cases} pi - qj = r + m, \\ pj - qk = r + l. \end{cases} \tag{5.3}$$

Let us consider the first and the second equations separately. Since p and q are relatively prime, there exist integers i_0, j_0 that solve the first equation. Moreover, any set of pairs (i_n, j_n): $i_n = i_0 + nq$, $j_n = j_0 + np$ is a solution also. In an analogous way we obtain the solutions of the second equation in the form $\tilde{j}_n = j_0 + nq$, $k_n = k_0 + np$. Since p and q are relatively prime, there exist integers m', l' such that $j_{m'} = \tilde{j}_{l'}$. Then the triple $(i_{m'}, j_{m'}, k_{l'})$ is a solution of (5.2). Consider the new sequences $i_n = i_{m'} + nq^2$, $j_n = j_{m'} + npq$, $k_n = k_{m'} + np^2$. It is obvious that any triple (i_n, j_n, k_n) is a solution of (5.2) as well.

Let us consider now the nonlinear system

$$\begin{cases} j_n = \rho i_n - \tau - \tau_1(\rho) + f_1(\rho, \tau, i_n, j_n, k_n), \\ k_n = \rho j_n - \tau - \tau_2(\rho) + f_2(\rho, \tau, i_n, j_n, k_n). \end{cases} \tag{5.4}$$

Assume that $\|f_l\|_{C^1} < \delta$ in the ϵ-neighborhood under consideration. This can be achieved by choosing n sufficiently large.

Since each triple (i_n, j_n, k_n) is a solution of the linear system (5.3), we can rewrite the system (5.4) in the following form:

$$\begin{cases} \left(\rho - \dfrac{p}{q}\right) i_n - \left(\tau - \dfrac{\rho}{q}\right) - \left(\tau_1(\rho) - \dfrac{m}{q}\right) + f_1(\rho, \tau, i_n, j_n, k_n) = 0, \\ \left(\rho - \dfrac{p}{q}\right) j_n - \left(\tau - \dfrac{\rho}{q}\right) - \left(\tau_2(\rho) - \dfrac{m}{q}\right) + f_2(\rho, \tau, i_n, j_n, k_n) = 0, \end{cases}$$

or, which is the same,

$$\begin{cases} \rho = \dfrac{p}{q} + \dfrac{\tau_1(\rho) - mq^{-1} - f_1(\rho) + lq^{-1} + f_2}{i_n - j_n}, \\ \tau = \dfrac{r}{q} + \dfrac{(\tau_1(\rho) - mq^{-1} - f_1) j_n - (\tau_2(\rho) - lq^{-1} + f_2) i_n}{i_n - j_n}. \end{cases} \tag{5.5}$$

Finding ρ from the first equation, we get $\rho_n = pq^{-1} + \nu(\tau, i_n, j_n, k_n)$, where the function ν satisfies the following conditions:

$$\left| \frac{\partial^t \nu(\tau, i_n, j_n, k_n)}{\partial \tau^t} \right| \leqslant M \frac{\delta}{i_n - j_n}, \qquad t = 0, 1. \tag{5.6}$$

Substituting this expression into the second equation of (5.5) we obtain

$$\tau = \frac{r}{q} + \frac{(\tilde{\tau}_1(\tau, i_n, j_n, k_n) - mq^{-1}) j_n - (\tilde{\tau}_2(\tau, i_n, j_n, k_n) - lq^{-1}) i_n}{i_n - j_n} + \alpha(\tau, i_n, j_n, k_n),$$

$$\lim_{n \to \infty} \|\alpha(\tau, i_n, j_n, k_n)\|_{C^1} = 0, \qquad \lim_{n \to \infty} \frac{\partial \tilde{\tau}_l(\tau, i_n, j_n, k_n)}{\partial \tau} = 0, \qquad l = 1, 2.$$

By applying the method of successive approximations, we find τ_n. Now we show that our solutions (ρ_n, τ_n) lie in the ϵ-neighborhood under consideration. This is can be achieved by choosing a sufficiently large n. It is obviously for ρ_n, see (5.6). As for τ_n, we can write the following sequence of inequalities:

$$\left|\tau_n - \frac{r}{q}\right| \leqslant \frac{2\delta i_n}{i_n - j_n} + \beta_1(n) \leqslant \frac{2\delta q}{q-p} + \beta_2(n) \leqslant \frac{2\delta}{1-\rho} + \beta_3(n),$$
$$\lim_{n\to\infty} \beta_l(n) = 0, \qquad l = 1, 2, 3.$$

Since $\delta(\epsilon) < \epsilon(1-\rho)/4$, we have $|\tau_n - r/q| < \epsilon$ for sufficiently large n. Then for (ρ^*, τ^*) we can take (ρ_n, τ_n). The lemma is proved. □

Theorem 2.2 follows immediately from the lemma.

Acknowledgments. We would like to thank professor Nicolis for useful discussions about the cusp-bifurcations problem of periodic orbits in systems with a saddle-focus homoclinic curve and systems close to them.

References

[1] L. P. Shilnikov, *A case of the existence of a countable set of periodic motions*, Dokl. Akad. Nauk SSSR **160** (1965), 558–561; English transl., Soviet Math. Dokl. **6** (1965), 163–166.

[2] ———, *A contribution to the problem of the structure of an extended neighborhood of a rough state of saddle-focus type*, Mat. Sb. **81 (123)** (1970), 92–103; English transl., Math. USSR-Sb. **10** (1970), 91–102.

[3] I. M. Ovsyannikov and L. P. Shilnikov, *On systems with a saddle-focus homoclinic curve*, Mat. Sb. **130 (172)** (1986), no. 4, 552–570; English transl., Math. USSR-Sb. **58** (1987), 557–574.

[4] S. V. Gonchenko, D. V. Turaev, and L. P. Shilnikov, *On models with nonrough Poincaré homoclinic curves*, Physica D **62** (1993), 1–14.

[5] N. K. Gavrilov and L. P. Shilnikov, *On three-dimensional dynamical systems close to systems with a structurally unstable homoclinic curve*, I, Mat. Sb. **88 (130)** (1972), 475–492; II, **90 (132)** (1973), 139–156; English transl., Math. USSR-Sb. **17** (1972), 467–485; **19** (1973), 139–156.

[6] S. V. Gonchenko, D. V. Turaev, and L. P. Shilnikov, *Dynamical phenomena in multidimensional systems with a structurally unstable homoclinic Poincaré curve*, Dokl. Akad. Nauk **330** (1993), no. 2, 144–147; English transl., Russian Acad. Sci. Dokl. Math. **47** (1993), no. 3, 410–415.

[7] S. A. Alekseeva and L. P. Shilnikov, *On the bifurcations of periodic orbits in systems with a saddle-focus homoclinic curve*, Differentsial′nye Uravneniya **33** (1997), no. 4, 440–447; English transl. in Differential Equations **33** (1997).

[8] L. P. Shilnikov and O. V. Stenkin, *On the bifurcations of triple circuit periodic orbits in systems with a nonrough Poincaré homoclinic orbit*, Differentsial′nye Uravneniya **33** (1997), no. 3, 377–384; English transl. in Differential Equations **33** (1997).

Nizhny Novgorod State Pedagogical University, Minin Square, 7A, Nizhny Novgorod 603 005, Russia

Department of Differential Equations, Institute for Applied Mathematics and Cybernetics, 10 Ul′yanova st., Nizhny Novgorod 603005, Russia
E-mail address: shilnikov@focus.nnov.ru

Translated by S. A. ALEKSEEVA

Amer. Math. Soc. Transl.
(2) Vol. **200**, 2000

Collapse and Continuity of Geodesic Frameworks of Surface Foliations

S. Aranson, V. Medvedev, and E. Zhuzhoma

Dedicated to the memory of E. A. Andronova-Leontovich

ABSTRACT. The continuity of a complete topological invariant (the geodesic framework) of irrational foliations on a closed orientable hyperbolic surface is proved. At the same time, it is shown that the geodesic framework of irrational foliations has a variability which is similar to the variability of the irrational Poincaré rotation number. Some bifurcations of geodesic frameworks are considered and the "dynamical" properties of foliations supporting the special bifurcation called a collapse of the geodesic framework are studied.

Introduction

For a large class of flows (or dynamical systems with continuous time) on the 2-torus Poincaré [**24**] introduced the notion of rotation number, which is a topological invariant (up to a transformation by a square integer unimodular matrix). For a minimal flow on the torus, the rotation number is irrational and is a complete topological invariant. This invariant describes the asymptotic properties of trajectories of a torus flow and shows how trajectories "rotate" in the first homology group (or, which is the same, the first homotopy group) of the 2-torus.

It is well known that the Poincaré rotation number of a minimal or transitive torus flow depends continuously on the flow in the space of C^r-flows, $r \geqslant 1$ (see for example [**10**], [**13**, Proposition 4.2.1], [**16**, Theorem 3.3.2]). On the other hand, this rotation number possesses a variability in the following sense. Given a transitive torus flow, there is an arbitrarily close (in the C^1-topology) torus flow with a rotation number different from the rotation number of the original transitive flow.

There is a deep theory of the dependence of the Poincaré rotation number on perturbations of the torus flow or on the parameter of the family of torus flows ([**10, 13**], see the references of [**7**]). An extension of this theory to flows on orientable closed surfaces of genus $g \geqslant 2$ was carried out in [**8**].

2000 *Mathematics Subject Classification*. Primary 57R30; Secondary 37C15, 37C20.
The research was partially supported by INTAS (grant No. 97-1843) and RFBR (grant No. 99-01-00230).

©2000 American Mathematical Society

Surface foliations with singularities are natural generalizations of surface flows. Their investigation is interesting not only by itself, but is also necessary in the study of homeomorphisms with invariant foliations. Important classes of such homeomorphisms are Anosov and pseudo-Anosov homeomorphisms, A-diffeomorphisms etc. So it is natural to consider similar questions for surface foliations. The main difference between flows and foliations is the nonorientability of singularities as well as global nonorientability of the foliation. Therefore the investigation of foliations is more difficult than the study of flows and displays new phenomena.

A complete topological invariant of irrational foliations on an orientable closed surface of genus $g \geqslant 2$ is the geodesic framework [**4, 5**] (up to the action of the group of covering transformations). We recall that the geodesic framework is a special geodesic lamination consisting of geodesics with the same asymptotic directions as the leaves of the foliation (see also [**6, 17**]). This invariant shows how the leaves of the foliation wind in the fundamental group (first homotopy group) of the surface. We can consider the geodesic framework as a skeleton of the surface foliation which encodes complete information on the asymptotic behavior of all the leaves.

On the set of geodesic laminations we introduce the structure of a topological space (note that this topology is different from a similar one introduced in the case of flows, because there are no natural directions on the corresponding geodesics). So we can study the dependence of geodesic frameworks on bifurcations of the foliation.

This paper is organized as follows. In Section 1 we introduce basic definitions and notations. In Section 2 we prove the continuity of the geodesic framework of irrational foliations on the closed orientable hyperbolic surface. The main result of this section is the following theorem.

THEOREM 2.1. *Let $F \in HTS_g$ be a highly transitive C^1-foliation on a closed orientable hyperbolic surface M of genus $g \geqslant 2$ and let U be a neighborhood of the geodesic framework $Gf(F)$ of F. Then there is a neighborhood $O^1(F)$ of F in the space $\mathcal{F}^1(M)$ such that any foliation $H \in O^1(F)$ has a nonempty geodesic framework $Gf(H)$ belonging to U.*

This result is similar to the fact that the Poincaré rotation number of a minimal torus flow depends continuously on perturbations of the flow in the space of C^1-flows. Theorem 2.1 shows that from the "physical" viewpoint the geodesic framework is a "good" characteristic of a highly transitive C^1-foliation.

To prove Theorem 2.1, we need the following lemma on the "stability" of global sections of a surface foliation with structurally stable singularities, which is of interest by itself.

LEMMA 2.1. *Let a C^1-foliation $F \in S_g$ have a global section C. Then there is a neighborhood $U(F)$ of F in the space $\mathcal{F}^1(M)$ of C^1-foliations such that every $H \in U(F)$ has the global section C as well.*

In Section 3 we prove that a geodesic framework of an irrational foliation has some variability. This means that given an irrational foliation, there is an arbitrarily close (in the C^1-topology) foliation with nonempty geodesic framework different from the geodesic framework of the original foliation. This result is similar to the variability of the irrational Poincaré rotation number.

In Section 4 we consider some bifurcations of geodesic frameworks that correspond to bifurcations of surface foliations. Special attention is paid to the bifurcation called collapse of the geodesic framework. This bifurcation reminds one of the "blue sky catastrophe" in the space of geodesic laminations (see Problem 37 in [**25**]). The main theorem of this section is the following.

THEOREM 4.4. *Let F_μ, $\mu \in [0;1]$, be a continuous family of C^∞-foliations on a closed orientable hyperbolic surface M of genus $g \geqslant 2$. Suppose that*

1) *for all $\mu \in [0;1)$ the geodesic framework $Gf(F_\mu) \neq \varnothing$ is rational;*
2) *lengths of closed geodesics of $Gf(F_\mu)$ tend to infinity uniformly as $\mu \to 1$;*
3) *$Gf(F_1) = \varnothing$;*
4) *every F_μ, $\mu \in [0;1)$, is widely disposed (in the sense of Definition 1.1 below).*

Then the foliation F_1 has infinitely many singularities.

We would like to thank D. Anosov, G. Hector, and I. Nikolaev for valuable discussions.

1. Preliminaries

1.1. Surface foliations. We recall some definitions and results in the theory of surface foliations.

Let S be a finite subset of a closed surface M. A *topological foliation F of M with singularity set* $\mathrm{Sing}(F) = S$ is a partition of M into subsets, each called a *leaf* of F, such that every point of $\mathrm{Sing}(F)$ is a singularity and the restriction of F to $M - \mathrm{Sing}(F)$ is a codimension-one topological foliation in the usual sense. This means that for each point of $M - \mathrm{Sing}(F)$ there exists a homeomorphism $\varphi\colon U \to \mathbb{R}^2$ of a neighborhood U of this point such that the image of each component of the intersection of any leaf with U is a horizontal interval in $\mathbb{R}^2$.

As for the behavior of leaves in neighborhoods of singular points, it must be specially described and different assumptions may be considered. Let us recall that the standard topological saddle with ν separatrices (for integer $\nu \geqslant 1$) is a foliation F_ν of the punctured complex plane $\mathbb{C} \setminus \{0\}$ defined as its partition into subsets of the form $\{z : \operatorname{Im} z^{\nu/2} = \mathrm{const}\}$. Then the point $x \in \mathrm{Sing}(F) = S$ is called a *topological saddle with ν separatrices* if it is locally topologically equivalent to F_ν in some punctured neighborhood U of x.

The saddle with one, two, and three separatrices is called a *thorn*, *fake saddle*, and *tripod* respectively.

DEFINITION 1.1. A foliation F is called *widely disposed* if no transversal segment together with an arc of some leaf of F bound a simply connected domain.

If a foliation has only singularities of saddle type, then the foliation is widely disposed if and only if it has no thorns.

Let l be a nonsingular nonclosed leaf of F. Any point $x \in l$ divides l into two semileaves l^+, l^-. A semileaf $l^{(\cdot)}$ is called *nontrivially recurrent* if its intrinsic topology does not coincide with the topology of $l^{(\cdot)}$ considered as a subset of M. The leaf l is said to be *nontrivially recurrent* if both its semileaves are nontrivially recurrent.

DEFINITION 1.2. The topological closure of a nontrivially recurrent semileaf is called a *quasiminimal set* of the foliation.

DEFINITION 1.3. By an *irrational foliation* we mean a foliation without fake saddles and separatrix connections such that the whole surface is a quasiminimal set of the foliation.

In other words, any irrational foliation is a highly transitive foliation (that is, every one-dimensional leaf is dense in M) without fake saddles.

1.2. Differentiability of foliations. The differentiability class of a foliation F at a nonsingular point is well known. This class is C^r at the point $m \in M - \mathrm{Sing}(M)$ if there is a C^r-diffeomorphism bringing the foliated neighborhood $F|U$, $m \in U \subset M - \mathrm{Sing}(M)$, to a family of parallel lines of $\mathbb{R}^2$.

There are many ways to extend the definition of differentiability to the set $\mathrm{Sing}(F)$, see [**15, 18**]. In our opinion, the most natural way is the one due to I. Bronshtein, I. Nikolaev [**11, 12**], who used two-fold branched coverings. The following theorem, proved in [**14, 11**].

THEOREM 1.1. *Let F be a foliation on a compact orientable surface M with $2k$ nonorientable singularities $p_1, \dots, p_{2k}$ and q orientable singularities. Then there is a unique two-fold branched covering $p\colon \widetilde{M} \to M$ with the following properties:*

1. *$W := \{p_1, \dots, p_{2k}\}$ is the set of branch points, each of index 2.*
2. *$\widetilde{M}$ is a compact orientable surface of genus $g = 2g_M + k - 1$, where g_M is the genus of M.*
3. *There is an involution $\theta\colon \widetilde{M} \to \widetilde{M}$, $\theta^2 = id$, which fixes the points of $p^{-1}(W)$ and only these points.*
4. *The natural projection $\widetilde{M} \to \widetilde{M}/\theta = M$ is p.*
5. *There is a flow $f^t(F)$ on $\widetilde{M}$ such that the integral curves of $f^t(F)$ generate a covering foliation $\widetilde{F}$ for F under the covering p (i.e., $\widetilde{F}/\theta = F$). In particular, θ preserves trajectories of $f^t(F)$.*
6. *The flow $f^t(F)$ has $2(p+q)$ fixed points.*

DEFINITION 1.4. Let F be a foliation on a compact orientable surface M. Let $f^t(F)$ be a flow of maximal smoothness class on the compact orientable surface $\widetilde{M}$ which satisfies the assumptions of Theorem 1.1. Then F is said to be *C^r-smooth* ($r \geqslant 0$) if the flow $f^t(F)$ is of class C^r.

Denote by $\mathcal{F}^r(M)$ the space of all C^r-smooth foliations on M endowed with the uniform C^r-topology.

1.3. Bronshtein–Nikolaev theory. Let us fix a set $W = \{p_1, \dots, p_{2k}\}$ of $2k$ points on a compact orientable surface M. Then it fixes an involution θ satisfying Theorem 1.1.

Consider the subclass $\mathcal{F}^r_W(M) \subset \mathcal{F}^r(M)$ of foliations with $\mathrm{Sing}_-(F) = W$, where $\mathrm{Sing}_-(F)$ denotes the set of nonorientable singularities of F.

Let $f^t(F_1)$, $f^t(F_2)$ be C^r-flows that correspond to foliations $F_1, F_2 \in \mathcal{F}^r_W(M)$ and satisfy the assumptions of Theorem 1.1.

DEFINITION 1.5. F_1, F_2 are said to be *topologically equivalent with fixed set W* if there exists a homeomorphism $h\colon M \to M$ such that $h \circ \theta = \theta \circ h$ and the flows $f^t(F_1)$, $f^t(F_2)$ are topologically equivalent via the homeomorphism h.

Take any two C^r-foliations $F_1', F_2' \in \mathcal{F}^r(M)$ such that $\mathrm{Sing}_-(F_i')$ consists of $2k$ points, $i = 1, 2$.

DEFINITION 1.6. The foliations $F_1', F_2' \in \mathcal{F}^r(M)$ are said to be *topologically equivalent* if there exist C^r-diffeomorphisms $\varphi_i\colon M \to M$ $(i = 1,2)$ and C^r-foliations F_1, $F_2 \in \mathcal{F}^r_W(M)$ such that

1) $\varphi_i(\mathrm{Sing}_-(F_i')) = W$, $i = 1,2$;
2) φ_i takes leaves of F_i' into leaves of F_i, $i = 1,2$;
3) F_1, F_2 are topologically equivalent with fixed set W.

DEFINITION 1.7. A C^r-foliation $F \in \mathcal{F}^r(M)$ is called *structurally stable* if there is a neighborhood $U(F) \subset \mathcal{F}^r(M)$ of F such that any $H \in U(F)$ is topologically equivalent to F.

I. Bronshtein and I. Nikolaev [**11, 12**] studied structurally stable nonorientable singularities and proved that the only structurally stable nonorientable singularity with negative index is the so-called *hyperbolic tripod* (which is defined by its special normal form; we do not need it explicitly, so we omit details).

Denote by S_g a class of foliations with structurally stable singularities of negative index on a closed orientable surface of genus $g \geqslant 2$. So S_g is a class of foliations with structurally stable saddles and hyperbolic tripods.

Denote by HTS_g the subclass of S_g consisting of highly transitive foliations. It is easy to see that HTS_g consists of irrational foliations. Moreover, any irrational foliation with structurally stable singularities is of the class HTS_g.

1.4. Poincaré model. Let Δ be the hyperbolic (or Lobachevsky) plane, which is a simply connected Riemannian 2-manifold of constant negative curvature -1. We regard Δ as the Poincaré realization as the unit disk $|z| < 1$ of the complex z-plane with the metric $ds = 2|dz|/(1 - |z|^2)$. The circle $S_\infty = \partial\Delta = (|z| = 1)$ is called the *absolute* or *circle at infinity*. The geodesics of Δ are the circular arcs orthogonal to S_∞. We shall suppose that endpoints of any geodesic of Δ belong to the circle at infinity.

Any closed orientable surface M of genus $g \geqslant 2$ can be thought of as the orbit space Δ/Γ, where Γ is a discrete group of orientation-preserving isometries of Δ acting freely on Δ (i.e., the elements of Γ have no fixed points in Δ). The group Γ is isomorphic to the fundamental group of M.

The natural projection $\pi\colon \Delta \to \Delta/\Gamma \cong M$ is the universal covering map.

Two subsets of $\Delta \cup S_\infty$ are called *congruent* if there is a transformation from Γ taking one subset into the other.

1.5. The topology of the set of geodesics. Let $L \subset M$ be a nonclosed simple (i.e., without self-intersections) geodesic. L is called *nontrivial recurrent* if its intrinsic topology does not agree with the topology of L as a subset of M.

We denote by $\mathrm{NGeo}(M)$ the set of all nondirected geodesics on M. Our goal is to introduce a topological space structure on $\mathrm{NGeo}(M)$, which would describe the asymptotic behavior of the geodesics.

Let L be a nondirected geodesic through some point $m \in M$. We denote by $K(m, L, \epsilon)$ the set of geodesics $L_\alpha \in \mathrm{NGeo}(M)$ such that L_α passes through the same point $m \in M$ and the smallest angle between L_α, L is less than $\epsilon > 0$.

Let $A \subset L$ be a compact arc. We denote

$$U(L, A, \epsilon) = \bigcup_{m \in A} K(m, L, \epsilon).$$

One can prove that the family of all $U(L, A, \epsilon)$-sets satisfies the axioms of a base of a topological space and, as a consequence, defines a topological space structure on $\mathrm{NGeo}(M)$.

1.6. Geodesic framework. We recall that a *geodesic lamination* on M is a foliation of a closed subset of M by simple (without self-intersections) geodesics. In other words, a geodesic lamination is a nonempty collection of disjoint simple geodesics whose union is closed in M.

If some leaves of a foliation have asymptotic directions, then for such a foliation one can construct a special geodesic lamination, called its geodesic framework. First of all we construct a geodesic framework for a highly transitive foliation with finite set of singularities.

Consider a leaf l both semileaves of which have asymptotic directions. Let $\bar{l}$ be any lift of l to Δ. Suppose that $\alpha(\bar{l}) \neq \omega(\bar{l})$, where $\alpha(\bar{l}), \omega(\bar{l}) \in S_\infty$ are endpoints of $\bar{l}$. Let $\overline{L}(\bar{l})$ be the geodesic with endpoints $\alpha(\bar{l}), \omega(\bar{l})$ connecting $\alpha(\bar{l})$ and $\omega(\bar{l})$.

The geodesic $\pi(\overline{L}(\bar{l})) \stackrel{\text{def}}{=} g(l)$ does not depend on the lift $\bar{l}$ and is called the *geodesic corresponding to the leaf l.*

Let Q be a quasiminimal set of F. Let l be a nontrivial recurrent leaf, and let l be dense in Q. Then both positive and negative semileaves of $\bar{l}$ have asymptotic directions and $\alpha(\bar{l}) \neq \omega(\bar{l})$. Then there is a geodesic $g(l)$ corresponding to l.

The topological closure $\mathrm{clos}[g(l)]$ of $g(l)$ is a geodesic lamination. This geodesic lamination is independent of the choice of l because any nontrivial recurrent leaf of Q is dense in Q [**19, 9**]. Hence the following notion is well defined.

DEFINITION 1.8. The geodesic lamination

$$\mathrm{clos}[g(l)] \stackrel{\text{def}}{=} Gf(Q)$$

will be called the *geodesic framework of the quasiminimal set Q.*

If F is a highly transitive foliation, then the quasiminimal set of F is the whole surface M. In this case

$$Gf(M) \stackrel{\text{def}}{=} Gf(F)$$

will be called the *geodesic framework of F.*

In the general case, we must construct geodesic frameworks of foliations more carefully. We must use extensions of a separatrix in the sense of Bendixson [**8**].

2. The continuity of a geodesic framework

We recall that a closed transversal C of a foliation F is said to be a *global section* if every one-dimensional leaf of F intersects C.

LEMMA 2.1. *Let a C^1-foliation $F \in S_g$ have a global section C. Then there is a neighborhood $U(F)$ of F in the space $\mathcal{F}^1(M)$ of C^1-foliations such that every $H \in U(F)$ has the global section C as well.*

PROOF. It is obvious that there exists a neighborhood $U_1(F)$ of F such that any foliation $H \in U_1(F)$ has the curve C as a closed transversal curve (not necessarily global). Since the singularities of F are structurally stable saddles (including nonorientable ones) there is a neighborhood $U_2(F)$ of F and a number $\delta_0 > 0$ such that all singularities of any $H \in U_2(F)$ are structurally stable saddles belonging to the δ_0-neighborhood U_0 of the set $\mathrm{Sing}(F)$ of singularities of F. Moreover, we can assume that the δ_0-neighborhood of every singularity of F contains the unique saddle of H.

Let us assume the contrary. Then there are a sequence of foliations $H_n \in U_1(F) \cap U_2(F)$ converging to F as $n \to \infty$ and a sequence of points $x_n \in M$ such that the one-dimensional leaf l_n through x_n of the foliation H_n does not intersect C. Since all the singularities of H_n are saddles, every one-dimensional leaf of H_n leaves U_0. In particular, l_n does. So we can assume that $x_n \notin U_0$.

Since M is compact, without loss of generality we can assume that the sequence x_n converges to some point x_0. Since $x_n \notin U_0$, the point x_0 belongs to a one-dimensional leaf l_0 of F. By the conditions of the lemma, $l_0 \cap C \neq \varnothing$. This result implies $l_n \cap C \neq \varnothing$ for large enough n. This contradiction proves the lemma. □

Two geodesics $\overline{L}_1, \overline{L}_2 \subset \Delta$ are called *ε-close in the Euclidean sense* if the corresponding endpoints are ε-close in the Euclidean metric on S_∞.

LEMMA 2.2. *Let $L \subset M$ be a geodesic and $V(L)$ a neighborhood of L in* NGeo(M). *Let $\overline{L} \subset \Delta$ be a lift of L. Then there is $\delta > 0$ such that if a geodesic $\overline{L}_1$ is δ-close to $\overline{L}$ in the Euclidean sense, then $\pi(\overline{L}_1) \subset V(L)$.*

PROOF. By the definition of the topological structure on NGeo(M), there is a geodesic segment $\Sigma \subset L$ and a number $\varepsilon > 0$ such that the $U(L, \Sigma, \varepsilon)$-neighborhood of L belongs to $V(L)$. By geometric considerations, there exists $\delta > 0$ such that if $\overline{L}_1$ is δ-close to $\overline{L}$ in the Euclidean sense, then $\overline{L}_1 \in \overline{U}(\overline{L}, \overline{\Sigma}, \varepsilon)$-neighborhood of $\overline{L}$. Moreover, if δ is small enough, then $\pi(\overline{L}_1) \in U(L, \Sigma, \varepsilon)$, since the geodesics of Δ are circular arcs orthogonal to S_∞. □

Recall that HTS_g is the subclass of S_g consisting of highly transitive foliations.

THEOREM 2.1. *Let $F \in HTS_g$ be a C^1-foliation on the closed orientable hyperbolic surface M of genus $g \geqslant 2$ and let U be a neighborhood of the geodesic framework $Gf(F)$ of F. Then there is a neighborhood $O^1(F)$ of F in the space $\mathcal{F}^1(M)$ such that any foliation $H \in O^1(F)$ has a nonempty geodesic framework $Gf(H)$ belonging to U.*

PROOF. For the sake of simplicity, we divide our proof into several steps. The end of the proof of each step will be denoted by $\diamond$.

STEP 1. *The foliation $F \in HTS_g$ has a global section, say C, which is a simple closed curve nonhomotopic to zero.*

PROOF. Indeed, since F has a nontrivial recurrent leaf then, according to [**9**], there is a simple closed transversal C. Since every one-dimensional leaf of F is dense on M, it follows that C is a global section of F. Since the singularities of F are of negative index, we see that C is nonhomotopic to zero. $\diamond$

STEP 2. *Let $\overline{F}$ be a lift of F. The preimage $\pi^{-1}(C)$ consists of a countable set of disjoint simple curves $\overline{C}_n$, $n \in \mathbb{N}$, each with endpoints on S_∞. Moreover, any curve $\overline{C}_n$ is a transversal of the covering foliation $\overline{F}$.*

The proof is obvious. $\diamond$

Let $\overline{C} \subset \Delta$ be a transversal curve of $\overline{F}$ without self-intersections whose endpoints belong to S_∞. These endpoints divide S_∞ into two intervals. If one of these intervals, say I, is of Euclidean length less than ε, then the domain of the hyperbolic plane restricted to I and $\overline{C}$ will be called an *ε-trap bounded by the curve $\overline{C}$*. The union of $\overline{C}$, I, and the ε-trap will be called a *closed ε-trap*.

Let $\bar{l}$ be a leaf of $\overline{F}$. We recall that if $\bar{l}$ is a separatrix, then we denote by $\mathrm{Ext}(\bar{l})$ the extensions of $\bar{l}$ in the Bendixson sense to the left and to the right (both in the positive and in the negative directions). If $\bar{l}$ is not a separatrix, then we assume $\mathrm{Ext}(\bar{l}) = \bar{l}$.

It is obvious that $\mathrm{Ext}(\bar{l})$ has at least two and at most three endpoints belonging to the circle at infinity.

STEP 3. *Let $\bar{l}$ be a leaf of $\overline{F}$. Then for any $\varepsilon > 0$ there are closed ε-traps bounded by curves from $\pi^{-1}(C)$ and containing the endpoints of* $\mathrm{Ext}(\bar{l})$.

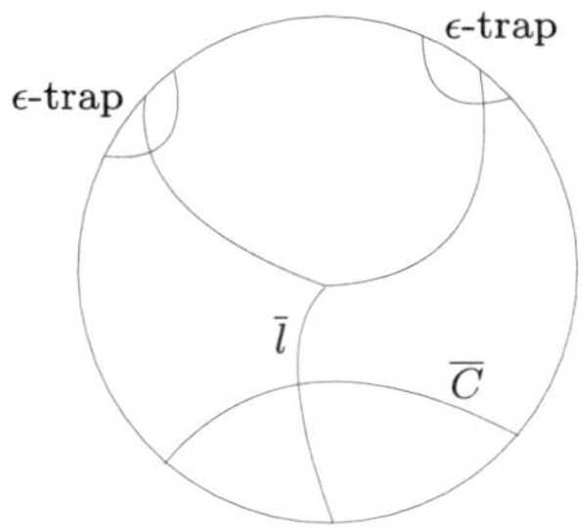

FIGURE 1

PROOF. It suffices to prove the following assertion. Let $\bar{l}$ be an oriented leaf of $\overline{F}$; suppose $\bar{l}$ is not an ω-separatrix. Then there is a closed ε-trap bounded by a component of $\pi^{-1}(C)$ and containing the ω-limit set of $\bar{l}$, which is a point of S_∞. Indeed, since $\pi(\bar{l})$ intersects C infinitely many times, it follows that $\bar{l}$ intersects some subsequence of curves $\overline{C}_{n_k}$. Since the surface M is the quotient-space of the hyperbolic plane by a discontinuous group of isometries, we see that the topological limit of the sequence of curves $\overline{C}_{n_k}$ is a unique point of S_∞, this point being the ω-limit set of $\bar{l}$. The result follows. ◇

STEP 4. *There is a neighborhood $O_0^1(F)$ of F such that any $H \in O_0^1(F)$ belongs to the class S_g. Moreover, we can assume that any $H \in O_0^1(F)$ has the global section C.*

PROOF. Immediately follows from Lemma 2.1 and the condition that the singularity of F is locally structurally stable. ◇

STEP 5. *Let l be an oriented leaf of $H \in O_0^1(F)$. Suppose l is an $\omega(\alpha)$-separatrix. Then we have one of the following cases.*

1. $\mathrm{Ext}_r^{\pm}(l)$ *consists of a finite number of leaves last of them has an asymptotic direction and is not an ω-separatrix.*
2. $\mathrm{Ext}_r^{\pm}(l)$ *is a homotopically nontrivial one-sided circuit consisting of separatrix connections including l.*

PROOF. Indeed, if $\mathrm{Ext}_r^{\pm}(l)$ contains leaf l_0, which is not an ω-separatrix, then l_0 is the last leaf of $\mathrm{Ext}_r^{\pm}(l)$. Since l_0 intersects the global section C, we see that l_0 has an asymptotic direction. If $\mathrm{Ext}_r^{\pm}(l)$ consists only of separatrix connections, then $\mathrm{Ext}_r^{\pm}(l)$ is a nontrivially homotopic one-sided circuit containing l_0 because $H \in O_0^1(F)$ has only singularities of negative index only. ◇

Similar assertions hold for another (e.g., left) global extension of l in the Bendixson sense.

STEP 6. *Let Σ_0, Σ_1, Σ_2 be transversal segments of the foliation F. Let l be an oriented leaf of F which is an ω-separatrix of some singularity s_0. Suppose the ω-separatrix l passes through an interior point $x(l)$ of Σ_0. Let l_1, l_2 be separatrices which are extensions of l in the Bendixson sense in the positive direction. Suppose l_1, l_2 pass through interior points of Σ_1, Σ_2 respectively. Then there exist a neighborhood $O_1^1(F) \subset O_0^1(F)$ of F in the space $\mathcal{F}^1(M)$ and an interval $(\alpha(l), \beta(l)) \subset \Sigma_0$ containing the point $x(l)$ such that for any leaf l' through $(\alpha(l), \beta(l))$ of a foliation $H \in O_1^1(F)$ the global positive extension* $\mathrm{Ext}^+(l')$ *of l' in the Bendixson sense intersects some of the segments Σ_1, Σ_2 (possibly, both).*

PROOF. Indeed, according to Step 4, there is a neighborhood $O_0^1(F)$ small enough so that any foliation $H \in O_0^1(F)$ has singularities which are only structurally stable tripods and saddles. Then the existence of $C_1^1(F)$ is an immediate consequence of the theorem on the stability of stable and unstable manifolds of a hyperbolic singularity ([**23**, Theorem 6.2, Chapter 2], [**11**]) and the theorem on the continuous dependence of leaves on initial conditions [**1**]. $\diamond$

A similar result holds for an α-separatrix l and global negative extensions in the Bendixson sense.

Let us choose some curve $\overline{C}$ from $\pi^{-1}(C)$, where C is the global section of F (see Step 1). By Step 2, we can take two points $a, b \in \overline{C}$ such that an interval I bounded by a, b does not contain congruent points except the points a, b, which are congruent. Denote by $\mathrm{Tr}(I)$ the set of leaves of $\overline{F}$ intersecting I. Without loss of generality we can assume that the leaf $\bar{l}(a)$ of $\overline{F}$ through the point a (as a consequence, through the point b) is not a separatrix.

By Step 1, there is a closed simple geodesic C_0 homotopic to the global section C. Choose a lift $\overline{C}_0$ of C_0 having the same endpoints with $\overline{C}$.

STEP 7. *Let $\overline{g}(\bar{l}(a))$, $\overline{g}(\bar{l}(b))$ be geodesics having the same endpoints with $\bar{l}(a)$, $\bar{l}(b)$ respectively. Then the geodesic $\overline{g}(\bar{l}(a))$ (respectively $\overline{g}(\bar{l}(b))$) intersects $\overline{C}_0$ at some point, say $\overline{a}_0$ (respectively $\overline{b}_0$). Moreover, every geodesic of $Gf(H)$ of a foliation $H \in O_0^1(F)$ has a lift that intersects $\overline{I}_0$, where $\overline{I}_0 \subset \overline{C}_0$ is the interval with endpoints $\overline{a}_0$, $\overline{b}_0$.*

PROOF. Since $H \in O_0^1(F)$, we get that C is the global section of H (see Step 4). Let L be any geodesic of $Gf(H)$ and $\overline{L}$ be its lift on Δ. By the definition of $Gf(H)$ and Step 5, L is the geodesic corresponding to either a leaf l having an asymptotic direction, or a separatrix l' which is not a separatrix connection or homotopically nontrivial one-sided circuit K. In any case, according to the construction of $Gf(H)$, there is a curve $T \subset \Delta$ consisting of leaves and singularities of $\overline{H}$ such that T and $\overline{L}$ have the same endpoints. Since C is the global section of H, we can assume that $\overline{C} \cap T \neq \varnothing$. Hence $C_0 \cap L \neq \varnothing$. $\diamond$

STEP 8. *Given a neighborhood U, there is $\delta > 0$ such that if a geodesic $\overline{L}_1 \subset \Delta$ is δ-close in the Euclidean sense to some geodesic $\overline{L} \in \overline{Gf}(\overline{F})$ with $\overline{L} \cap \overline{I}_0 \neq \varnothing$, then $\pi(\overline{L}_1) \in U$.*

PROOF. By the definition of the topological structure on the space of geodesics, for any geodesic $\overline{L} \in \overline{Gf(F)}$, $\overline{L} \cap \overline{I}_0 \neq \varnothing$, there is $\delta(\overline{L}) > 0$ such that if $\overline{L}_1$ is $\delta(\overline{L})$-close to $\overline{L}$ in the Euclidean sense, then $\pi(\overline{L}_1) \in U$. Let $\delta_{\max}(\overline{L})$ be a supremum of such $\delta(\overline{L})$.

Let us assume that there exists a sequence of geodesics $\overline{L}_i \in \overline{Gf(F)}$, $\overline{L}_i \cap \overline{I}_0 \neq \varnothing$, such that $\delta_{\max}(\overline{L}_i) \to 0$ as $i \to \infty$. Since $\overline{I}_0$ is compact, we see that the set of the points $\overline{m}_i = \overline{L}_i \cap \overline{I}_0$ has at least one point $\overline{m}_0$ of an accumulation. Without loss of generality we can assume that $\overline{m}_i \to \overline{m}_0$ as $i \to \infty$. Because a geodesic lamination is a closed set, there exist a geodesic $\overline{L}_0 \in \overline{Gf(F)}$ through $\overline{m}_0$.

Since $\overline{Gf(F)}$ is the lift of $Gf(F)$, we see that there is an i_0 such that $\overline{L}_i$ is $(\delta(\overline{L}_0)/2)$-close in the Euclidean sense to $\overline{L}_0$ for $i \geqslant i_0$. Therefore any geodesic $\widetilde{L} \subset \Delta$, which is $(\delta(\overline{L}_0)/2)$-close in the Euclidean sense to $\overline{L}_i$, $i \geqslant i_0$, is $\delta(\overline{L}_0)$-close in the Euclidean sense to $\overline{L}_0$. As a consequence, $\pi(\widetilde{L}) \in V$. This is contradiction with the assumption $\delta_{\max}(\overline{L}_i) \to 0$. Hence,

$$\min_{\overline{L} \in \overline{Gf(F)}, \overline{L} \cap I \neq \varnothing} \delta_{\max}(\overline{L}) = \alpha > 0.$$

We can take $\delta = \alpha/2$. ◇

Now we are ready to complete the proof of the theorem.

For a leaf $\overline{l} \in Tr(I)$ we denote by $x(\overline{l})$ the unique point of the intersection of $\overline{l}$ with I. By Step 3, there are closed δ-traps bounded by curves $\overline{C}_\nu(\overline{l})$, $1 \leqslant \nu \leqslant 3$, from $\pi^{-1}(C)$ and containing endpoints of $\mathrm{Ext}(\overline{l})$. According to Step 6, for any leaf $\overline{l} \in Tr(I)$ there exists a neighborhood, say $O^1(F, \overline{l}) \subset O_0^1(F)$, of F in the space $\mathcal{F}^1(M)$ and an interval $(\alpha(\overline{l}), \beta(\overline{l})) \subset \overline{C}$ containing the point $x(\overline{l})$ such that for any leaf $\overline{l}'$ through $(\alpha(\overline{l}), \beta(\overline{l}))$ of a lift $\overline{H} \in O^1(F, \overline{l})$ the global extensions $\mathrm{Ext}(\overline{l}')$ of $\overline{l}'$ in the Bendixson sense intersect some (possibly all) of the curves $\overline{C}_\nu(\overline{l})$, $1 \leqslant \nu \leqslant 3$.

Since I is compact, we see that there is a finite collection of leaves $\overline{l}_1, \dots, \overline{l}_k$ of $\mathrm{Tr}(I)$ such that intervals $(\alpha(\overline{l}_1), \beta(\overline{l}_1)), \dots, (\alpha(\overline{l}_k), \beta(\overline{l}_k))$ cover I.

Let us prove that the neighborhood

$$O^1(F) = O^1(F, \overline{l}_1) \cap \dots \cap O^1(F, \overline{l}_k) \cap O_0^1(F)$$

satisfies the conditions of the theorem. Take $H \in O^1(F)$. Since $O^1(F) \subset O_0^1(F)$, by Step 4 H belongs to S_g, i.e., all the singularities of H are structurally stable tripods and saddles. Moreover, H has the global section C.

Since all singularities of H are of negative index, $Gf(H) \neq \varnothing$.

Take a geodesic $L' \in Gf(H)$. We must prove $L' \in U$. According to the definition of the geodesic framework and Step 5, L' is the geodesic corresponding to the leaf l', which is either a leaf having an asymptotic direction, or a separatrix which is not a separatrix connection or a homotopically nontrivial one-sided circuit of the foliation H.

Since C is a global section of H, l' intersects C. As a consequence, there is a lift $\overline{l}'$ of l' intersecting the interval $I \subset \overline{C}$ at some point $x'(\overline{l}')$. By the definition of a geodesic framework, there is a lift of the geodesic L', denoted by $\overline{L}'$, that has endpoints belonging to the set of endpoints of the extension $\mathrm{Ext}(\overline{l}')$.

Since the intervals $(\alpha(\overline{l}_1), \beta(\overline{l}_1)), \dots, (\alpha(\overline{l}_k), \beta(\overline{l}_k))$ cover I, it follows that $x'(\overline{l}') \in (\alpha(\overline{l}_i), \beta(\overline{l}_i))$ for some $1 \leqslant i \leqslant k$. By Step 3, $\mathrm{Ext}(\overline{l}')$ intersects some (possibly all) the curves $\overline{C}_\nu(\overline{l}_i)$, $1 \leqslant \nu \leqslant 3$. Since the curves $\overline{C}_\nu(\overline{l}_i)$, $1 \leqslant \nu \leqslant 3$, bound δ-traps, there is a geodesic $\overline{L} \in \pi^{-1}(Gf(F))$ which is δ-close to $\overline{L}'$. According to the construction of the interval I_0 (see Step 7), the geodesic $\overline{L}$ intersects I_0. By Step 8, $L' = \pi(\overline{L}') \in U$. The theorem is proved. □

3. Variability of geodesic frameworks

The following theorem means that a geodesic framework of an irrational surface foliation has a variability property that is similar to the variability property of the irrational Poincaré rotation number of a flow on the torus or a circle diffeomorphism without periodic orbits. It can be regarded as a "geodesic" closing lemma as well.

THEOREM 3.1. *Let F be a highly transitive C^1-foliation on the closed orientable hyperbolic surface M of genus $g \geqslant 2$. Suppose that all the singularities of F are structurally stable. Then for any neighborhood U of the geodesic framework $Gf(F)$ of F and any neighborhood $O^1(F)$ of F in the space $\mathcal{F}^1(M)$ of C^1-foliations there is a foliation $H \in O^1(F)$ such that the geodesic framework $Gf(H)$ has a closed geodesic belonging to U. Moreover, $Gf(H)$ has no nontrivial recurrent geodesics.*

PROOF. It suffices to prove the existence of a foliation $H \in O^r(F)$ such that H has a homotopically nontrivial closed leaf and has no quasiminimal sets. We prove that one can find such a foliation in the set of Morse–Smale foliations.

We recall that a foliation F on the compact surface is called a *Morse–Smale foliation* if it satisfies the following conditions:

1) F has finitely many singularities and periodic leaves, each of which is hyperbolic;
2) the $\omega(\alpha)$-limit set of any leaf of F is either one singularity or one periodic leaf;
3) there are no separatrix connections.

It is known [**11, 12, 22**], that Morse–Smale C^r-foliations are dense in the space $\mathcal{F}^r(M)$ of C^r-foliations for any $r \geqslant 1$ on any orientable compact surface M. Let us take a Morse–Smale foliation $H \in O^r(F)$ in a neighborhood so small that singularities of H are hyperbolic tripods and saddles only. Moreover, according to Theorem 2.1, we can assume that $Gf(H) \subset U$. Since H has finitely many singularities and all of them are structurally stable, it follows that there is a directed (positive) semileaf l' of H such that the ω-limit set of l' is not a singularity. Since any Morse–Smale foliation has no quasiminimal sets and one-sided circuits, we see that the ω-limit set of l' is a limit cycle C_0, which is homotopically nontrivial because all the singularities of H have negative index. As a consequence, the geodesic framework $Gf(H)$ has a closed geodesic belonging to U and $Gf(H)$ has no nontrivial recurrent geodesics. □

4. Collapse of geodesic frameworks

Now we consider a family of surface foliations and the corresponding bifurcations of geodesic frameworks.

THEOREM 4.1. *On any closed orientable hyperbolic surface M of genus $g \geqslant 2$ there is a continuous family of C^∞-foliations F_μ, $\mu \in [0;1]$, such that*

1) *for all $\mu \in [0;1)$ the geodesic framework $Gf(F_\mu) \neq \varnothing$ is irrational and does not depend on μ;*
2) *the foliation F_1 has a rational geodesic framework $Gf(F_1)$;*
3) *there is a neighborhood U of $Gf(F_1)$ such that $G(f_\mu^t) \notin U$ if $\mu \in [0;1)$.*

THEOREM 4.2. *On any closed orientable hyperbolic surface M of genus $g \geqslant 2$ there is a continuous family of C^∞-foliations F_μ, $\mu \in [0;1]$, such that*

1) *for all $\mu \in [0;1)$ the geodesic framework $Gf(F_\mu) \neq \varnothing$ is rational and does not depend on μ;*
2) *there is a neighborhood U of $Gf(F_1)$ such that $G(F_\mu) \notin U$ if $\mu \in [0;1)$.*

PROOF OF THEOREMS 4.1, 4.2. The crucial step in proving both theorems is the following result.

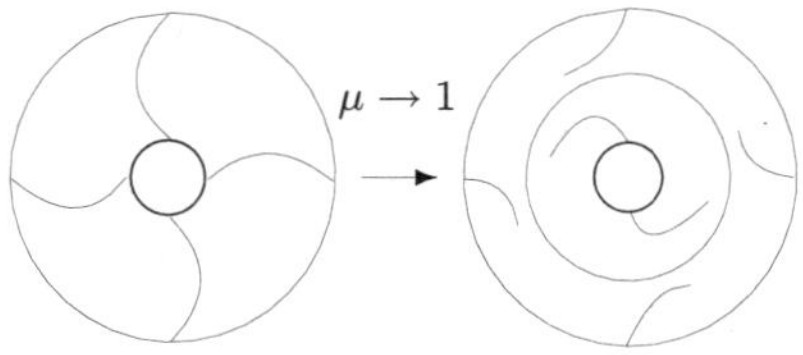

FIGURE 2

On the closed annulus K there is a family of C^∞-foliations f_μ depending continuously on the parameter $\mu \in [0;1]$ and satisfying the following conditions (see Figure 2):

(a) for all $\mu \in [0;1]$ the foliation f_μ is transversal to the boundary of K;
(b) for all $\mu \in [0;1)$ the foliation f_μ is a foliation-product whose leaves are closed;
(c) the foliation f_1 has a closed leaf nonhomotopic to zero.

A similar family of flows was constructed by Medvedev [**20, 21**] to demonstrate the bifurcation called "blue sky catastrophe" (see [**25**, Problem 37]).

Let us prove Theorem 4.1. Take an irrational foliation F_0 on the closed orientable hyperbolic surface M of genus $g \geqslant 2$. Such a foliation has a closed global section denoted by C. Let N be a tubular neighborhood of C homeomorphic to the annulus K. It is obvious that the restriction $F_0|N$ is topologically equivalent to the foliation f_0 considered above. Now we include the family of C^∞-foliations f_μ into N. As a result we get a family, which we denote by F_μ, of surface foliations. Since each F_μ, $\mu \in [0;1)$, is topologically equivalent to F_0 it follows that the geodesic framework $Gf(F_\mu) \neq \varnothing$ is irrational and does not depend on μ. But because C is a global section of F_0, we see that the limit set of any leaf of the foliation F_1 consists of singularities and one closed leaf, say C_0. Therefore F_1 has a rational geodesic framework $Gf(F_1)$. Since, according to construction, C_0 is homotopic to C, there is a neighborhood U of $Gf(F_1)$ such that $G(f_\mu^t) \notin U$ whenever $\mu \in [0;1)$.

The proof of Theorem 4.2 is similar, with one obvious modification: we must start with a foliation F_0 having nonempty rational geodesic framework and a global section nonhomotopic to zero. □

There are continuous families of surface foliations that correspond to another type of bifurcation of geodesic frameworks.

THEOREM 4.3. *On a closed orientable hyperbolic surface M of genus $g \geqslant 2$ there is a continuous family of C^∞-foliations F_μ, $\mu \in [0;1)$, such that*

1) *for all $\mu \in [0;1)$ the geodesic framework $Gf(F_\mu) \neq \varnothing$ is rational;*
2) *the lengths of the closed geodesics of $Gf(F_\mu)$ tend to infinity uniformly as $\mu \to 1$.*

PROOF. The construction of the family is similar to the construction in Theorems 4.1, 4.2, with the use of [**3**, Lemma 7]. □

REMARK. We can construct the family F_μ required by Theorem 4.3 above so that all singularities of any F_μ are structurally stable (saddles and nodes) and are situated at fixed points of the surface. Moreover, one can construct the family F_μ so that it is widely disposed [**8**].

We can extend the family F_μ, $\mu \in [0;1)$, to a foliation F_1 in different ways. For example, we can obtain a continuous family F_μ, $\mu \in [0;1]$, for which the geodesic framework $Gf(F_1)$ is either rational, or irrational, or in the empty set. In the last case $Gf(F_1) = \varnothing$, we shall say that there is a *collapse of the geodesic framework*.

Let us consider properties of a foliation F_1 that supports the collapse of a geodesic framework.

THEOREM 4.4. *Let F_μ, $\mu \in [0;1]$, be a continuous family of C^∞-foliations on a closed orientable hyperbolic surface M of genus $g \geqslant 2$. Suppose that*

1) *for each $\mu \in [0;1)$ the geodesic framework $Gf(F_\mu) \neq \varnothing$ is rational*;
2) *the lengths of the closed geodesics of $Gf(F_\mu)$ tend to infinity uniformly as $\mu \to 1$*;
3) $Gf(F_1) = \varnothing$;
4) *every F_μ, $\mu \in [0;1)$, is widely disposed.*

Then the foliation F_1 has infinitely many singularities.

PROOF. By the conditions, the foliation F_μ, $\mu \in [0;1)$, has either a homotopically nontrivial circuit K_μ or a homotopically nontrivial closed leaf denoted by the same letter K_μ. Since the lengths of the corresponding closed geodesics (which are freely homotopic to K_μ) of $Gf(F_\mu)$ tend to infinity uniformly as $\mu \to 1$ it follows that the lengths of the leaves K_μ tend to infinity as well.

Let us assume the contrary. Since the leaves K_μ are homotopically nontrivial curves, the topological limit $\mathrm{TLim}(K_\mu)$ of K_μ has the cardinality of the continuum. Therefore there is a point $m_0 \in M$, belonging to the topological limit $\mathrm{TLim}(K_\mu)$ of K_μ, such that a one-dimensional directed semileaf $l^+(m_0)$ passes through m_0. According to the theorem on the continuous dependence of leaves on the initial conditions [**2**, Theorem 8], $l^+(m_0)$ belongs to $\mathrm{TLim}(K_\mu)$.

Since $Gf(F_1) = \varnothing$, the limit set $\lim(l^+(m_0))$ of $l^+(m_0)$ is either a closed leaf l_0 homotopic to zero, or a closed circuit homotopic to zero, or a singularity O.

First we consider the case $\lim(l^+(m_0)) = l_0$. Since l_0 homotopic to zero, $l^+(m_0) \neq l_0$. Let Σ be a transversal segment intersecting l_0. Denote by $a, b \in \Sigma$ two consecutive intersection points of $l^+(m_0)$ with Σ. Then the union of the segment $ab \subset \Sigma$ and the arc of $l^+(m_0)$ with endpoints a, b form a closed curve C that bounds a simply connected domain D.

Without loss of generality, we can assume that Σ are transversal to F_μ and $l^+(m_0)$ approaches spirally l_0 after passing through the points a and b. Since the curve bounds a simply connected domain, any foliation F_μ has a region of the index $1/2$. We obtain a contradiction with the assumption that F_μ, $\mu \in [0;1)$, is widely disposed.

A similar proof works in the case where $\lim(l^+(m_0))$ is a closed circuit homotopic to zero.

Let us consider the case where the limit set $\lim(l^+(m_0)) = O$ is the singularity. There is a neighborhood U of O containing the unique singularity O because we suppose that F_1 has a finite number of singularities. We can assume that U is simply connected and $m_0 \notin U$. Since K_μ are homotopically nontrivial, the curves K_μ leave O. Hence, according to [**1**, Theorem 60, Chapter 8], the semileaf $l^+(m_0)$ has a Bendixson extension, say l^1, which is a separatrix of O. By repeating these arguments, we get a sequence of leaves $l^1, \dots, l^{k+1} = l^1$ such that l^{i+1} is the Bendixson extension of l^i, $i = 1, \dots, k$, and the union of all the leaves l^i form a homotopically nontrivial circuit K. This is a contradiction with $Gf(F_1) = \varnothing$. □

References

[1] A. A. Andronov, E. A. Leontovich, I. I. Gordon, and A. G. Maier, *Qualitative theory of dynamical systems of the second order.* "Nauka", Moscow, 1966; English transl. Israel Program of Scientific Translations, Jerusalem–London, 1973.

[2] ———, *Theory of bifurcations of dynamical systems on the plane*, "Nauka", Moscow, 1967; English transl., Israel Program of Scientific Translations, Jerusalem and London, 1973.

[3] S. Aranson, *The absence of nonclosed Poisson stable semi-trajectories and trajectories doubly asymptotic to a double cycle for dynamical systems of the first degree of structural unstability on orientable two-dimensional manifolds*, Mat. Sb. **76 (118)** (1968), 214–230; English transl., Math. USSR-Sb. **5** (1968), 205–219.

[4] ———, *On topological equivalence of the foliations with singularities and homeomorphisms with invariant foliations on two-dimensional manifolds*, Uspekhi Mat. Nauk **41** (1986), no. 3, 197–198; English transl. in Russian Math. Surveys **41** (1986).

[5] ———, *Topology of vector fields, foliations with singularities and homeomorphisms with invariant foliations on closed surfaces*, Trudy Mat. Inst. Steklov. **193** (1992), 15–21; English transl., Proc. Steklov Inst. Math. (1993), no. 3, 13–18.

[6] S. Aranson and V. Grines, *On the representation of minimal sets of flows on 2-manifolds by geodesic lines*, Izv. Akad. Nauk SSSR Ser. Mat. **42** (1978), no. 1, 104–129; English transl., Math. USSR-Izv. **12** (1978), 103–124.

[7] S. Aranson, M. Malkin, and E. Zhuzhoma, *On relation between smooth and topological properties of circle transformations (Denjoy like theorem)*, Dep. VINITI, no. 3052-84, VINITI, Moscow, 1984. (Russian)

[8] S. Aranson, V. Medvedev, and E. Zhuzhoma. *On continuity of geodesic frameworks of surface flows*, Mat. Sb. **188** (1997), no. 7, 3–22; English transl. in Russian Ac. Sci. Math.-Sb.

[9] S. Aranson and E. Zhuzhoma, *On the structure of quasiminimal sets of surface foliations*, Mat. Sb. **185** (1994), no. 8, 31–62; English transl. in Russian Ac. Sci. Math.-Sb. **82** (1995)

[10] V. I. Arnold, *Small denominators* I; *on the mapping of a circle itself*, Izv. Akad. Nauk SSSR Ser. Mat. **25** (1961), 21–86. (Russian)

[11] I. U. Bronstein and I. V. Nikolaev, *Structurally stable fields of line elements on surfaces*, J. Nonlinear Analysis. Theory, Methods and Appl. (to appear)

[12] ———, *On Peixoto graphs of Morse–Smale foliations*, CRM-2214, Univ. de Montréal, Preprint, 1994.

[13] M. Herman, *Sur la conjugaison différentiable des difféomorphismes du circle et des rotations*, IHES Publ. Math. **49** (1979), 5–234.

[14] A. Hurwitz, *Ueber Riemann'sche flachen mit gegebenen verzweigungspunkten*, Math. Ann. **39** (1891), 1–61.

[15] C. Gutierrez, *Structural stability for line fields with singularities on two-dimensional manifolds*, Lecture Notes in Math., vol. 468, Springer-Verlag, Heidelberg, 1975, pp. 17–19.

[16] I. P. Kornfeld, Y. G. Sinai, and S. V. Fomin, *Ergodic Theory*, "Nauka", Moscow, 1980. (Russian)

[17] G. Levitt, *Foliations and laminations on hyperbolic surfaces*, Topology **22** (1983), no. 2, 119–135.

[18] G. Levitt and H. Rosenberg, *Differentiability and topology of labyrinths in the disc and annulus*, Topology **26** (1987), no. 2, 173–186.

[19] A. Maier, *On trajectories on closed orientable surfaces* Mat. Sb. **12** (1943), 71–84. (Russian)

[20] V. Medvedev, *On new type of bifurcations on manifolds*, Mat. Sb. **113** (1980), no. 3, 487–492; English transl. in Math. USSR-Sb. **41** (1982).

[21] ———, *About the bifurcation "blue sky catastrophe" on two-dimensional manifolds*, Mat. Zametki **51** (1992), no. 1, 118–125; English transl., Math. Notes **51** (1992), no. 1–2, 76–81.
[22] I. V. Nikolaev, *Foliations with singularities of semi-integer index*, CRM-2206, Université de Montréal, Preprint, 1994.
[23] J. Palis and J. W. de Melo, *Geometric theory of dynamical systems. An introduction*, Springer-Verlag, New York, Heidelberg, and Berlin, 1982.
[24] H. Poincaré, *Sur les courbes définies par les équations différentielles*, J. Math. Pures Appl. 1886, no. 2, 151–217.
[25] J. Palis and C. Pugh, *Fifty problems in dynamical systems*, Lecture Notes in Math., vol. 486, Springer-Verlag, Heidelberg, 1975, pp. 345–353.

Nizhny Novgorod State Technical University, Minin Str., 24, Nizhny Novgorod 603155, Russia
E-mail address: ara@uic.nnov.ru

Department of Differential Equations, Institute for Applied Mathematics and Cybernetics, 10 Ul'yanova st., Nizhny Novgorod 603005, Russia
E-mail address: medvedev@unn.ac.ru

Nizhny Novgorod State Technical University, Minin Str., 24, Nizhny Novgorod 603155, Russia

Translated by E. ZHUZHOMA

Amer. Math. Soc. Transl.
(2) Vol. **200**, 2000

Homoclinic and Heteroclinic Linkages in Concrete Systems: Nonlocal Analysis and Model Maps

Vladimir N. Belykh

To the memory of Professor E. A. Leontovich-Andronova

ABSTRACT. The qualitative analysis of several multidimensional dynamical systems using phase pictures of systems in the plane is presented. The existence problem of various homoclinic and heteroclinic linkages illustrating complex and chaotic behavior is discussed. Geometric models expressed in terms of explicitly defined maps are introduced. Some new strange attractors of these maps are exhibited.

Introduction

This paper deals with the existence problem of various homoclinic and heteroclinic linkages in dynamical systems. This problem, which was first raised and considered by A. A. Andronov and E. A. Leontovich for flows in the plane [**1**] more than 50 years ago and is intimately related to the general theory of complex basins and attractors, is now considered as one of the basic contemporary problems in the theory of dynamical systems.

We study the sufficient conditions for the existence of homoclinic and heteroclinic orbits for certain multidimensional nonintegrable systems; the existence of such orbits serves as criteria of complicated dynamics. For this purpose, we use the methods of comparison systems and Lyapunov functions [**2, 3, 4, 5, 6, 7, 8**], integral manifolds, small parameters, etc. Here we present some theorems on the existence of homoclinic and heteroclinic linkages in multidimensional systems based on bifurcations of certain reduced systems in the plane. Namely, we consider

1) the equations of an automatic control system with one nonlinear function (Section 1);
2) a three-dimensional oscillatory system (Section 2) that becomes the Lorenz system in a particular case;
3) a system containing a slow variable (Section 3).

2000 *Mathematics Subject Classification.* Primary 37C29.
This work was supported by the Russian Foundation for Basic Research (grant 96-01-01428) and by INTAS (grant 93-0570 (ext.)).

©2000 American Mathematical Society

Finally, in Section 4 we present geometric models for different linkages and describe the explicit mappings conforming to the appropriate flow—defined Poincaré return maps. Those mappings possess a set of new strange attractors.

1. Systems with one nonlinear function

We consider the system

$$\frac{dy}{dt} = Ay + r\Phi(x), \qquad x = C^* y, \tag{1.1}$$

where y is the n-vector of variables, A is an $n \times n$ matrix, r, c are n-vectors of constant parameters, and the symbol "$*$" stands for transposition; $\Phi\colon \mathbb{R}^1 \to \mathbb{R}^1$ is a nonlinear function.

In contrast with the traditional problems on global stability and dichotomy, we study the chaotic dynamics in the system (1.1) related to the existence of Poincaré homoclinic and heteroclinic orbits, which are the initial cause of the limiting trajectories complex behavior.

Applying to the system (1.1) with a nonsingular matrix A the change of variables $y \to Hy$ with nonsingular constant matrix H we get a system of the same form such that now $C^* = (1, 0, \ldots, 0)$, the first column of the new matrix A is equal to zero and the function Φ is replaced by $m_0 x + \Phi(x)$ with some constant m_0. Separating the first and, say, second equations in the transformed system (1.1) and denoting $x = y_1$, $y = y_2$, $v_i = y_{i+2}$, $i = 1, \ldots, n-2$, we obtain the system

$$\begin{aligned} \dot{x} &= -a\Phi(x) + y + d_1^* v, \\ \dot{y} &= -b\Phi(x) - \lambda y + d_2^* v, \\ \dot{v} &= c_1 \Phi(x) + c_2 y + Bv, \end{aligned} \tag{1.2}$$

where $x, y \in \mathbb{R}^1$, $v \in \mathbb{R}^{n-2}$, Φ is a smooth continuous function admitting discontinuities of the derivatives at certain critical points, $a \in \mathbb{R}^1$, $b > 0$, $\lambda \geqslant 0$ are scalars, $\beta \equiv b + \lambda a > 0$, $c_{1,2}$, $d_{1,2}$ are $(n-2)$-vectors, B is a Hurwitz $(n-2) \times (n-2)$ matrix. Denote $c = (c_1, c_2)$, $d = (d_1, d_2)$.

It is assumed that the function $\Phi(x)$ behaves like a polynomial of odd degree, in that it has $2m$ critical points $x = x_i^c$, $i = 1, \ldots, 2m$, and $2m+1$ zeros $x = x_i^0$, $i = 1, \ldots, 2m+1$, that alternate with the points x_i^c. (The choice of an odd number of intervals of monotonicity for $\Phi(x)$ is made for convenience and is not a real constraint.)

To be definite we assume that

$$\Phi(x_1^c) > 0, \qquad \Phi'(x) \neq 0, \quad x \neq x_i^c, \qquad \lim_{|x| \to \infty} |\Phi(x)| = \infty. \tag{1.3}$$

System (1.2) has $2m+1$ equilibrium points $O_k(x_{2k}^0, 0)$, $k = 1, \ldots, m$ and $P_k(x_{2k-1}^0, 0)$, $k = 1, \ldots, m+1$.

For the region $G_u\colon \{\|v\| < u = \text{const}\}$ we define the values

$$\gamma = \max_{\|v\| < u} (|a^{-1} d_1^* v|, |b^{-1} d_2^* v|) = d_0 u, \qquad d_0 = \max(|a|^{-1} \|d_1\|, b^{-1} \|d_2\|),$$

and introduce the two-dimensional comparison system $A^{\pm}$

$$\begin{aligned} \dot{x} &= y - a\Phi(x) \pm a\gamma, \\ \dot{y} &= -\lambda y - b\Phi(x) \pm b\gamma, \end{aligned}$$

which has the following property. The projection of the vector field (1.2) to the plane (x, y) for any $v \in G_u$ rotates clockwise in the region $y > 0$ $(y < 0)$ and counterclockwise in the region $y < 0$ $(y > 0)$ with respect to the trajectories of the system A^+ (A^- respectively) for $\beta = \text{const}$.

Denoting $G_\kappa \colon \{|x| < \kappa, |y| < \eta\}$, we introduce the truncated system $(1.2)_0$ such that $(1.2)_0$ is the system (1.2) for $(x, y) \in G_\kappa$ and is the zero vector field outside G_κ. We put $x^0_{2k} = 0$ for simplicity and consider (1.2) in the region $G_{sd} = I_k \times \mathbb{R}^{n-1}$, where $I_k \colon \{x^c_{2k-1} < x < x^c_{2k}\}$, so that $O_k \in I_k$, $\Phi'(x) < 0$ for $x \in I_k$.

Since the matrix B is stable, the linear system $\dot{v} = Bv + \widetilde{w}(t)$, where $\widetilde{w}(t)$ is any bounded function $\|\widetilde{w}(t)\| < \omega = \text{const}$, $t \in \mathbb{R}^1$, has the absorbing domain $\|v\| < S\omega$, where $S > 0$ is defined by the spectrum of B. The following assertion holds [**4**].

THEOREM 1. *Let the following inequality hold:*

$$q \equiv d_0 S(\|c_1\| + 2b\lambda^{-1}\|c_2\|) < 1. \tag{1.4}$$

Then all equilibrium points O_k are of the saddle (saddle-focus) type with unstable and stable manifolds $W^{u(s)}_k$, $\dim W^u_k = 1$, $\dim W^s_k = n - 1$. All trajectories of the system (1.2) leave the regions $G_{sd} \setminus W^s_k$ $(G_{sd} \setminus W^u_k)$ as $t \to +\infty$ $(t \to -\infty$ respectively).

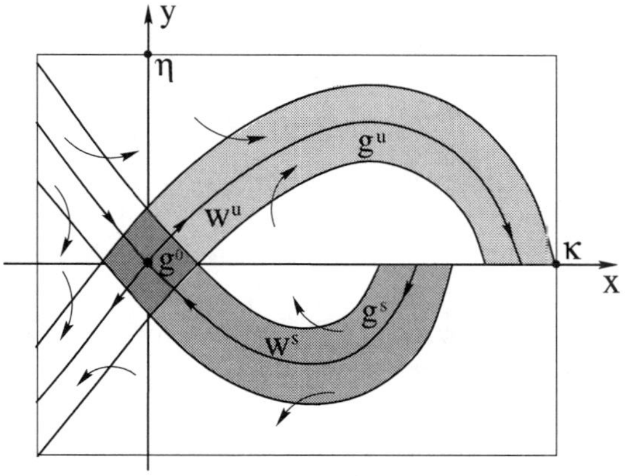

FIGURE 1. The saddle cell g^0 and the separatrix channels g^u and g^s formed by the separatrices of the comparison systems $A^\pm$. The saddle equilibrium point $O \in g^0$, and parts of manifolds W^u and W^s lie in g^u and g^s. The arrows indicate the orientation of the vector field (1.2) at the boundaries of g^u and g^s.

PROOF. Let κ, u be constants such that $G_\kappa \times G_u \subset G_{sd}$. The systems $A^\pm$ have the equilibria $O^\pm_k(x^\pm, 0)$, which are saddles because $\Phi' < 0$. The separatrices of $O^\pm_k$ form the *separatrices channels* $g^{s(u)}_k$ and a *saddle cell* $g^0_k = g^s_k \cap g^u_k$ [**2, 3, 4, 5**] (see Figure 1); the orientation of the vector field $(1.2)_0$ at its boundaries was described above. It is easy to verify that $g^0_k \subset \{x^+ \leqslant x \leqslant x^-, |y| < 2\lambda^{-1}b\gamma\} \subset G_\kappa$ provided that

$$\gamma < \overline{\Phi} = \min(\Phi(-\kappa), |\Phi(\kappa)|), \qquad \eta > 2\lambda^{-1}b\gamma. \tag{1.5}$$

Putting $u = S\omega = S(\|c_1\|\overline{\Phi} + \|c_2\|\eta)$, we see that inequalities (1.5) imply (1.4) and thus for $q < 1$ we have $g_k^0 \subset G_\kappa$. Hence, choosing (κ, η), we construct the saddle cell $G_k^0(1) = g_k^0 \times G_u$ for $(1.2)_0$. Since $g_k^0 \subset G_\kappa$, $G_k^0(1)$ is the cell for (1.2). Now choosing new values

$$\kappa = \max(|x^+|, x^-), \qquad \eta = \max |y|, \quad y \in g_k^0$$

and repeating the same procedure, we obtain the saddle cell $G_k^0(2) \subset G_k^0(1)$. Hence, under conditions (1.4) we have constructed a contracting mapping $G_k^0(i) \to G_k^0(i+1)$ such that $G_k^0(i) \to O_k$ as $i \to \infty$. □

Note that the equilibrium points P_k can be stable as well as unstable. Moreover, the loss of stability is related to the Andronov–Hopf bifurcation.

Obviously, for $c = 0$ the system (1.2) has a globally stable 2-dimensional manifold J: $\{v = 0\}$ with the following system in it (A^0 being $A^\pm$ at $\gamma = 0$)

$$\dot{x} = -a\Phi(x) + y, \qquad \dot{y} = -b\Phi(x) - \lambda y. \tag{1.6}$$

For $\lambda = a = 0$ this system has the integral

$$V(x, y) \equiv \frac{y^2}{2} + b\int_0^x \Phi(\xi)\, d\xi = \text{const}\,. \tag{1.7}$$

By (1.3), all the curves of the family (1.7) are closed when Φ is not periodic and may be nonclosed in the case of a 2π-periodic Φ with $\int_0^{2\pi} \Phi(x)dx \neq 0$. Hence, the separatrix loops of a saddle O_{k_1} and separatrix connections of the saddles O_{k_1} and O_{k_2} of system (1.2) for $c = \lambda = a = 0$ defined by the curves

$$V(x, y) = V(x_{2k}^0, 0), \qquad k = 1, \dots, m, \tag{1.8}$$

are homoclinic and heteroclinic orbits lying in the central manifold J.

In the case $c = 0$, $\lambda \neq 0$, we obtain the existence of bifurcations of the system (1.2), which are bifurcations of (1.6), in the following way. Taking the expression $V(x, y)$ in (1.7) as the Lyapunov function with derivative $\dot{V} = -\lambda y^2 - ab\Phi^2$, we see that system (1.6) is globally asymptotically stable in the parameter range

$$p^s = (\lambda \geqslant 0,\, a \geqslant 0) \setminus (0, 0)$$

and globally unstable for

$$p^u = (\lambda = 0,\, a < 0).$$

For $\beta \equiv b + a\lambda = \text{const}$, as λ increases along any line $a = k_1\lambda + k_2$ connecting p^u and p^s, the vector field of (1.6) rotates uniformly clockwise. Hence, the correspondence $p^u \to p^s$ leads to a sequence of bifurcations including that of the homoclinic orbit asymptotic to O_{k_1} for $\lambda = \lambda_1^0(a, \beta)$ and to a heteroclinic linkage of O_{k_1} and O_{k_2} for $\lambda = \lambda_{1,2}^0(a, \beta)$, $\lambda_1^0(0, b) = 0$, $\lambda_{1,2}^0(0, b) = 0$.

This assertion is valid for the systems $A^\pm$ if the functions $\Phi \pm \gamma$ have the same properties as Φ, i.e., $\gamma < \overline{\gamma}$, and that is what we assume below.

For a rather large range $\|c\| < c^*$, we shall prove the next assertion [**4, 5**].

THEOREM 2. *Let $a < 0$ and condition* (1.4) *of Theorem* 1 *holds.*

1) *Let* (1.8) *be a homoclinic curve of a saddle O_{k_1} and, for $\gamma < \overline{\gamma}$, $0 < \lambda < \overline{\lambda}$, $\overline{a} < a < 0$, let (κ, η) in system $(1.2)_0$ be minimal, so that the channels $g_k^{s(u)}$ connecting g_k^0 and a section $\{y = 0\}$ lie in G_κ (Figure* 1). *Let*

$$d_0 S(\overline{\Phi}\|c_1\| + \eta\|c_2\|) < \overline{\gamma}. \tag{1.9}$$

Then the system (1.2) *has the bifurcation surface* $H_1\colon (\lambda = \lambda_1(a,b,c,d))$ *corresponding to the bifurcation of the homoclinic orbit* $\Gamma_1 \subset (W^u_{k_1} \cap W^s_{k_1})$, *where* $\lambda_1(a,b,0,d) = \lambda_1(a,b,c,0) = \lambda_1^0(a,b)$.

2) *Let* (1.8) *be a heteroclinic linkage of the saddles* O_{k_1} *and let* O_{k_2} *and the previous condition for* $g_k^{s(u)}$ *connecting* g_k^0 *and the section* $\{x = \text{const}\}$ *hold. Then the system* (1.2) *has the bifurcation surface* H_{12} $(\lambda = \lambda_{12}(a,b,c,d))$, *corresponding to the bifurcation of the heteroclinic orbit* $\Gamma_{12} \subset (W^u_{k_1} \cap W^s_{k_2})$, *where* $\lambda_{1,2}(a,b,0,d) = \lambda_1(a,b,c,0) = \lambda_{1,2}^0(a,b)$.

PROOF. First we claim that, due to perturbation theory, the assertion of the theorem is true when $\|c\|$ or (and) $\|d\|$ are small parameters. For the continuation of the bifurcation set as $\|c\|$, $\|d\|$ increase, we apply the method of comparison systems as follows. According to condition 1) of the theorem, both A^+ and A^- have homoclinic orbits for $\lambda = \lambda_1^{\pm}(a,\beta)$. Then for $\beta = \text{const}$, as λ increases along the family of lines $a = k_1\lambda + k_2$ from the parameter range p^u to the range p^s, the channels g_k^u and g_k^s interchange their positions. Due to Theorem 1 the manifolds $W_k^{u(s)}$ leave $G_k^0 = g_k^s \times G_u$ and attain the cross-section $\{y = 0\}$ along the corresponding transitive parts of $g_k^{u(s)} \times G_u$. Hence at some value $\lambda = \lambda_1$ the manifold W_k^u intersects W_k^s, forming the homoclinic orbit asymptotic to O_k. Due to conditions (1.9), we have $W_k^u \cap W_k^s \subset G_\kappa \times G_u$ and therefore the homoclinic orbit of the truncated system $(1.2)_0$ is that of system (1.2). Assertion 2) of the theorem is proved in a similar manner. □

COROLLARY. *Any complex homo-heteroclinic linkage defined by the family* (1.8) *generates the corresponding system* (1.2) *of homo-heteroclinic orbit bifurcation. Let* O_{k_1} *be a saddle-focus with positive saddle value. Then in the neighborhood of* H_1 *there exists a nontrivial bifurcation set related to the loss of hyperbolicity of the infinite set of Smale horseshoes of the corresponding Poincaré mapping. Moreover, any homo-heteroclinic linkage, including saddle-foci, has a neighborhood containing a component of complex chaotic trajectories.*

Thus Theorem 2 serves as a criterion for a complicated dynamics. Note that the complexity of chaotic structures essentially depends on the number of saddles m. Namely, for $m = 1$ the generating curve (1.8) is one "figure-eight" linkage, for $m = 2$ the family (1.8) is either "three sausages" or two "figure-eight in figure-eight" linkages, and so on.

2. Three-dimensional oscillatory system

Consider a vector field $F(X)$ in $\mathbb{R}^3$, $X = (x,y,z)$, defined by the three-dimensional system

$$
\begin{cases}
\dot{x} = H_y + \lambda p(x,y,z) \equiv P,\\
\dot{y} = -H_x + \lambda q(x,y,z) \equiv Q,\\
\dot{z} = f(z) + \alpha r(x,y) \equiv R,
\end{cases}
\tag{2.1}
$$

where λ, α are nonnegative parameters,

$$
H(x,y,z) = h(x,y) + \beta r(x,y)z \tag{2.2}
$$

and the functions satisfy conditions

$$
(2.3)\qquad \begin{gathered} P(0,0,z)=0, \quad Q(0,0,z)=0, \quad r(0,0)=0, \quad h(0,0)=0, \\ f(0)=0, \quad f(z)z<0, \quad z\neq 0, \quad f'(0)<0. \end{gathered}
$$

From (2.1)–(2.3) it follows that

1) $r_x(0,0)=r_y(0,0)=0$, $p(0,0,z)=q(0,0,z)=0$;
2) $O(0,0,0)$ is an equilibrium point;
3) W_0^s $(x=0,\ y=0,\ z\in\mathbb{R}^1)$ is the integral line with the equation $\dot z = f(z)$ on it and hence W_0^s is a stable submanifold of O.

We state the following obvious proposition.

PROPOSITION 1. *For $\alpha=0$ the system (2.1) has the integral manifold I $(z=0,$ $x,y\in\mathbb{R}^2)$ globally stable in $\mathbb{R}^3$. The flow in I is defined by the two-dimensional system*

$$
(2.4)\qquad \begin{aligned} \dot x &= h_y + \lambda p(x,y,0), \\ \dot y &= -h_x + \lambda q(x,y,0). \end{aligned}
$$

From the linearization of the vector field F at O it follows that the type of O is determined by $f'(0)$ and the two eigenvalues of the linearization (2.4) at O. We assume that the equilibrium point $O_0(0,0)$ of (2.4) is a saddle whose unstable separatrix W_0^u is tangent to the one-dimensional unstable manifold W^u of the saddle point $O(0,0,0)$ of (1.1), the latter having a stable 2d-manifold $W^s\supset W_0^s$. For $\lambda=0$ this assumption implies that the integral $h(x,y)=0$ of the Hamiltonian system

$$
\dot x = h_y, \qquad \dot y = -h_x,
$$

encloses the saddle point O_0 and the separatrix W_0^u, which is assumed to be the homoclinic orbit asymptotic to O_0. Let us introduce the line L: $(x=a_1\xi,\ y=a_2\xi,$ $\xi>0,\ z=0)$.

ASSUMPTION 1. For $\lambda>0$ let

1) unstable separatrix W_0^s be asymptotic to an equilibrium point differing from O_0 or to a cycle of (2.4);
2) L is transversal to the flow of (2.4) and does not intersect W_0^u;
3) system (2.1) is dissipative, so that G: $(\|X\|<K)$ for some large enough K is an absorbing domain. Introducing the half-plane $S=L\times\mathbb{R}^1$ that has W_0^s as an edge, and the function

$$
(2.5)\qquad V = H(x,y,z) + \beta\alpha^{-1}\int_0^z f(\eta)\,d\eta,
$$

we assume the following

ASSUMPTION 2. 1) Let the half-plane S be a cross-section transversal to the flows of (2.1), i.e.,

$$
(a_1Q-a_2P)|_s \neq 0.
$$

2) Let the family of sections S' with arbitrary $a_1=a_1'$, $a_2=a'$ (covering the entire space $\mathbb{R}^3\setminus W_0^s$) be transversal to the flow of (2.1) for $\lambda=0$ in the region $V>c$ for some large $c>0$ (linear independence of (H_x,H_y)).
3) The function (2.5) is nonpositive in the equilibria.

THEOREM 3. *Let (2.3) and let Assumptions 1 and 2 be satisfied. Then in the parameter region* $(\alpha > 0, \lambda > 0)$ *as we pass from* $(\alpha > 0, \lambda = 0)$ *to* $(\alpha = 0, \lambda > 0)$, *the manifold* W^u *passes through a heteroclinic or a homoclinic bifurcation.*

PROOF. For $\lambda = 0$ the function (2.5) has the following derivative along the trajectory with respect to (2.1)

$$\dot{V} = \beta\alpha^{-1}\dot{z}^2.$$

Hence, for $\lambda = 0$ the system (2.1) has no cycles and the function V increases along W^u so that, due to Assumption 2, W^u reaches the cross-section S being transversal to S at the point $M = W^u \cap S$. On the other hand, according to Assumption 1, we have $W^u \cap S = \varnothing$ for $\alpha = 0$. Thus as $(\alpha > 0, \lambda = 0) \to (\alpha = 0, \lambda > 0)$, the point M disappears. But this may occur only when either M turns into the cross-point of another unstable manifold at the heteroclinic bifurcation, or merges with the point O at the homoclinic bifurcation. □

One may consider system (2.1) as the generalization of the Lorenz system written in oscillatory form [**6**], which is (2.1) in the case

$$h(x,y) = x^4/4 - x^2/2 + y^2/2, \qquad r(x,y) = x^2/2, \tag{2.6}$$
$$p = 0, \quad q = -y, \quad f = -kz.$$

It is easy to verify that all the assumptions used are valid for the particular case (2.1), (2.6), as well as for (2.6) with the Van-der-Pol term $q = -(x^2 - a^2)y$, and so on.

3. System with small parameter

Consider the system with a slow variable

$$\begin{aligned} \dot{x} &= P(x,y) + \mu p(x,y,\mu), \\ \dot{y} &= \mu Q(x,y,\mu), \end{aligned} \tag{3.1}$$

where $x \in \mathbb{R}^n$, $y \in \mathbb{R}^1$, P, p, and Q are assumed to be sufficiently smooth vectors and scalar functions, respectively, $\mu > 0$ is a small parameter.

For all $y \in \mathbb{R}^1$ the reduced system

$$\dot{x} = P(x,y), \qquad y = \text{const} \tag{3.2}$$

is assumed to have an equilibrium state $x = \varphi(y)$ $(P(\varphi(y), y) \equiv 0)$. We present the result [**7, 8**] related to the existence of a homoclinic orbit for (3.1).

THEOREM 4. *Let the Jacobian matrix* $DP(\varphi(y), y)$, $y \in \mathbb{R}^1$, *have* m *eigenvalues to the left and* $n - m$ *eigenvalues to the right of the imaginary axis. Then there exists a number* $\mu_0 > 0$ *such that, if* $\mu \in (0, \mu_0)$, (3.1) *has a unique one-dimensional integral manifold* $J(\mu)$: $\{x = \widetilde{\varphi}(y,\mu), y \in \mathbb{R}^1\}$ *close to the reduced manifold* $J(0)$: $\{x = \varphi(y)\}$ *which is the intersection of the stable* $W^s_{\text{loc}}(\mu)$ *and unstable* $W^u_{\text{loc}}(\mu)$ *local manifolds. The nonlocal parts of* $W^s(\mu)$ *and* $W^u(\mu)$ *are the continuations of* $W^{s(u)}_{\text{loc}}(\mu)$ *by the flow in finite time and are continuous in the parameter* μ.

Observe that the nonlocal behavior of (3.1) in the scenario of Theorem 4 depends both on the arrangement of the nonlocal manifolds and the equilibria of the equation

$$\dot{y} = \mu Q(\widetilde{\varphi}(y), \mu)$$

at the manifold $J(\mu)$.

COROLLARY. *Let the functions P, p, and Q be 2π-periodic in y and $Q(x, y, \mu) > 0$. Then $J(\mu)$ is a saddle periodic orbit in the cylindrical phase space $\mathbb{R}^n \times S^1$ of (3.1).*

THEOREM 5. *Let the saddle-type equilibrium point of the reduced system (3.2) $O = J(0)|_{y=y_0}$ have a homoclinic orbit Γ_0, which is the transversal intersection of the manifolds $W^s(0)$ and $W^u(0)$:*

$$\dim(W^s_M \cap W^u_M) = 1,$$

where $W^{s(u)}_M$ are the tangent manifolds to $W^{s(u)}(0)$ at the point $M \in \Gamma_0$. System (3.1) then has a homoclinic orbit Γ_μ asymptotic to the manifold $J(\mu)$.

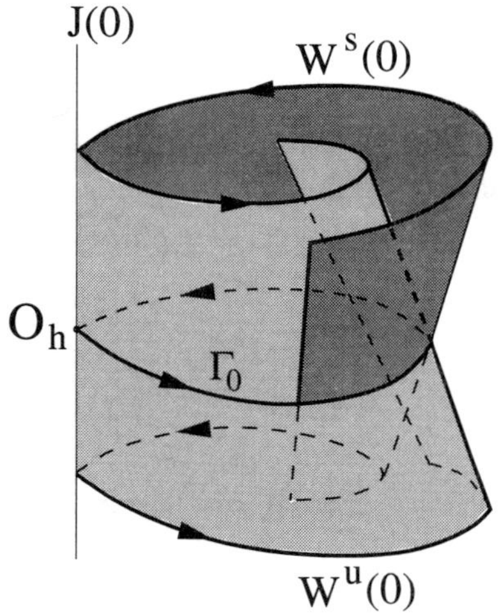

FIGURE 2. The arrangement of the stable $W^s(0)$ and unstable $W^u(0)$ manifolds of the reduced system in Theorem 5.

Note that the proof of this theorem [**8**] is based on the continuous dependence of the finite parts of the manifolds W^s, W^u on the parameter μ. Figure 2 illustrates the arrangement of the reduced manifolds $W^s(0)$ and $W^u(0)$ in Theorem 5.

COROLLARY 1. *If (3.1) has an equilibrium point O_μ: $(y = y_\mu) \subset J(\mu)$ passing through O as the function Q varies, then there exists a bifurcation of the homoclinic orbit Γ_μ asymptotic to O_μ.*

COROLLARY 2. *In the case where (3.1) is 2π-periodic in y and $Q > 0$, Theorem 5 implies that the homoclinic orbit Γ_μ is asymptotic to the periodic orbit $J(\mu)$.*

EXAMPLE 1. The nonautonomous pendulum system written in the form

$$\dot{x_1} = x_2, \quad \dot{x_2} = \sin x_1 - a \sin y, \quad \dot{y} = \mu, \tag{3.3}$$

where μ is a small frequency, for $a < 1 - \varepsilon$, where $\varepsilon(\mu) > 0$ is small, has a saddle periodic orbit close to the curve $x_1 = \arcsin(a \sin y)$, $x_2 = 0$. Since for $\mu = 0$ (3.3) has homoclinic orbits for $y = 0, \pi$, for small $\mu > 0$ system (3.3) has a saddle periodic orbit with two (and hence infinite number) homoclinic orbits asymptotic to it. Note that the well-known Mel′nikov formula gives the range $a \ll 1$, $\mu > 0$ for this example.

EXAMPLE 2. The three-dimensional system

$$\begin{aligned} \dot{x} &= y, \\ \dot{y} &= F(z) - E(x,z)x - D(x,y,z)y, \\ \dot{z} &= \mu C(x,y,z), \end{aligned} \tag{3.4}$$

represents most of the concrete systems exhibiting chaotic behavior. In terms of mechanics, the right-hand side of (3.4) may be interpreted as follows: F is the driving force, E is the coefficient of elasticity, D is the damping coefficient, and C is a control function that changes D, E, F via $z(t)$. If in the reduced system (3.4) for $\mu = 0$ (zero-rate control) a change of arrangement of the separatrices through a homoclinic orbit occurs as the parameter z increases, then according to Theorem 5 system (3.4) has a corresponding homoclinic orbit for $\mu > 0$.

Observe that Theorem 5 can be similarly applied to system (2.1) when the functions f and r have a small multiplier.

4. Geometric models

In order to analyze the detailed structures of chaotic sets of trajectories and bifurcations, we consider the following geometric models for the vicinities of homo-heteroclinic linkages [**5, 9**].

Let G_k, $k = 1, \dots, m$, be domains homeomorphic to balls in $\mathbb{R}^n$ containing saddles $O_k \in G_k \subset I_k$. Let g_k be global tubular neighborhoods (tubes) of the unstable manifolds W_k^u consisting of two parts: g_k^+ and g_k^- pasted to G_k at the opposite sides of the stable manifolds W_k^s. Let the endpoints of each tube $g_k^{+(-)}$ be pasted to the surface of the ball G_l, where $l = 1, \dots, m$ depends on the linkage: $l = k$ corresponds to a homoclinic and $l \neq k$ to a heteroclinic type of linkage. Assume that the balls G_k and tubes $g_k^{+(-)}$ are pairwise disjoint, and their union

$$G = \bigcup_{k=1}^{m} (G_k \cup g_k^+ \cup g_k^-)$$

is connected. Thus G is the connected geometric figure consisting of m balls and $2m$ handles.

Suppose the following conditions are valid.

1) G is an absorbing domain of the vector field F simulating system (1.2).
2) In each ball G_k, the vector field F has the saddle type equilibrium point O_k and may be considered as linear.
3) The trajectories of F in each tube $g_k^{+(-)}$ are transitive and generate the mappings $S_{lk}^{+(-)}\colon \partial G_k \to \partial G_l$.

STATEMENT 1. *The vector field F has an attractor $A \subset G$.*

Denote by $d_k^{+(-)} = \partial G_k \cap g_k^{+(-)}$ the disks of pasted points of ∂G_k and $g_k^{+(-)}$. Let $l_k = \partial G_k \cap W_k^s$, $M_k^{+(-)} = d_k^{+(-)} \cap W_k^u$, $D_k = \partial G_k \setminus (d_k^+ \cap d_k^-)$. The stable manifold W_k^s divides D_k into two parts $D_k^{+(-)}$ such that $D_k = D_k^+ \cup l_k \cup D_k^-$.

Due to these assumptions, we obtain

STATEMENT 2. *The vector field F generates discontinuous mappings $f_{lk} = S_{lk}^{+}T_k$, where*

$$(4.1)\qquad \begin{aligned} &T_k\colon D_k^{+(-)} \to d_k^{+(-)},\ l_k \to M_k^{+} \cup M_k^{-},\\ &S_{lk}^{+(-)}\colon d_k^{+(-)} \to D_l, \qquad k,l = 1,\dots,m. \end{aligned}$$

The limiting set of f_{lk} is the transversal cross-section of the attractor A.

In the case of (1.2) with $n = 3$ and a cubic-like $\Phi(x)$ $(m = 1)$, when O_1 is a saddle-focus, the geometric model was proposed in [**10**]; the simulating mapping (4.1) with local polar coordinates for the cylinder G_1 has the explicit form

$$(4.2)\qquad \begin{aligned} &\overline{\varphi} = \psi - \pi/2 + (\pi/2 + a|x|^{\nu}\cos(\varphi - \omega\ln|x|))\operatorname{sgn} x,\\ &\overline{x} = \mu\operatorname{sgn} x - a|x|^{\nu}\sin(\varphi - \omega\ln|x|), \qquad x \in D_1, \end{aligned}$$

where $D_1 = \{|x| < h,\ \varphi \in S^1\}$, μ, ν, ω, ψ are positive parameters, and $\nu < 1$. The chaotic attractor of (4.2) is characterized by the following assertion.

THEOREM 6. *In the parameter range*

$$\nu < 1, \quad a^{1/(1-\nu)} < h, \quad 0 \leqslant \mu \leqslant h - ah^{\nu}$$

the map (4.1) has a chaotic attractor A_1, which is a quasiattractor for $1/2 < \nu < 1$ (A_1 has a countable set of stable periodic orbits for a dense set of parameters) and a strange attractor having no stable trajectories for $0 < \nu < 1/2$.

REMARK. Within the interesting parameter range $0 < \nu < 1/2$, the simulating map (4.1) is noninvertible and can be related to the continuous system as a quotient map.

Now consider a similar model for the general case $n > 3$, assuming that O_1 is a saddle-focus and the linear system in the cell G_1 is given in the form

$$(4.3)\qquad \dot{x} = \gamma x, \quad \dot{y} = Ly, \quad \dot{u} = -\sigma u - \omega_0 v, \quad \dot{v} = \omega_0 u - \sigma v,$$

where $x \in \mathbb{R}^1$, $y \in \mathbb{R}^p$, $p = n-3$, $(u,v) \in \mathbb{R}^2$; the parameters satisfy the inequalities $\gamma > 2\sigma > 0$, $\omega_0 > 0$, and L is a Hurwitz matrix with eigenvalues α_i, $i = 1,\dots,p$, such that $\operatorname{Re}\alpha_i \leqslant \operatorname{Re}\alpha_1 < -\sigma$, $G_1 = \{|x| \leqslant h,\ u^2 + v^2 \leqslant r^2\}$. Using the polar coordinates $u = \rho\cos\varphi$, $v = \rho\sin\varphi$ and the elapse time $\tau = -\gamma^{-1}\ln|x(0)/h|$, we obtain the following formulas for the mapping $T_1\colon (\varphi(0), x(0), y(0)) \mapsto (u(\tau), v(\tau), y(\tau))$

$$(4.4)\qquad \begin{aligned} &u(\tau) = a|x|^{\nu}\cos(\varphi - \omega\ln|x|),\\ &v(\tau) = a|x|^{\nu}\sin(\varphi - \omega\ln|x|),\\ &y(\tau) = |x|^{\alpha}R(x)y, \end{aligned}$$

where $x(0) = x$, $y(0) = y$, $\varphi(0) + \psi = \varphi$, $\psi = \omega\ln h$, $\nu = \sigma/\gamma$, $\omega = \omega_0/\gamma$, $a = rh^{-\nu}$, $\alpha = |\operatorname{Re}\alpha_1|/\gamma$.

Matrix $R(x)$ generated by the solutions of (4.3) has the elements $R_{i,j}$, $i,j = 1,\dots,p$, which are bounded for $|x| < h$, $|R_{ij}| < K$, so that $\|R(x)\| < Kp^2$. The elapse linear mappings S_{11}^{+} and S_{11}^{-} simulating the "figure-eight" structure of G and returning the coordinates $(x,\varphi,y) \in D_1$ to $(\overline{x},\overline{\varphi},\overline{y}) \in D_1$ are given by

$$(4.5)\qquad \begin{aligned} &\overline{\varphi} = \psi - \pi/2 + (\pi/2 + u(\tau))\operatorname{sgn} x,\\ &\overline{x} = \mu\operatorname{sgn} x - v(\tau),\\ &\overline{y} = \Delta\operatorname{sgn} x + y(\tau), \end{aligned}$$

where $x > 0$ for S_{11}^{+} and $x < 0$ for S_{11}^{-}, the controlling scalar parameters are positive, $\mu > 0$ and $\delta > 0$, and $\Delta = \text{column}(\delta, \dots, \delta)$. The special form of (4.5) is needed to match the odd symmetry. Thus the resulting modeling map $f_{11} = S_{11}^{+(-)}T_1 : D_1 \to D_1$, due to (4.4), (4.5), can be written as the map (4.2) with additional third map equation

$$\overline{y} = \Delta \operatorname{sgn} x + |x|^{\alpha} R(x) y. \tag{4.6}$$

Since $\|R(x)\| < Kp^2$, $|x| < h$ and $\alpha > 0$, for $Q = Kp^2h^{\alpha} < 1$ the map (4.6) is contractive. Moreover, the following assertion is valid.

LEMMA 1. *Let $Q < 1/2$. Then for any orbit $x(k)$, $k = 0, 1, \dots,$ of the difference equations* (4.2) *such that $|x(k)| < h$, the map* (4.6) *has an absorbing domain consisting of two separate components $\Omega^{+} \subset \{y_i > 0\}$ and $\Omega^{-} \subset \{y_i < 0\}$, where $y = \text{column}(y_i)$, $i = 1, \dots, p$, and the trajectories of* (4.6) *leave $\Omega^{+(-)}$ for $\Omega^{-(+)}$ when $x(k)$ changes sign.*

PROOF. Consider the norm $\|y\| = p^{-1} \sum_{i=1}^{p} |y_i|$ and let $\operatorname{sgn} x(k) = 1$. Obviously, $\|\overline{y}\| < Q\|y\| + \delta$. Then $\|\overline{y}\| < \|y\|$ in the region $\|\overline{y}\| > \delta(1-Q)^{-1}$ and hence $\Omega^{+} \subset \{\|\overline{y}\| < \delta(1-Q)^{-1}\}$. For $\delta > Q\|\overline{y}\|$ and any $y_i > 0$, the image is positive, $\overline{y}_i > 0$, $i = 1, \dots, p$. Therefore, if $Q < 1/2$, then $\Omega^{+} \subset \{y_i > 0,\ i = 1, \dots, p\}$.

Similarly, for $\operatorname{sgn} x(k) = -1$ we see that the domain Ω^{-} satisfies

$$\Omega^{-} \subset \{\|\overline{y}\| < \delta(1-Q)^{-1}\} \cap \{y_i < 0,\ i = 1, \dots, p\}.$$

Finally, for $\operatorname{sgn} x(k) = -1$ and $y_i > 0$ the image is again positive, $\overline{y}_i > 0$, $i = 1, \dots, p$. □

Thus Theorem 6 and Lemma 1 imply the following result.

THEOREM 7. *In the parameter range*

$$0 < \nu < 1/2, \quad Q < 1/2, \quad \mu < \mu^*, \quad a < a^*,$$

the $(n-1)$-dimensional map (4.2), (4.6), *being a one-to-one map with singularity at $x = 0$, has a strange attractor A_2 without stable trajectories.*

Note that the addition of the supplementary dimension in order to avoid the noninvertibility of (4.2) was proposed by D. V. Turaev and L. P. Shilnikov.

In the case of n-dimensional oscillatory systems similar to (2.1), when the origin is the saddle point O_1, the linearization in G_1 may be written in a form similar to (4.3):

$$\dot{x} = \gamma x, \quad \dot{y} = Ly, \quad \dot{u} = -\sigma_1 u, \quad \dot{v} = -\sigma_2 v.$$

In order to simulate a structure with a symmetric homoclinic linkage, we assume the map to have central symmetry and for the sake of simplicity consider only the linear contribution to the maps $S_{11}^{+(-)}$. Then the model mapping $f_{11}^{+(-)} : D_1^{+(-)} \to d_1^{+(-)} \to D_1$, where $D_1 : (u = c)$, $d_1^{+(-)} : (x = \pm h)$ acquires the form

$$\overline{x} = (\mu + a|x|^{\nu}) \operatorname{sgn} x, \tag{4.7}$$

this equation being taken together with (4.6).

This map satisfies the conditions of the theorem from [**11**] and has a Lorenz-type strange attractor precisely for the range of parameters

$$1/2 < \nu < 1, \quad (a/2)^{1/(1-\nu)} \leqslant \mu \leqslant (a\nu)^{1/(1-\nu)}, \quad Q < 1/2.$$

Finally we note that the dynamics of the complex mapping f_{lk}, which can be constructed on the basis of (4.2) and (4.7) as its main constituents, reflects the complicated behavior of the flows in the neighborhood of the homo-heteroclinic linkages.

References

[1] A. A. Andronov, E. A. Leontovich, I. I. Gordon, and A. G. Maier, *Theory of bifurcations of dynamical systems on the plane*, "Nauka", Moscow, 1967; English transl., Israel Program of Scientific Translations, Jerusalem and London, 1973.

[2] V. N. Belykh, *Bifurcations and attractors of phase systems*, Nonlinear Dynamics and Quantum Dynamical Systems (Gaussig, 1990), Math. Res., vol. 59, Akademie-Verlag, Berlin, 1990, pp. 21–26.

[3] V. N. Belykh and V. I. Nekorkin, *On the qualitative investigation of the multidimensional phase system*, Sibirsk. Mat. Zh. **18** (1977), no. 4, 723–735; English transl. in Siberian Math. J. **18** (1977).

[4] V. N. Belykh, *Homoclinic and heteroclinic trajectories of a family of multidimensional dynamical systems*, Trudy Mat. Inst. Steklov. **216** (1997), 14–25; English transl. in Proc. Steklov. Inst. Math.

[5] ———, *Chaotic dynamics in multidimensional control systems*, Proceedings of the 1st International Conference Control of Oscillations and Chaos, St. Petersburg, Russia, vol. 1, 1997, pp. 105–108.

[6] ———, *On bifurcation saddle separatrices of the Lorenz system*, Differentsial′nye Uravneniya **20** (1984), no. 10, 1666–1674; English transl. in Differential Equations **20** (1984).

[7] L. N. Belyustina and V. N. Belykh, *On nonautonomous phase system of equation with a small parameter, which contains invariant tori and robust homoclinic curves*, Izv. Vyssh. Uchebn. Zaved. Radiofiz. **15** (1972), 1039–1048; English transl. in Radiophys. and Quantum Electronics, **15** (1972).

[8] V. N. Belykh, *Generation of a structurally stable doubly asymptotic trajectory in a system with slow varying parameter*, Differentsial′nye Uravneniya **11** (1975), no. 11, 2083–2085; English transl. in Differential Equations **11** (1975).

[9] ———, *New types of strange attractors and its models*, Abstracts of III National Conference on Nonlinear Oscillations of Mech. Systems, Ninzhny Novgorod, 1993, pp. 28–32.

[10] V. N. Belykh and L. O. Chua, *New type of strange attractor from a geometric model of Chua's circuit*, Internat. J. Bifur. Chaos Appl. Sci. Engrg. **2** (1992), no. 3, 697–704.

[11] V. S. Afraimovich, V. V. Bykov, and L. P. Shilnikov *On the Lorenz-attractor type structurally unstable attracting limit sets*, Trudy Moskov. Mat. Obshch. **44** (1982), 150–212; English transl. in Trans. Moscow Math. Soc. **1983**, issue 3.

Mathematics Department, Volga State Academy of Water Transportation, 5, Nesterova st., Nizhny Novgorod 603600, Russia

E-mail address: belykh@unn.ac.ru

Translated by THE AUTHOR

Amer. Math. Soc. Transl.
(2) Vol. **200**, 2000

On the Mutual Arrangement of a Conic and a Quintic in the Real Projective Plane

Anna A. Binstein and Grigoriĭ M. Polotovskiĭ

Dedicated to the memory of Professor E. A. Leontovich-Andronova

ABSTRACT. The problem of isotopy classification of arrangements of a real projective conic and a quintic is considered under natural conditions of general position and maximality. The main attention is given to obtaining restrictions.

Foreword of the second author. For more than 15 years I worked together with Evgeniya Aleksandrovna (E. A. below) Leontovich-Andronova in the Department of Differential Equations that she headed at the Institute for Applied Mathematics and Cybernetics. In fact, this happened mainly because of the efforts of E. A.: at the time, to hire someone with a "defective" application form (in my case the defect was being Jewish) was a highly nontrivial problem. Here I should say that I was not a student of E. A. in any formal sense, and soon after coming to the Institute and searching for "my own" problem, I began studying plane algebraic curves under the direction of D. A. Gudkov, with E. A.'s full approval. Of course, following D. Hilbert and A. A. Andronov, E. A. considered (see [**1, 2**]) the topology of real algebraic varieties ideologically and methodologically close to the qualitative theory of differential equations, the field of her permanent research interest. However, E. A. generally offered her collaborators a great deal of freedom in the choice of their field of research, and took interest in a wide range of mathematical topics. For example, when Yu. V. Matiyasevich concluded the proof of the algorithmic undecidability of Hilbert's 10th problem, E. A. decided to acquire a deeper understanding of this result, and organized a semester-long seminar of expository character in algorithmic problems, to which she invited the mathematical logician M. I. Lyogon'kiĭ.

E. A. followed my research regularly and in detail, constantly encouraged and assisted me in many ways. Her constant support was also often noted (see, e.g., [**5**]) by my teacher D. A. Gudkov.

I hope that this Foreword explains my gratitude to E. A. and justifies the inclusion of the article below in this volume dedicated to her memory.

2000 *Mathematics Subject Classification.* Primary 14P25, 14E99.

©2000 American Mathematical Society

Introduction

In this paper we continue studying the topology of real plane decomposable curves of degree 7 in the most interesting case of two cofactors under certain natural conditions of general position and maximality.

Let C_2 be the set of real points of a 2nd degree M-curve, i.e., C_2 is an oval in the real projective plane $\mathbb{R}P^2$, and let C_5 be the set of real points of an M-curve of degree 5, i.e., C_5 consists of an odd branch J_5 and six ovals each outside the others in $\mathbb{R}P^2$. As usual, by an *oval* (*odd branch*) we mean a two-sided (one-sided, respectively) circle embedded in $\mathbb{R}P^2$. We assume that C_2 intersects J_5 at 10 distinct points.

The problem we study is: *Which curves $C_2 \cup C_5$ are possible up to isotopy in $\mathbb{R}P^2$ under the above conditions?*

A general approach to solve this problem is the following:

(i) obtain the list of *admissible topological models*; here a topological model is a collection of eight circles in $\mathbb{R}P^2$ such that the union of these circles may pretend to represent some curve $C_2 \cup C_5$ up to a homeomorphism; the model is called *admissible* if it does not contradict the topological consequences of the Bezout theorem for intersections with a straight line or a conic;

(ii) prove that the given model from the list (i) cannot be realized by a degree 7 curve, or

(iii) present a realization of the given model by a degree 7 curve.

These items and the result of applying this approach were briefly described in [**3, 4**]. Here we will pay more attention to *restrictions*, i.e., consequences of the Bezout theorem and part (ii).

1. The list of admissible models

We shall depict our topological models by using the *Poincaré disk* as the model of the projective plane $\mathbb{R}P^2$: the interior of the disk is an affine chart and the boundary circle with identified diametrically opposite points is the infinite line l_∞ of this chart. Sometimes the line l_∞ will be shown in figures as a dotted circle, in other cases it will be omitted. It is obvious that one can always choose the affine chart so that C_2 is an ellipse E lying in this chart and the odd branch J_5 intersects l_∞ without tangency at s points. There are three different cases: $s = 1$, $s = 3$ or $s = 5$.

First, all admissible arrangements of $E \cup J_5$ are enumerated. The algorithm of this enumeration, based on gluings of possible collections of arcs of J_5 inside E with outer ones, was described in detail in [**4**]. Here we give results only as a list of pictures (see Figures 1, 2, 3 for the cases $s = 1$, $s = 3$, $s = 5$, respectively, where the ellipse E is shown as the standard circle). Of course, there exists a coding for these pictures, but here we prefer to give figures in order to show fragments of proofs.

The set $\mathbb{R}P^2 \setminus (E \cup J_5)$ always consists of 11 curvilinear polygons B_1–B_{11} (some of them are digons), among which s polygons intersect l_∞ and will be called *infinite polygons*. Our next goal is to find which of the polygons may contain ovals of the quintic. Such polygons will be called *admissible domains*.

1.1. Pencils of straight lines. The following lemmas are evident corollaries of the Brusotti theorem on the independence of perturbations of nondegenerate

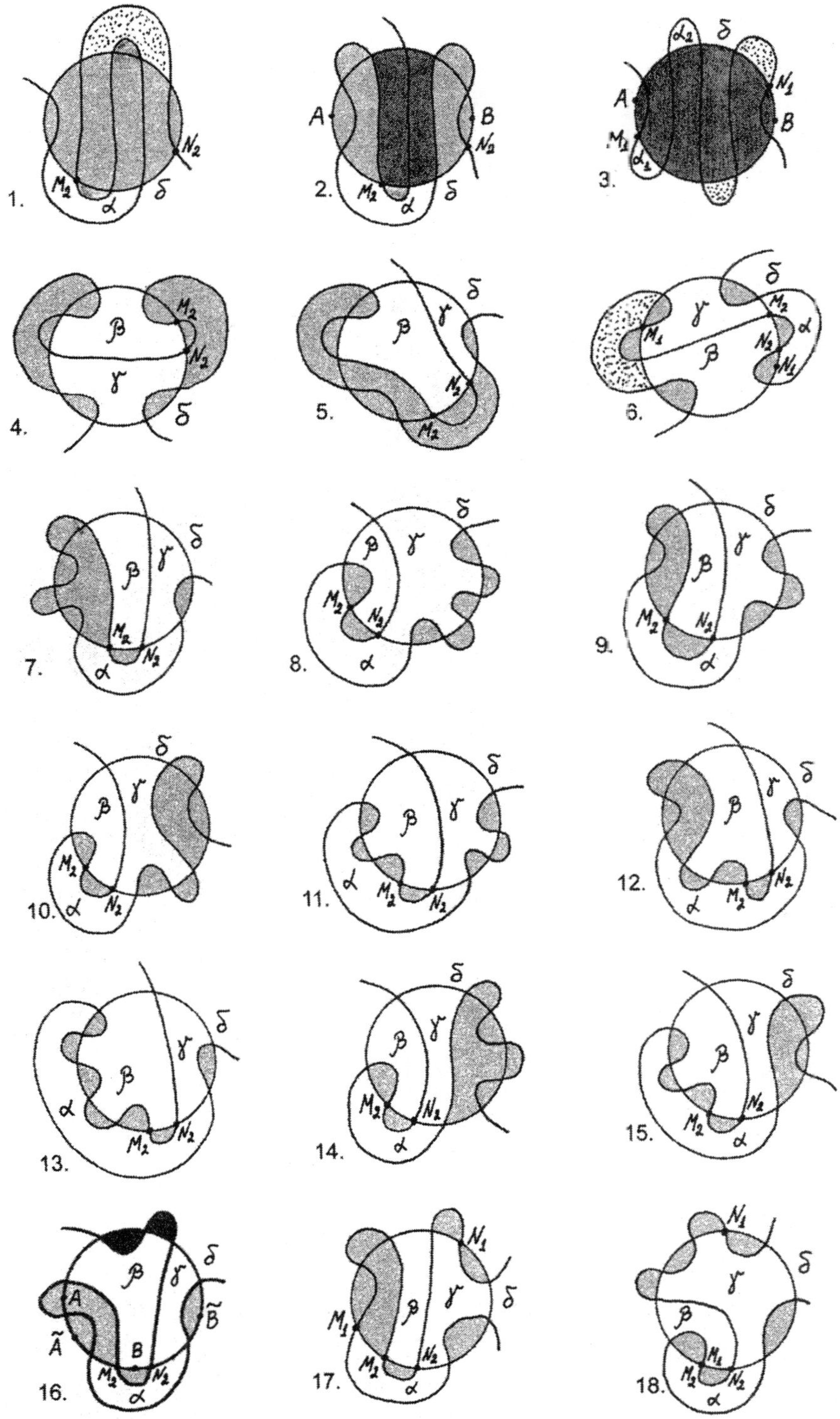

FIGURE 1

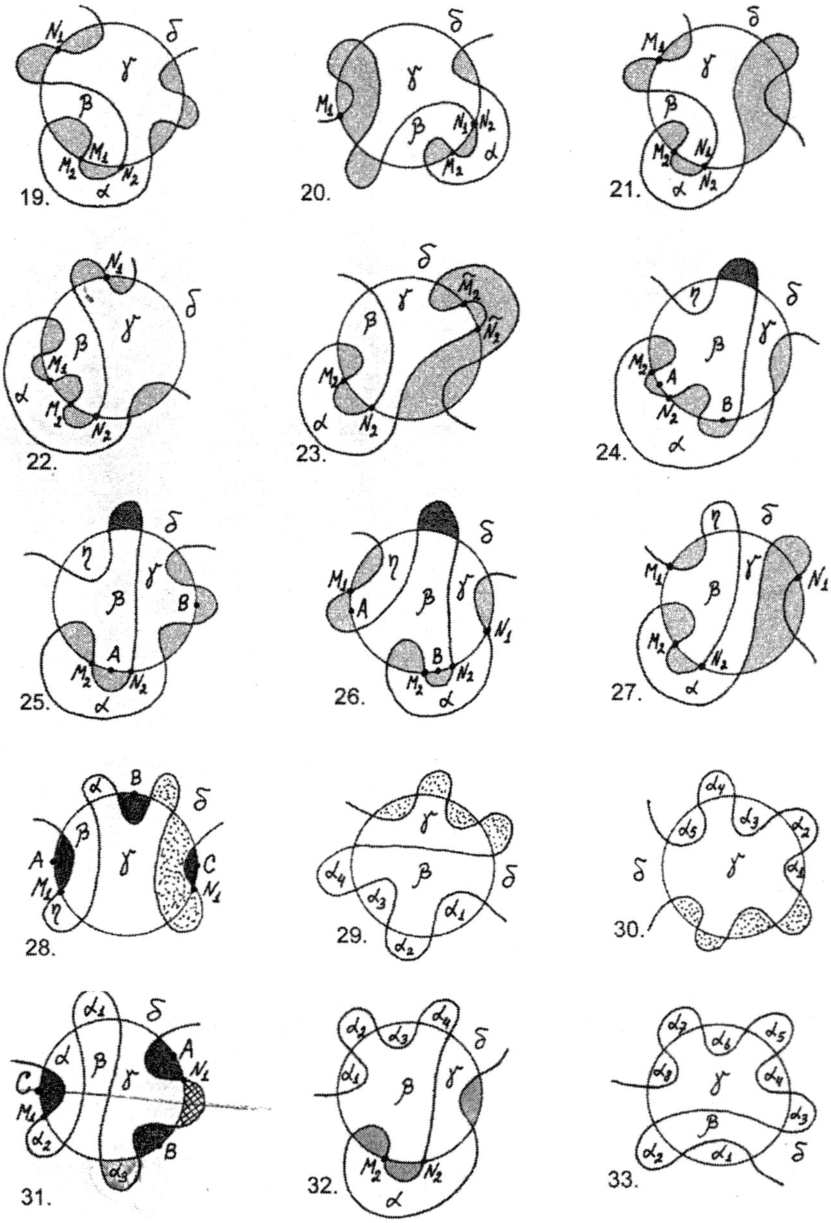

FIGURE 1 (CONTINUED)

double points (see [**5**]) and the Bezout theorem on the number of common points of two algebraic curves.

LEMMA 1. *Polygons B_i, B_j without a common side cannot contain ovals simultaneously.*

LEMMA 2. *If B_i is a finite polygon that has no common sides with any infinite polygon, then B_i is inadmissible.*

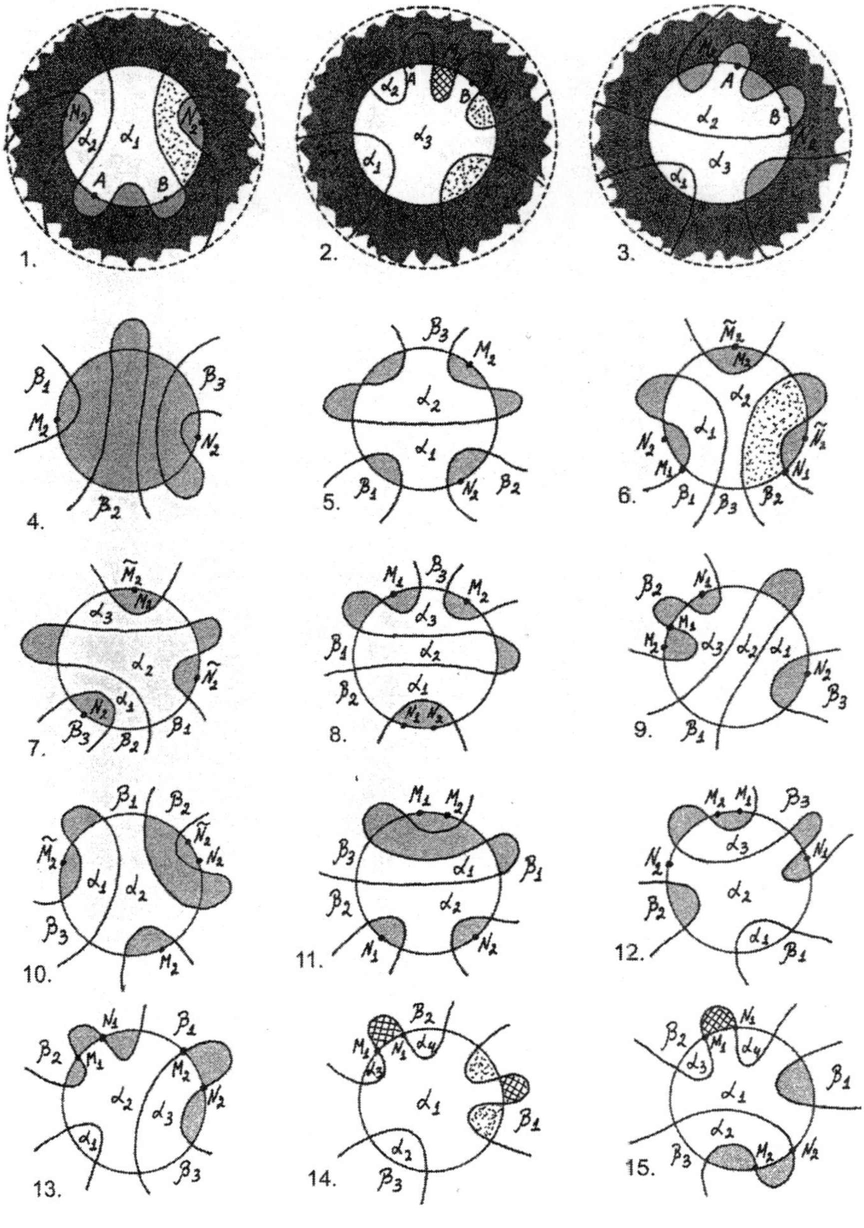

FIGURE 2

LEMMA 3. *Let B_i be an inadmissible domain by Lemma 2 and $\bar{B}_i \cap \bar{B}_j = \varnothing$. Then B_j is inadmissible.*

If two polygons satisfy the condition of Lemma 1 and are equivalent due to the symmetry of the model, then we arbitrarily choose one of them as admissible and mark the others by dotted shading. Polygons which are inadmissible by Lemmas 2, 3 are painted light-grey.

Besides, sometimes for some polygon B_i it is possible to find a pencil Pl of straight lines such that for every point $p \in B_i$ line $l \in Pl$ through the point p

intersects $E \cup J_5$ at 7 points. These B_i are inadmissible; we indicate them by cross-hatching in Figures 1 and 2; see, for example, the model 31 (Figure 1) where the center of Pl is the point C.

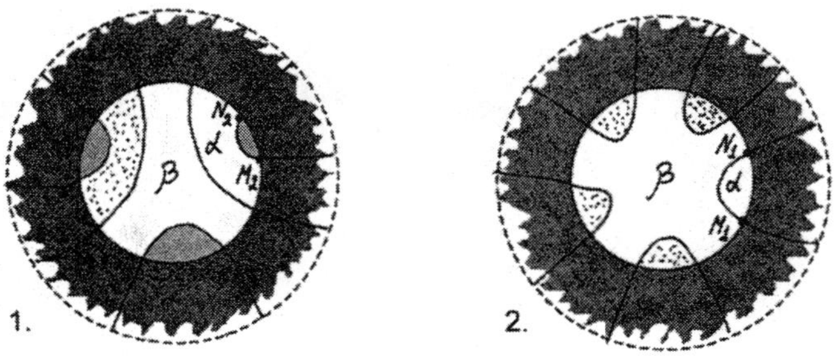

FIGURE 3

1.2. Pencils of conics. As well as the pencils Pl above, we shall use pencils Pc of conics: if conics from some "segments" of the "line" Pc sweep the polygon B_i and each conic from this "segment" intersects J_5 in 10 points, then B_i is inadmissible by the Bezout theorem (we use quotation marks for objects in the space $\mathbb{R}P^5$ of real conics to distinguish them from objects of the same name in $\mathbb{R}P^2$).

Let us consider the pencil Δ_{ab} of conics, where a, b are tangents to the ellipse E at the points A and B, respectively; $\Delta_{ab} = AB \cup E_{ab} \cup E \cup \Theta_{ab} \cup (a \cup b) \cup \Gamma_{ab}$, where the "point" AB is the double line AB, the "segment" E_{ab} sweeps the interior of the ellipse E, and so on, see Figure 4. It is clear that in order to prohibit a given polygon, a suitable choice of base points is necessary. These points A and B (or $\widetilde{A}$ and $\widetilde{B}$) are shown in Figures 1 and 2; in the models 28 and 31 in Figure 1 for the base points we take all pairs formed by the points A, B, C; the pencils of conics in Figure 3 are pencils of ellipses expanding through E.

The polygons B_i that are inadmissible by this method are painted dark grey.

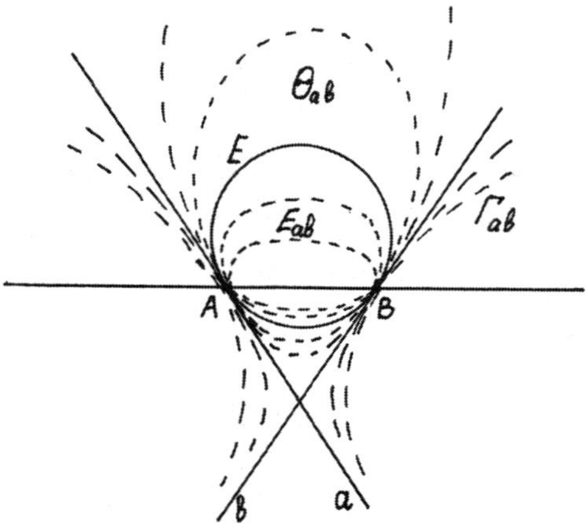

FIGURE 4

2. Restrictions for admissible models

Using considerations in Section 1, we obtained inadmissible domains marked in some way (sometimes some polygons are inadmissible for several reasons simultaneously but we show only one). In the remaining polygons we write Greek letters $\alpha, \beta, \gamma, \ldots$ (sometimes with indices) which indicate the numbers of ovals of C_5 in the related polygons. Since C_5 is an M-curve, it follows that

$$\alpha + \beta + \gamma + \cdots = 6. \tag{1}$$

This condition together with Lemma 2 and elementary arguments of the equivalence of polygons by symmetry give us 1956 pairwise distinct admissible models for $C_2 \cup C_5$ in the case $s = 1$, they yield 1194 in the case $s = 3$, and 14 in the case $s = 5$. Now we shall find some restrictions on the numbers $\alpha, \beta, \gamma, \ldots$.

2.1. Application of complex orientation theory. Let us recall some notions and facts (for the details, see [**6**]).

If the set $\mathbb{R}F$ of real points of the real n-degree curve F divides the set $\mathbb{C}F$ of its complex points, then $\mathbb{R}F$ obtains two opposite boundary orientations coming from the two halves of this partition. These orientations are called *complex orientations.* Any pair of ovals with one of lying inside the other, is called an *injective pair.* An injective pair O, O' is called *positive* if the orientations of O, O' are induced by some orientation of the annulus bounded by them in $\mathbb{R}P^2$ and *negative* in the opposite case. The number of positive (negative) pairs is denoted by Π^+ (Π^-, respectively). If n is odd, an oval O is called *positive* if $[O] = -2[J]$, where $[c]$ is the integral one-dimensional homology class of $\mathbb{R}P^2 \setminus \operatorname{Int} O$ determined by the cycle c, and *negative* in the opposite case. The number of positive (negative) ovals is denoted by Λ^+ (Λ^-, respectively). Complex orientations of an n-degree curve having l ovals satisfy the following relations:

$$2(\Pi^+ - \Pi^-) = l - k^2, \qquad \text{if } n = 2k, \tag{2}$$

$$\Lambda^+ - \Lambda^- + 2(\Pi^+ - \Pi^-) = l - k(k+1), \qquad \text{if } n = 2k+1. \tag{3}$$

We use the complex orientations to obtain restrictions in the following way. Let some model of $C_2 \cup C_5$ with values $\alpha = \alpha_*$, $\beta = \beta_*$, $\gamma = \gamma_*$, $\ldots$ be realized by some curve of degree 7. Then C_2 and C_5 have some complex orientations as M-curves. Let an orientation of E be fixed. Orient J_5 in an arbitrary way and remove all the double points of $E \cup J_5$ in accordance with the chosen orientations so that the 6 ovals of C_5 are only slightly moved. By the Brusotti theorem, it is possible to do this by a small perturbation of the coefficients of the curve. Then the 11 polygons B_i are transformed into the new odd branch J_7 and m new ovals for the some choice of the orientation of J_5 or $11 - s - m$ ovals for the opposite choice, where $m \equiv 1 \pmod 2$ and $m \leqslant (11 - s)/2$ (the value of m depends on the initial model). The induced orientation of the obtained nonsingular curve (6 "old" ovals preserve their orientations) is its complex orientation by the Fiedler theorem (see [**6, 7**]).

Denote by α^+ (α^-), β^+ (β^-) and so on, the numbers of positive (respectively, negative) ovals of the curve C_5 in the polygons marked by the corresponding letters, so that

$$\alpha^+ + \alpha^- = \alpha_*, \quad \beta^+ + \beta^- = \beta_*, \ldots. \tag{4}$$

Applying (3) to the curve C_5 we obtain

$$\Delta\alpha + \Delta\beta + \Delta\gamma + \cdots = 0, \tag{5}$$

where $\Delta\alpha = \alpha^+ - \alpha^-$ and so on. Now writing (3) for the obtained nonsingular 7-degree curve, we get one more linear relation for $\Delta\alpha, \Delta\beta, \ldots$. Hence this system of linear equations must have a nonnegative integer solution, this gives us additional restrictions on $\alpha_*, \beta_*, \gamma_*, \ldots$.

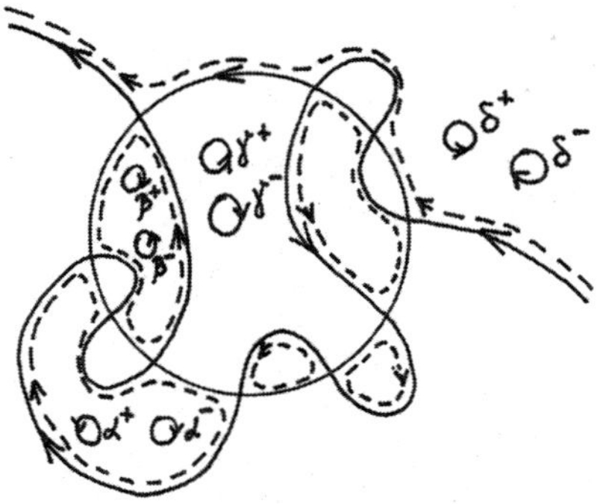

FIGURE 5

For example, for the model 10 of Figure 1, we have (see Figure 5)

$$\Delta\alpha + 3\Delta\beta - \Delta\gamma + \Delta\delta = 0. \tag{6}$$

Subtracting (5) from (6) we obtain $\Delta\beta - \Delta\gamma = 0$. Since $\beta_* \equiv \Delta\beta \pmod 2$ and the same for α_*, γ_* and so on, the last equality gives the necessary condition

$$\beta_* + \gamma_* \equiv 0 \pmod 2.$$

2.2. Restrictions based on the classification of affine quintics. Recall that the classification of real M-decomposable 6th degree curves was obtained in [**8**]. In particular, in [**8**] it was proved that every disposition of an affine M-quintic and a straight line is isotopic to one of the models of Figure 6, where P, Q, R are the numbers of ovals in the corresponding domains,

$$(P, Q) \in \{(1,5), (5,1)\} \tag{7}$$

for the case of Figure 6.1 (all digons are equivalent here) and

$$(P, Q, R) \in \{(2,0,4), (0,2,4), (6,0,0), (0,6,0)\} \tag{8}$$

for the case of Figure 6.2 (quadrangles P and Q are equivalent here).

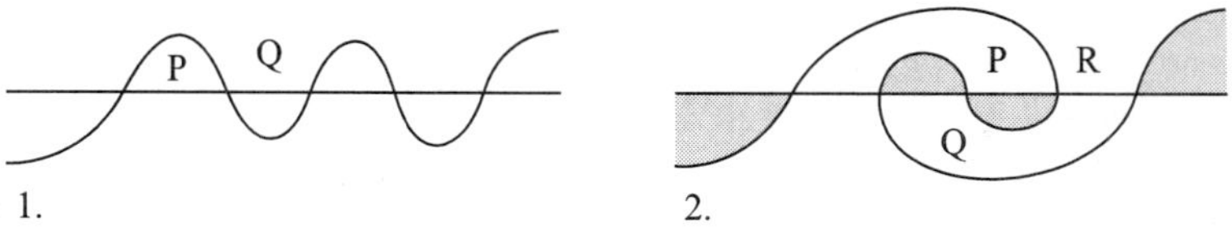

FIGURE 6

Taking into account the convexity of C_2, it is easy to choose straight lines intersecting J_5 in five points for each model from the list of Figures 1–3 except for 29, 30, 33 in Figure 1. Then conditions (7) and/or (8) give some restrictions on the values of $\alpha, \beta, \ldots$.

Let us consider as an example the model 17 in Figure 1. Draw the straight line $L^{(1)}$ through the points M_1, N_1. It is easy to see that one obtains a disposition of $L^{(1)} \cup J_5$ of the type shown in Figure 6.1, see Figure 7.1. Here the shaded zones cannot contain ovals by the Bezout theorem (consider rotations of $L^{(1)}$ around points M_1 and N_1), hence only one digon (marked by the letter δ', $\delta' + \delta'' = \delta$) may contain ovals. Condition (7) gives $\delta' \in \{1, 5\}$, hence $\delta \neq 0$.

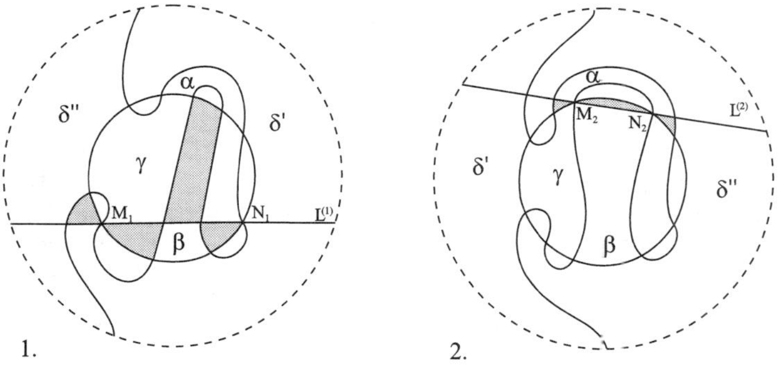

FIGURE 7

Further, if we draw the straight line $L^{(2)}$ through the points M_2, N_2, then we obtain a disposition of $L^{(2)} \cup J_5$ of the type shown in Figure 6.2, where $P = \beta + \delta''$, $Q = \alpha$, $R = \gamma + \delta'$ ($\delta' + \delta'' = \delta$), see Figure 7.2. Here condition (8) gives $\alpha \in \{0, 2, 6\}$ and $\gamma \leqslant 4$, and also if $\alpha = 0$ and $\beta \geqslant 3$, then $\gamma = 0$; if $\alpha = 2$, then $\beta = 0$.

Another example (for the model 10 in Figure 2) was given in [**4**].

Our choice of $L^{(1)}$ and $L^{(2)}$ is shown in Figures 1–3 by specifying the points M_1, N_1 and M_2, N_2 (or $\widetilde{M}_2$, $\widetilde{N}_2$), respectively, of their intersection with E.

3. Final remarks

3.1. Applying the methods of Section 2, as a result we have "only" 587, 166, and 4 unrestricted models of $C_2 \cup C_5$ in the cases $s = 1$, $s = 3$, and $s = 5$, respectively.

3.2. Some models of $C_2 \cup C_5$ can be realized by elementary methods. For example, it is not difficult to realize 66 such models (including all 4 in the case $s = 5$) by the small parameter method from dispositions $C_1' \cup C_1 \cup C_5$, where $C_1 \cup C_5$ is shown in Figure 6 and C_1' is a straight line intersecting C_5 in five real points different from $C_1 \cap C_5$.

3.3. Recently (1997) S. Yu. Orevkov suggested some new methods of constructions and restrictions for real algebraic curves, see [**9**, **10**]. These methods appear very effective in the problem of classifying decomposable curves. In particular, all 10 admissible (by Section 2) distributions of ovals of C_5 for the model 4 in Figure 1 were prohibited in [**9**], and in [**11**] the classifications of M-cubics and M-quartics in general position with the maximum number of intersection points of the pair of ovals is completed.

Thus in fact the number of open questions in the classification of $C_2 \cup C_5$ is less than is indicated in 3.1. We hope that the application of Orevkov's methods may complete the classification of degree 7 decomposable curves, or, at least, this problem will be substantially advanced.

The authors thank Professor L. Lerman for reading the manuscript and making valuable remarks.

References

[1] E. A. Leontovich-Andronova, *Some mathematical works of the Gor′kiĭ School by A. A. Andronov*, Proc. of 3d All-Union Mathematical Congress, USSR Academy of Science, vol. 3, 1958, 116–125. (Russian)

[2] ———, *On some analogies between plane algebraic curves and plane algebraic dynamical systems*, Dokl. Akad. Nauk SSSR **129** (1959), no. 3, 503–506. (Russian)

[3] G. M. Polotovskiĭ, *On the classification of decomposing plane real algebraic curves*, Lecture Notes in Math., vol. 1524, 1992, pp. 52–74.

[4] ———, *On the classification of decomposable 7th degree curves*, Contemp. Math. (2000). (to appear)

[5] D. A. Gudkov, *Topology of real projective algebraic varieties*, Uspekhi Mat. Nauk **29** (1974), no. 4, 3–79; English transl., Russian Math. Surveys **29** (1974), no. 4, 1–79.

[6] V. A. Rokhlin, *Complex topological characteristics of real algebraic curves*, Uspekhi Mat. Nauk **33** (1978), no. 5, 77–90; English transl., Russian Math. Surveys **33** (1978), no. 5, 85–98.

[7] Th. Fiedler, *Eine Beschränkung für die Lage von reellen ebenen algebraischen Kurven*, Beiträge Algebra Geom. **11** (1981), 7–19.

[8] G. M. Polotovskiĭ, *Catalog of M-decomposed curves of 6th degree*, Dokl. Akad. Nauk SSSR **236** (1977), no. 3, 548–551; English transl., Soviet Math. Dokl. **18** (1977), no. 5, 1241–1245.

[9] S. Yu. Orevkov, *Link theory and oval arrangements of real algebraic curves*, Topology **38** (1998), no. 4, 779–810.

[10] ———, *New affine M-sextic.* I, Funktsional. Anal. i Prilozhen. **32** (1998), no. 2, 91–94; English transl., Functional Anal. Appl. **32** (1998), no. 2, 141–143; II, Uspekhi Mat. Nauk **53** (1998), no. 5, 243–244; English transl. in Russian Math. Surveys **53** (1998), no. 5.

[11] S. Yu. Orevkov and G. M. Polotovskiĭ, *Projective M-cubics and M-quartics in general position with maximum intersecting pair of ovals*, Algebra i Analiz **11** (1999), no. 5, 166–184; English transl. in St. Petersburg Math. J. **11** (2000).

Department of Mech. and Math., Nizhny Novgorod University, Nizhny Novgorod, Russia 603600

E-mail address: `polot@curve.unn.ac.ru`

Translated by G. M. POLOTOVSKIĬ

Amer. Math. Soc. Transl.
(2) Vol. **200**, 2000

The Structure of the Boundary Curve for Planar Controllability Domains

N. N. Butenina

ABSTRACT. For a wide class of controllable dynamical systems with bounded control we prove that, under the local controllability condition at a point K, the boundary of the K-controllable domain contains only ω-orbiunstable semi-trajectories of auxiliary sewed systems.

We study controllable dynamical systems (CDS) of the form

$$x' = P(x) + u(t)Q(x), \tag{1}$$

where x belongs to a domain $\Omega \subset \mathbb{R}^2$, $P = (P_1, P_2)$, $Q = (Q_1, Q_2)$ are C^k-(vector-values) functions ($k \geqslant 3$), $u(t)$ ($m \leqslant u(t) \leqslant n$) is a bounded function piecewise continuous on any finite segment.

We use the following notation.

- U is the *set of controllability* to the point K;
- N is the complement to U in Ω (the *set of noncontrollability*);
- $F = \det(P(x), Q(x))$; the curve F_0 given by the equation $F = 0$ is the *contact curve*;
- F^+ (respectively, F^-) is the part of Ω, where $F > 0$ (respectively, $F < 0$);
- if $u(t) \equiv \mu$, $\mu = \text{const}$, $m \leqslant \mu \leqslant n$, then (1) is called the *μ-system*;
- if $u(t) = m$ (respectively, $u(t) = n$) in F^+, $u(t) = n$ (respectively, $u(t) = m$) in F^-, $m \leqslant u(t) \leqslant n$ in F_0, then (1) is called the *mn-system* (respectively, the *nm-system*);
- $O_\varepsilon(M)$ is the ε-neighborhood of the point M;
- $O_\varepsilon(L^+)$ is the ε-neighborhood of the semitrajectory L^+.

An *m-trajectory* (respectively, an *m-arc*) is a trajectory of the m-system (respectively, an arc of an m-trajectory). n- (respectively, mn-, nm-) trajectories and arcs are defined in a similar way. An m- (respectively, n-, mn-, nm-) arc is denoted by $(AB)_{m(\text{resp.},n,mn,nm)}$; B is the end of the arc corresponding to the larger value of the parameter t; an open segment of the contact curve is denoted by (AB), A is the left endpoint of this segment; the notions of "to the left" and "to the right" are defined with respect to $\overline{\operatorname{grad} F}$.

According to [**3**, p. 95], a trajectory of the sewed mn- (respectively, nm-) system belongs to one of the following types:

2000 *Mathematics Subject Classification.* Primary 37C99; Secondary 37N35.

©2000 American Mathematical Society

1) an *equilibrium state*; these include all equilibrium states of the m- and of the n-system as well as those points of the contact curve where the trajectories of the m- and of the n-system are of opposite directions;

2) closed trajectories (not identically constant);

3) nonclosed trajectories without selfintersections;

4) trajectories entering an equilibrium state in finite time (as t either grows or decreases); in this case we say that the ω- (respectively, α-) limit set of the trajectory is an *improper equilibrium state.*

If a trajectory L of the mn- (respectively, of the nm-) system contains a semi-trajectory of the m- or of the n-system tending to an equilibrium point O, then O is called a *proper equilibrium state* of the trajectory L.

Main definitions.

DEFINITION 1. A point M^* of the contact curve is called an *a-point* if

1) $|F'_x(M^*)| + |F'_y(M^*)| \neq 0$;

2) for $\|P(M^*) + \mu_0 Q(M^*)\| \neq 0$, $\mu_0 \in [mn]$, we have $(\overline{\operatorname{grad}} F(M^*) \cdot (P(M^*) + \mu_0 Q(M^*)) = 0$.

(In [**6**] these points are called α-points.)

A point x of the contact curve, where $\|Q(x)\| \neq 0$, is an a-point if and only if $(\overline{\operatorname{grad}} F \cdot Q) = 0$ at this point (see [**6**]). It can be also easily shown that if an a-point does not coincide with an equilibrium point of either the m- or the n-system, then the order of tangency with F_0 is odd or even for both the m- and the n-trajectory at this point simultaneously.

DEFINITION 2. An a-point is said to be of even (respectively, of odd) order if trajectories of at least one of the m- and the n-system are tangent to F_0 with even (respectively, odd) order.

DEFINITION 3. An a-point of odd order is called a γ-point ([**5**]) if 1) this point is not an equilibrium point of either the m-, or the n-system; 2) the m- and the n-trajectory have opposite directions at this point.

DEFINITION 4. A point $M_0 \in F$ is called a *b-point* if $\|Q(M_0)\| = 0$.

If an equilibrium point of a μ-system ($m \leqslant \mu \leqslant n$) is a b-point, then the coordinates of this equilibrium point do not depend on μ (it is a stationary equilibrium) and $\overline{\operatorname{grad}} F(M_0) = 0$.

DEFINITION 5. An open segment of the contact curve that contains neither a-points nor singular points of the curve F_0 is called singular (respectively, nonsingular) if any of the two strict inequalities holds at each point of this segment: $(P_1 + mQ_1)(P_1 + nQ_1) < 0$ (respectively, > 0), $(P_2 + mQ_2)(P_2 + nQ_2) < 0$ (respectively, > 0).

From the geometric point of view the last definition means that m- and n-trajectories intersect F_0 at points of a singular (respectively, a nonsingular) segment in opposite (respectively, in coinciding) directions (in [**5**] this set is called the set of regular points).

Recall the following properties of a singular segment:

1) each point of a singular segment is an equilibrium point of some μ-system, $m < \mu < n$ [**6**];

2) if a point of a singular segment I belongs to U, then $I \subset U$ [**5**];

3) if $\|P + \mu Q\| \neq 0$ ($\mu = m, n$) at an a-point of even order adjacent to singular segments from both sides, then this point cannot belong to the boundary of the controllability set.

DEFINITION 6. A trajectory L of the mn- or of the nm-system is called ω- (respectively, α-) orbistable at a point M_0 if for any $\varepsilon > 0$ there is $\delta > 0$ such that for any trajectory Γ of the same system passing through a point $M \in O_\delta(M_0)$ at $t = \tau$ we have $\Gamma^+(M) \in O_\varepsilon(L^+(M_0))$ (respectively, $\Gamma^-(M) \in O_\varepsilon(L^-(M_0))$).

Here $\Gamma^+(M)$ and $\Gamma^-(M)$ are respectively the positive ($t \geqslant \tau$) and the negative ($t \leqslant \tau$) semitrajectories of Γ [**1**, p. 257].

DEFINITION 7. A trajectory L of the mn- or of the nm-system is called ω- (respectively, α-) orbistable if

1) L is ω- (respectively, α-) orbistable at each of its points;

2) L does not contain nonorbistable positive (respectively, negative) nontrivial semitrajectories neither of the m-, nor of the n-system. (Compare with Definition XV [**1**, p. 257].)

A positive (respectively, negative) semitrajectory of a ω- (respectively, of an α-) orbistable trajectory of the mn- or of the nm-system is called an ω (respectively, α-) *orbistable* semitrajectory.

REMARK 1. In an autonomous dynamical system with a smooth right-hand side the ω- (respectively, the α-) orbistability of a trajectory L at some point implies the ω- (respectively, α-) orbistability of L ([**1**, p. 257]). In sewed systems the situation is different.

EXAMPLE 1. Suppose an equilibrium state O of the mn-system that is not an equilibrium state either for the m- or for the n-system coincides with the node singular point of the contact curve. If a trajectory of the m-system passes through O from F^+ to F^+, then O is an ω-orbiunstable trajectory of the mn-system. Therefore, the trajectory of the mn-system tending to O is ω-orbiunstable as well (see Figure 1). (Here and below the m-segments (respectively, the n-segments) are shown by solid (respectively, by dashed) lines, while F_0 is shown by a dashed-dotted line.)

REMARK 2. Each equilibrium state of a sewed system not coinciding with the singular point of the contact curve is an ω-orbiunstable trajectory of either the mn-, or the nm-system.

EXAMPLE 2. Let O_m be a stable node of the m-system not coinciding with the a-point of odd order, and suppose trajectories of the n-system passing from F^- to F^+ intersect F_0 at some neighborhood of O_m (Figure 2a). Then O_m is an ω-orbistable trajectory of the mn-system (Figure 2b), and it is an ω-orbiunstable trajectory of the nm-system (Figure 2c). In this case the semitrajectory of the m-system tending to O from F^- along the exceptional direction ([**1**, p. 189]) is an orbiunstable trajectory of the nm-system, i.e., it is the ω-separatrix of the sewed equilibrium state.

REMARK 3. A trajectory of a sewed system that is ω-orbistable at each of its points can contain an ω-orbiunstable trajectory of the m- or of the n-system.

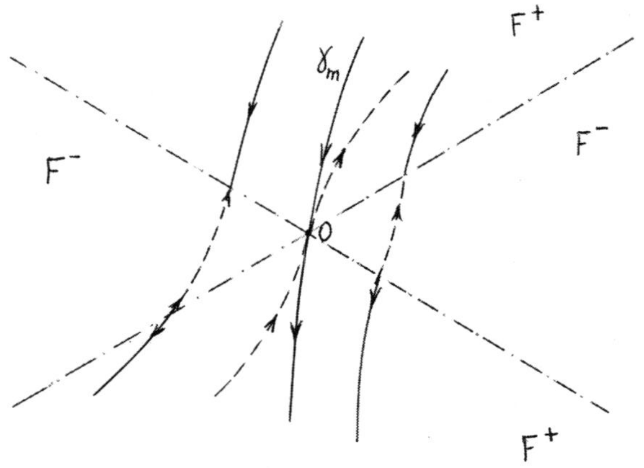

FIGURE 1

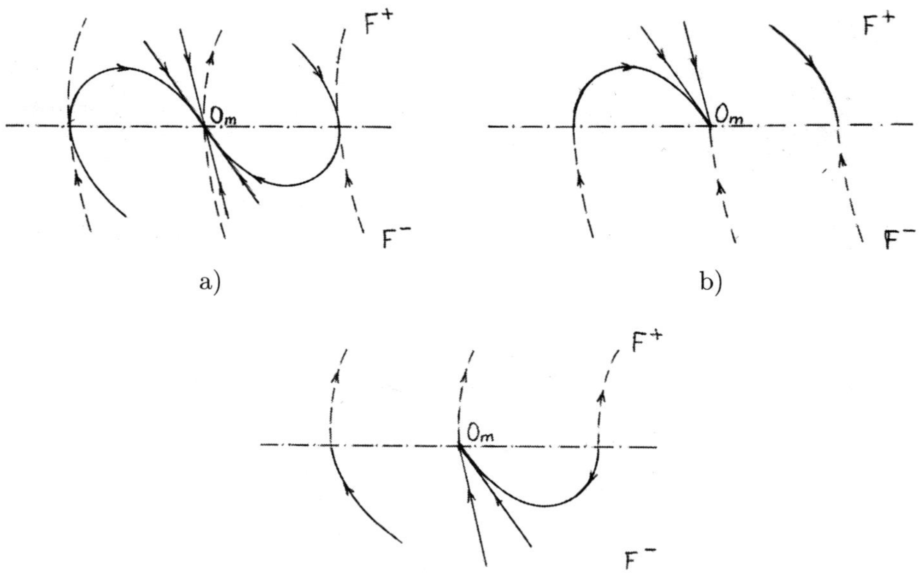

FIGURE 2

EXAMPLE 3. Let O_m be a saddle-node of the m-system with a stable parabolic sector [**2**, p. 92]; only one of its ω-separatrices, namely Γ_m^+, tends to O_m from F^+; n-trajectories in the neighborhood of O_m intersect F_0 from F^- to F^+. Being a semitrajectory of an mn-system, the separatrix Γ_m^+ is ω-orbistable at each of its points, but it contains an orbiunstable semitrajectory of the m-system. Γ_m^+ is an ω-orbiunstable semitrajectory of the mn-system (Figure 3).

Formulation of the problem. In what follows we suppose that the following assumptions are satisfied:

I) $F^+ \cap \Omega \neq \varnothing$, $F^- \cap \Omega \neq \varnothing$;

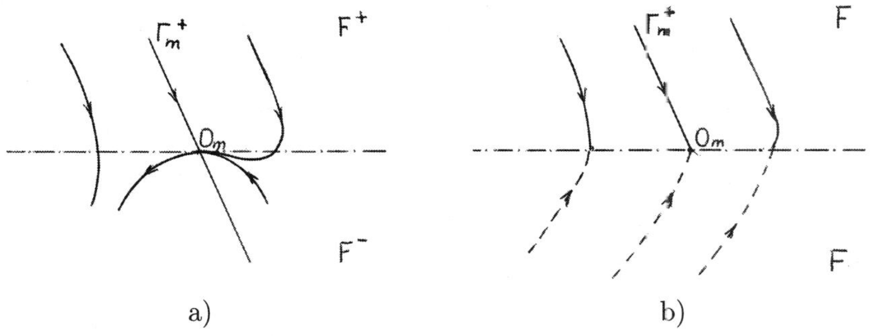

FIGURE 3

II) the contact curve contains only a finite set of singular points (where $F = F'_x = F'_y = 0$), and $|F''_{x^2}| + |F''_{y^2}| + |F''_{xy}| \neq 0$;

III) F_0 contains only a finite set of a-points;

IV) the m- and the n-system have only isolated equilibrium states not coinciding with the singular points of the contact curve;

V) the positive semitrajectories of the mn- and the nm-system are bounded;

VI) if a state K is the goal of the control, then the system is locally controllable at the point K, i.e., U is a domain [**5**], $\partial U \subset N$.

The set of points where the system (1) is locally controllable, contains points of singular segments and points of closed trajectories of the m- and the n-system, when these trajectories have nonempty intersections with F^+ as well as F^-.

It is known [**5**] that the boundary of the controllability domain contains only mn- and nm-semitrajectories since only these semitrajectories possess the property of monodirectional intersection. We are going to show that, under assumptions I)–VI), ∂U can contain only ω-orbiunstable semitrajectories of sewed systems.

Main lemmas and theorems.

LEMMA 1. *Let $\widehat{AB}$ be a simple arc of an mn- (respectively, nm-) system; and let $L^+(A)$ be a positive semitrajectory of the same system. If $\widehat{AB} \subset \partial U$, then $L^+(A) \subset \partial U$.*

PROOF. Since the system is locally controllable at the point K, we have $K \notin \partial U$. If M is an arbitrary point of the semitrajectory $L^+(A)$, then $M \notin U$ since otherwise $\widehat{AB} \subset U$.

Draw the nm- (respectively, the mn-) arcs l_A and l_M through the points A and M. Consider the ε-neighborhood $O_\varepsilon(\widehat{AM})$ of the arc $\widehat{AM}$ not containing the point K and draw in $O_\varepsilon(\widehat{AM})$ an mn- (respectively, an nm-) arc to the right (respectively, to the left) of $\widehat{AM}$ meeting the arcs l_A and l_M at points A_1 and M_1 respectively. Denote by G the domain bounded by the arcs $\widehat{AM}$, $\widehat{A_1M_1}$, l_A, l_M. By construction, $G \cap U \neq \varnothing$. It is possible to leave G along admissible trajectories either through $\widehat{AM}$ to N, either through l_M. Therefore, $l_M \cap U \neq \varnothing$, i.e., there are points belonging to U in each neighborhood of the point M ($M \in N$). Thus, $M \in \partial U$; $L^+(A) \subset \partial U$. Since ∂U is closed, the ω-limit set of the semitrajectory $L^+(A)$ is a subset of ∂U as well (see Figure 4). □

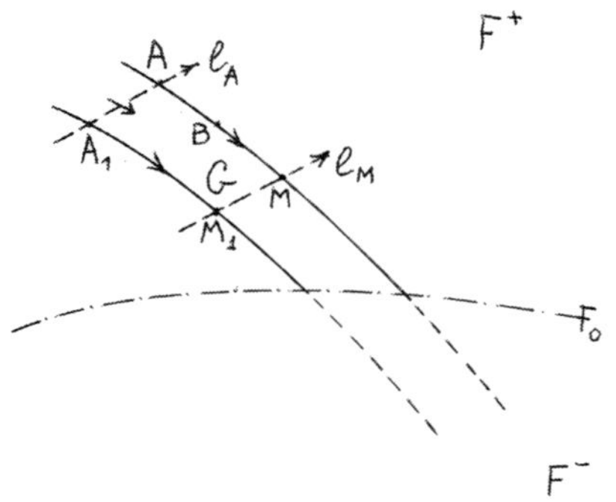

FIGURE 4

THEOREM 1. *If a nonclosed mn- (respectively, nm-) semitrajectory* L^+ *has at least one limit trajectory that is not an equilibrium point, then* $L^+ \cap \partial U = \varnothing$.

PROOF. Let M^* be the limit point of the semitrajectory L^+ that is not an equilibrium point and does not belong to F_0. Draw an nm- (respectively, mn-) arc l_0 through M^* having no contact with mn- (respectively, with nm-) trajectories. Let M_1 and M_2 be two successive intersection points of L^+ with l_0 belonging to an arbitrarily small neighborhood of M^*. If $L^+ \subset \partial U$, then there are points of l_0 in neighborhoods of M_1, M_2 belonging both to U and to N. Therefore, the admissible trajectory L ($l_0 \in L$) passes through N to U which is impossible. The theorem is proved (see Figure 5). □

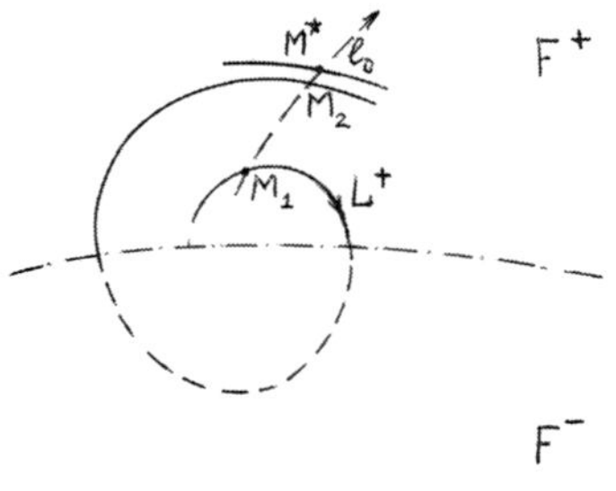

FIGURE 5

THEOREM 2. *The boundary of the controllability domain cannot contain a stable closed trajectory* L *of the mn- or of the nm-system.*

In order to prove the theorem we construct a subdomain in N containing L (Figure 6).

LEMMA 2. *Suppose that the limit set of a semitrajectory* L^+ *of the mn- or of the nm-system is a proper (respectively, improper) equilibrium state* O^*. *Then there is a semitrajectory* l^+ *such that* $l^+ \subset L^+$, $l^+ \cap F_0 = \varnothing$ *(respectively,* $l^+ \cap F_0 = O^*$*).*

PROOF. Assumption IV) guarantees that equilibrium states of the m- and of the n-system are distinct. Therefore, for any equilibrium state of the sewed system there exists a neighborhood where either m- or n-trajectories (or both) intersect F_0 in one direction (from F^- to F^+ or backwards). No trajectory of the sewed system

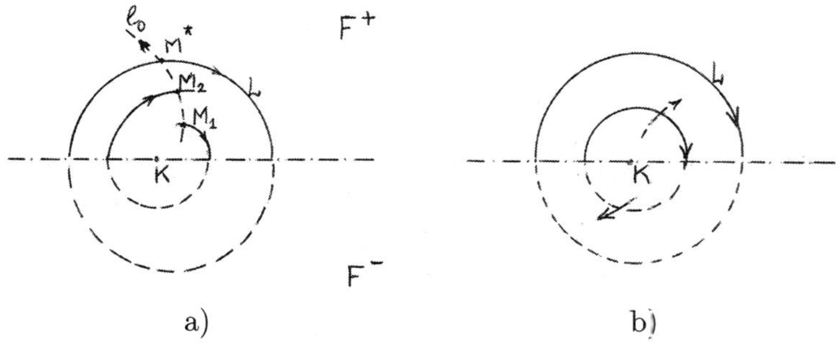

FIGURE 6

intersects F_0 in this neighborhood twice. Also if the equilibrium state O^* coincides with an a-point of odd order, then there is an arc l_0 of an m- or of an n-trajectory entirely (except the point O^* itself) contained either in F^+ or in F^-. A semitrajectory of one of the sewed system tends to O^* along l_0. For the other sewed system, l_0 is an arc without contact, and trajectories of the last system cannot intersect this arc twice. The lemma is proved. □

LEMMA 3. *If both eigenvalues of the linear approximation of the m- or of the n-system at an equilibrium point O^* are equal to zero, then O^* cannot coincide with an a-point of the contact curve.*

PROOF. Suppose for $\mu = m$ or n we have $\overline{\text{grad}}\,(P_1 + \mu Q_1) = (a, b)$, $\overline{\text{grad}}\,(P_2 + \mu Q_2) = (c, d)$ at O^*, $Q = (Q_1^0, Q_2^0)$. Then $F_x = aQ_2^0 - cQ_1^0$, $F_y = bQ_2^0 - dQ_1^0$ at this point. Note that $|a| + |b| + |c| + |d| \neq 0$, otherwise, O^* would coincide with a singular point of the contact curve.

Under the assumptions of the lemma we have $a + d = 0$, $a^2 + bc = 0$.

1) Suppose that $a = d = b = 0$, $c \neq 0$. Then $F_x = -c_1 Q_1^0$, $F_y = 0$, $Q_1^0 \neq 0$ (since $|F_x| + |F_y| \neq 0$ at O^*), $F_x Q_1^0 + F_y Q_2^0 = -c(Q_1^0)^2 \neq 0$, i.e., O^* does not coincide with an a-point. The case $a = d = c = 0$, $b \neq 0$ is treated similarly.

2) If $a \neq 0$, then

$$b \neq 0, \quad c \neq 0, \quad d \neq 0, \qquad \frac{c}{a} = -\frac{a}{b},$$

$$F_x = a\Big(Q_2^0 + \frac{a}{b}\,Q_1^0\Big), \quad F_y = b\Big(Q_2^0 + \frac{a}{b}\,Q_1^0\Big), \quad Q_2^0 + \frac{a}{b}\,Q_1^0 \neq 0,$$

$$(\overline{\text{grad}}\,F \cdot Q) = b\Big(Q_2^0 + \frac{a}{b}\,Q_1^0\Big)^2 \neq 0,$$

i.e., O^* does not coincide with an a-point of the contact curve. The lemma is proved. □

LEMMA 4. *Let γ^+ be a semitrajectory of the mn- (respectively, of the nm-) system not meeting F_0, whose ω-limit set is a proper equilibrium state O^*, and let l_0 be an arc of the nm- (respectively, of the mn-) trajectory meeting γ^+ at a point M and not intersecting F_0. If there is a point M_1 on l_0 to the right (respectively, to the left) of γ^+ such that the semitrajectory $\tilde{\gamma}^+(M_1)$ of the m-system outgoing from M_1 tends to O^* not intersecting F_0, then γ^+ cannot be a part of the boundary of the controllability domain.*

PROOF. For the sake of definiteness suppose that γ^+ is a semitrajectory of the mn-system tending to O^* from F^+ (other cases are treated similarly). Suppose that $\gamma^+ \subset \partial U$. Assumption VI) implies that $K \notin \partial U$. It is always possible to choose $\varepsilon > 0$ such that $K \notin O_\varepsilon(\gamma^+)$. Under the assumptions of the lemma any mn- or nm-trajectory intersecting the arc $(M_1M)_{nm}$ tends to O^* without meeting F_0. Choose a point M^* on $(M_1M)_{nm}$ such that $\gamma_m^+(M^*) \subset O_\varepsilon(\gamma^+)$ and consider the curvilinear triangle Δ with the boundary consisting of the arcs $(MO^*)_{mn}$, $(M^*O^*)_{mn}$, $[M^*M]_{nm}$. The only way out of Δ along admissible trajectories is through $[MO^*)_{mn} \cup O^*$. However, $[MO^*)_{mn} \cup O^* \subset N$. Therefore, $\Delta \subset N$, i.e., $\gamma^+ \not\subset \partial U$ (Figure 7). □

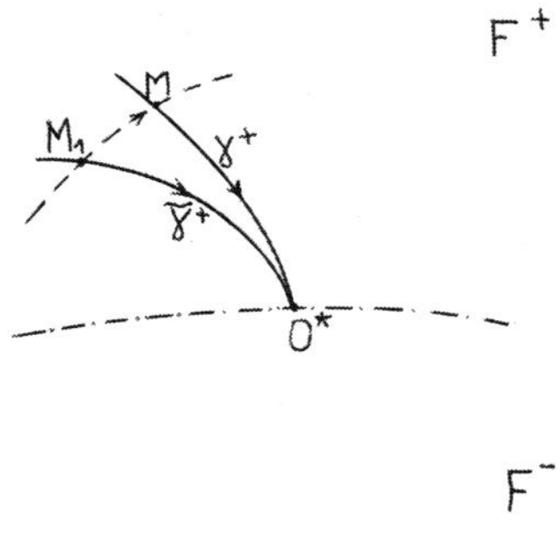

FIGURE 7

LEMMA 5. *Suppose that* 1) $\gamma^+(M)$ *is a semitrajectory of the* m- (*respectively, of the* n-) *system whose* ω-*limit set is an equilibrium state* O^*, $\gamma^+(M) \subset F^+$ (*respectively,* $\gamma^+(M) \subset F^-$); 2) *there is* $\delta > 0$ *such that* $O_\delta(M) \cap F_0 = \varnothing$ *and any trajectory of the* m- (*respectively, of the* n-) *system passing through* $O_\delta(M)$ *to the right* (*to the left*) *of* $\gamma^+(M)$ (*respectively, to the left* (*to the right*) *of* $\gamma^+(M)$) *intersects* F_0.

If a singular open segment is adjacent to O^* *on* F_0 *from the left* (*respectively, from the right*), *then* $\gamma^+(M)$ *is an orbiunstable semitrajectory of the* m- (*respectively, of the* n-) *system.*

PROOF. For the sake of definiteness consider the semitrajectory $\gamma^+(M)$ of the m-system (other cases are treated similarly). Since, by assumption of the lemma, a singular open segment I is adjacent O^* on F_0 from the left, we conclude that trajectories of the n-system intersect F_0 at points of this interval from F^- to F^+.

1. Suppose O^* does not coincide with an a-point of odd order. Then there is a neighborhood of O^*, where trajectories of the n-system intersect F_0 from F^- to F^+ (with possible tangency at O^*).

Consider a neighborhood such that its closure contains no equilibrium states of the m- and of the n-system, no a-points and no singular points of the contact curve different from O^*. In F^- trajectories of the m-system intersect trajectories of the n-system from the right to the left. Therefore, a trajectory of the m-system outgoing from a point of the singular segment I cannot tend to O^*, and it leaves the chosen neighborhood of O^* as t grows (Figure 8).

2. Suppose O^* is an a-point of odd order. Denote by l_n^0 the n-arc tangent to F_0 at O^* and having no other common points with F_0. We call the direction of this trajectory at the equilibrium state along the common tangent vector with F_0 *basic.* If $\gamma^+(M)$ tends to O^* along a direction different from the basic one, then $l_n^0 \subset F^-$

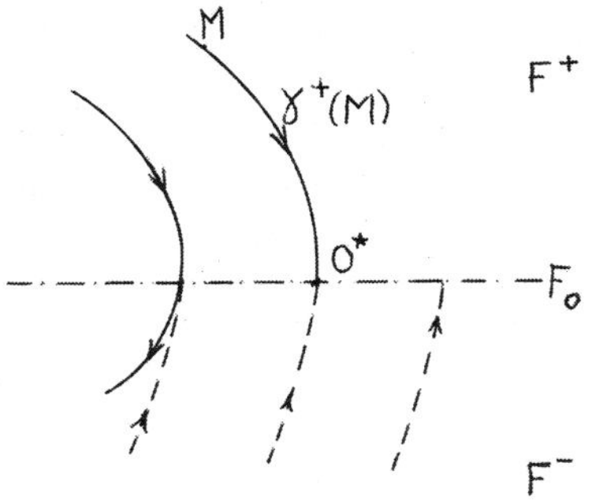

FIGURE 8

(otherwise the segment to the left of O^* is nonsingular) (see Figure 9). Here the situation is the same as in the first case, $\gamma^+(M)$ is an orbiunstable semitrajectory of the m-system.

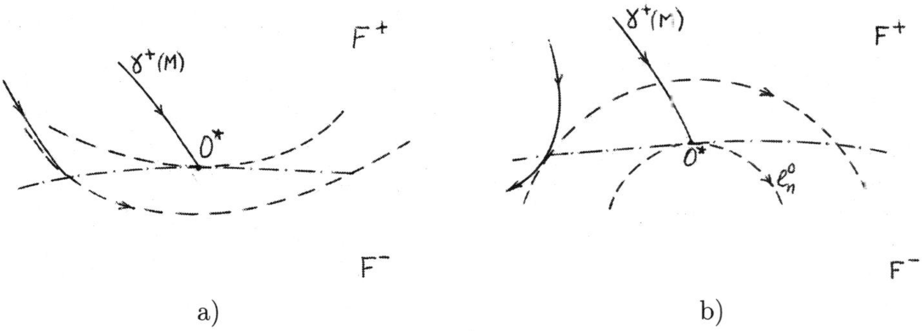

FIGURE 9

Suppose now that $\gamma^+(M)$ tends to O^* along the basic direction. If $l_n^0 \subset F^-$, then the situation coincides with that of the previous case (Figure 10a)).

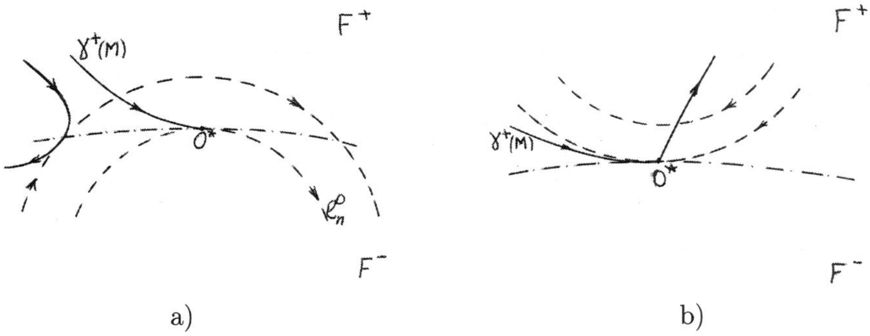

FIGURE 10

Suppose $l_n^0 \subset F^+$. Consider an ε-neighborhood of O^* satisfying the same assumptions as in the first case, and add the requirement that l_n^0 divides this neighborhood into two parts; denote the part entirely belonging to F^+ by $\widetilde{O}_\varepsilon(O^*)$.

Trajectories of the m-system intersecting l_n^0 go away from O^*, i.e., the trajectory coming into $\widetilde{O}_\varepsilon(O^*)$ has O^* as the α-limit set. Hence, m-trajectories can tend to O^* from F^+ only along the basic direction. Looking over all possible types of equilibrium states at an a-point (Lemma 3) we conclude that $\gamma^+(M)$ is an ω-separatrix of the equilibrium state O^* [**1**, pp. 372–385] (Figure 10b). □

COROLLARY 1. *If O^* is a simple equilibrium state of the m- (respectively, of the n-) system, then, under the assumptions of Lemma 5, $\gamma^+(M)$ is an ω-orbiunstable semitrajectory both for the m- (respectively, for the n-) and for the mn- (respectively, for the nm-) system.*

PROOF. Under the assumptions of the corollary the point O^* is a saddle equilibrium point and $\gamma^+(M)$ is an ω-separatrix of this saddle, and one of the α-separatrices leaves the saddle for F^-, while the other one leaves it for F^+ [**6**]. □

THEOREM 3. *If L is an ω-orbiunstable trajectory of the mn- or of the nm-system whose ω-limit set is the improper equilibrium point O^*, then the boundary of the controllability domain contains no positive semitrajectory of L.*

PROOF. For the sake of definiteness suppose, without loss of generality, that L is the trajectory of the mn-system tending to O^* from F^+.

Suppose ∂U contains a semitrajectory L^+. An improper equilibrium state of the semitrajectory L^+ can be of one of the following types: a point of the singular segment; a γ-point; an equilibrium point of the mn-system coinciding with a singular point of the contact curve; an equilibrium point of the n-system.

a) Let O^* be a point of the singular segment. In this case L is ω-orbistable. If $L \subset \partial U$, then $O^* \in \partial U$. But ∂U can contain no point of the singular segment. Hence, $L^+ \not\subset \partial U$.

b) If O^* is a γ-point, then O^* and L^+ are orbiunstable trajectories of the mn-system.

c) Suppose O^* is a singular point of the contact curve, and let γ_m be an m-arc containing O^*. In this case trajectories of the m- and of the n-system have opposite directions at O^*. If γ_m passes through O^* from F^- to F^+, then O^* is an orbiunstable trajectory of the mn-system. Suppose γ_m passes through O^* from F^+ to F^-. Then the n-arc l_n passing through O^* has an odd order of tangency with γ_m at O^*, and γ_m is to the left of l_n in a neighborhood $O_\delta(O^*)$ of O^*. Any trajectory of the n-system passing to the left of l_n sufficiently close to l_n intersects γ_m at two points, and one of these points belongs to $O_\delta(O^*) \cap F^-$, while the other one belongs to $O_\delta(O^*) \cap F^+$. Let l_n^1 be such a trajectory meeting γ_m at points M_1 ($M_1 \in F^-$) and M_2 ($M_2 \in F^+$). If $L^+ \subset \partial U$, then U is adjacent to γ_m in F^+ from the right. But then $M_1 \in U$, $(M_1, M_2)_m \in U$, i.e., $L^+ \subset U$, therefore $L^+ \not\subset \partial U$ (Figure 11).

d) Suppose O^* is an equilibrium state of the n-system. If this equilibrium state coincides with an a-point of odd order, then L^+ is an ω-orbiunstable semitrajectory of the mn-system. Therefore, there is a neighborhood $O_\varepsilon(O^*)$ of the point O^*, where trajectories of the m-system intersect F_0 from F^+ to F^- (with possible tangency at O^*). If a nonsingular interval I is adjacent to O^* from the left, then the first assumption of Definition 7 is not satisfied (trajectories of the n-system passing through I intersect trajectories of the m-system in F^- from the left to the right and cannot, therefore, tend to O^* from $O_\varepsilon(O^*)$) (see Figure 12a). Thus, under the assumptions of the theorem the singular segment J is adjacent to O^* from the

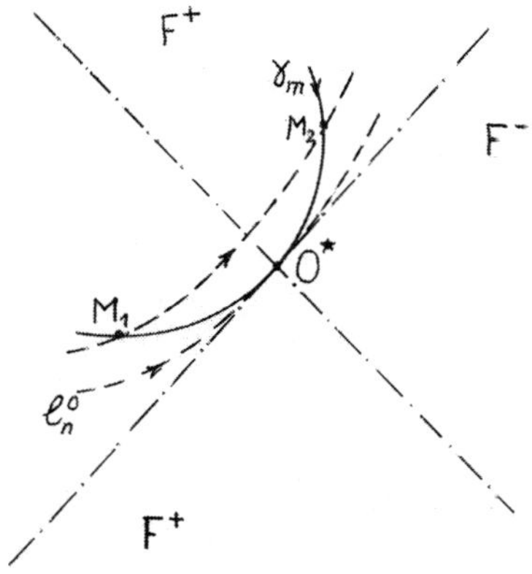

FIGURE 11

left. Since O^* is an ω-orbistable trajectory of the mn-system, the ω-limit set for a trajectory of the n-system intersecting γ_m in F^- closely enough to O^* is a point of the singular segment J. If $J \subset U$, then $\gamma_m \cap F^+ \in U$. If $J \subset N$, then ∂U contains a semitrajectory of the n-system tending to O^* from F^+ as well as L^+ (Figure 12b). However, it is easy to point out a domain in F^+ adjacent to $\gamma_m \cap F^+$ from the left and not contained in U (this domain is shaded in the figure). Hence, $L^+ \not\subset \partial U$. The theorem is proved. □

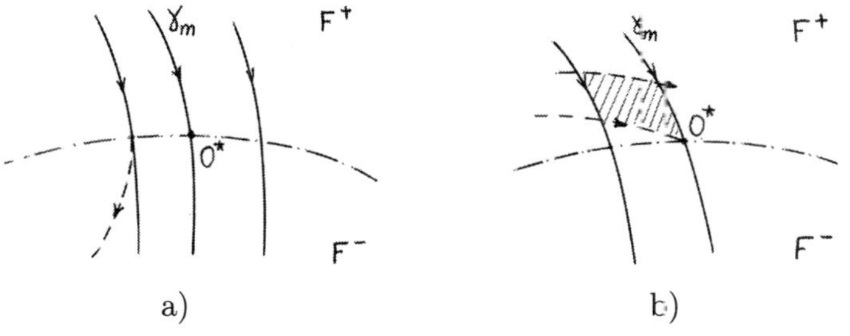

a) b)

FIGURE 12

COROLLARY 2. *An equilibrium state of the m- or of the n-system that is a center or a focus cannot belong to the boundary of the controllability domain.*

PROOF. An equilibrium state of the m- or of the n-system can be only an improper ω-limit set for trajectories of the mn- or of the nm-system. It can be easily shown that an nm- or an mn-trajectory tending to such an equilibrium state is ω-orbistable (independently of whether the focus is stable or not). □

THEOREM 4. *The positive semitrajectory L^+ of an ω-orbistable trajectory of the mn- or of the nm-system whose ω-limit set is a proper equilibrium state O^* cannot be a part of the boundary of the controllability domain.*

PROOF. It is sufficient to study the case where L^+ is a semitrajectory of the mn-system containing a semitrajectory γ_m of the m-system, i.e., O^* is an equilibrium state of the m-system, and $\gamma_m^+ \subset F^+$. In what follows, we make use of the notation introduced in the proof of Lemma 5.

1. Suppose O^* is not an a-point of odd order.

Through a point $M \in \gamma_m^+$ we draw an arc l_0 of the n-system not intersecting F_0. Suppose that any m-trajectory intersecting l_0 to the right of γ_m^+ closely enough to M intersects F_0. If a nonsingular segment is adjacent to O^* from the left in F_0, then the first requirement of Definition 7 is not satisfied; if it is a singular segment, then, by Lemma 5, γ_m^+ is an orbiunstable semitrajectory of the m-system, and the second requirement of Definition 7 is not satisfied. Since the semitrajectory L^+ is ω-orbistable, there is a point M_1 on l_0 to the right of γ_m^+ such that the m-trajectory passing through M_1 tends to O^* not intersecting F_0. This means that the assumptions of Lemma 4 are satisfied, and in this case $\gamma_m^+ \not\subset \partial U$, $L^+ \not\subset \partial U$.

2. Now let O^* be an a-point of odd order. If the assumptions of Lemma 4 are satisfied for γ_m^+, then $\gamma_m^+ \not\subset \partial U$, $L^+ \not\subset \partial U$. Consider the case, where the assumptions of Lemma 4 are not satisfied. If γ_m^+ is ω-orbistable, then a nonsingular segment is adjacent to O^* in F_0 from the left and $l_n^0 \subset F^+$. If a nonsingular segment is adjacent to O^* in F_0 from the right as well, then L^+ and γ_m^+ are ω-orbiunstable semitrajectories of the mn-system (Figure 13).

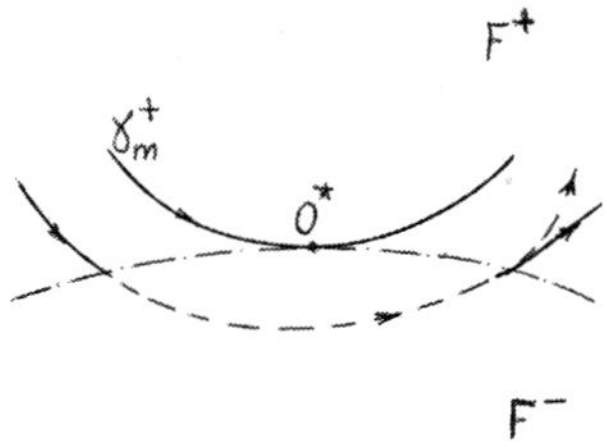

FIGURE 13

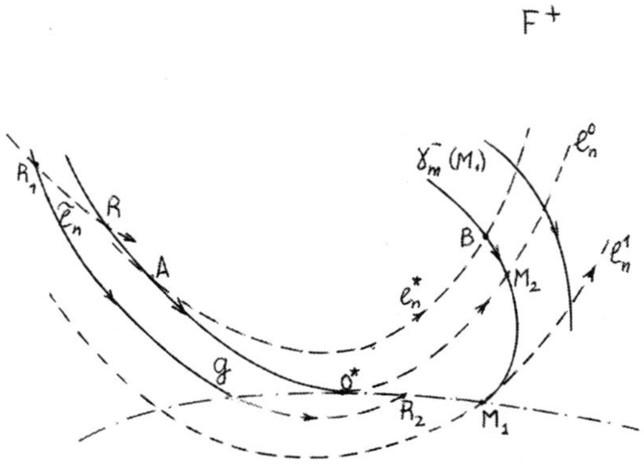

FIGURE 14

Therefore, a singular segment adjoins O^* in F_0 from the right; we denote this segment by J. Let us show that in this case O^* is a stable equilibrium state of the m-system. A trajectory of the n-system intersecting F_0 both from the left and from the right of O^* separates a closed domain such that all trajectories of the m-system in this domain tend to O^*. All trajectories of the m-system passing in F^+ to the right of γ_m^+ intersect F_0 from F^+ to F^-; a trajectory of the n-system intersecting γ_m^+ and close enough to l_n^0 separate a domain in $\widetilde{O}_\varepsilon(O^*)$ such that each trajectory of the m-system in this domain either tends to O^*, or intersects l_n^0. A trajectory of the m-system intersecting l_n^0 intersects also other trajectories of the n-system sufficiently close to l_n^0. Denote by l_n^1 such trajectory passing to the right of l_n^0 and intersecting the singular segment J at M_1. The semitrajectory $\gamma_m^-(M_1)$ (Figure 14) intersects in F^+ neither l_n^1 nor a trajectory of the m-system meeting both l_n^1 and l_n^0. Therefore, $\gamma_m^-(M_1)$ intersects l_n^0; denote the intersection point by M_2. Obviously, all m-trajectories intersecting $(O^*M_2)_n$ meet J as well. Thus, O^* is a stable equilibrium state of the m-system. By Lemma 3, it can be only a node (either simple or multiple). In this case L^+ is an ω-orbistable semitrajectory. Suppose that $\gamma_m^+ \subset \partial U$. Let $\varepsilon > 0$ be such that $O_\varepsilon(\gamma_m^+)$ does not contain K, and all trajectories of the m-system in $O_\varepsilon(O^*)$ tend to O^*. Draw an arc $\tilde{l}_n$ in $O_\varepsilon(\gamma_m^+)$ through a point R, $R \in \gamma_m^+$. A trajectory of the mn-system passing through a point R_1 lying on $\tilde{l}_n$ to the right of γ_m^+ and sufficiently close to R has a point $R_2 \in J$ as an ω-limit set. Consider the domain g bounded by the arcs $(RO^*)_m$, $[R_1R]_n$, $(R_1R_2)_{mn}$, and $[O^*R_2]$ ($[O^*R_2] \subset F_0$). Since, by assumption, $\gamma_m^+ \subset \partial U$, U adjoins γ_m^+ from the right. Since $K \notin O_\varepsilon(\gamma_m^+)$, there is an admissible trajectory passing from g to the point K and intersecting (O^*R_2). Hence, $J \subset U$. Let $R^* \in (O^*R_2)$ and suppose $\gamma_m^-(R^*)$ intersects l_n^0 in $O_\varepsilon(O^*)$. Then $\gamma_m^-(R^*)$ also intersects other trajectories of the n-system close to l_n^0. However, γ_m^+ also intersects trajectories of the n-system passing to the left of l_n^0 and sufficiently close to l_n^0. Denote the intersection points of l_n^* with γ_m^+ and $\gamma_m^-(R^*)$ by A and B correspondingly. Then the admissible semitrajectory consisting of $(AB)_n$ and $\gamma_m^+(B)$ intersects J. Since $J \subset U$, we have $\gamma_m^+ \subset U$. The theorem is proved. □

The following theorem combines the statements of Theorems 1–4 taking into account the behavior of the equilibrium states of sewed systems.

THEOREM 5. *If a semitrajectory L^+ of the mn- or of the nm-system is a part of the boundary of the controllability domain, then L^+ is an ω-orbiunstable semitrajectory of this system.*

COROLLARY 3. *If the m- and the n-system has only simple equilibrium states, then only positive semitrajectories of the mn- and of the nm-system ω-orbiunstable at each of their points enter the boundary of the controllability domain (see Definition 6).*

PROOF. In this case an ω-orbiunstable semitrajectory of the m- or of the n-system simultaneously is an ω-orbiunstable semitrajectory of the sewed system (see Corollary 1). □

A similar theorem is proved in [**7**] for CDS of special form:

$$x' = y, \qquad y' = f(x,y) + u(t), \tag{1}$$

$m \leqslant u(t) \leqslant n$, under the assumption that the equilibrium states of the m- and of the n-system are structurally stable. An algorithm constructing the controllability domain for system (1) is presented in [**8**].

References

[1] A. A. Andronov, E. A. Leontovich, I. I. Gordon, and A. G. Maier, *Qualitative theory of second-order dynamic systems*, "Nauka", Moscow, 1966; English transl., Israel Program of Scientific Translations, Jerusalem–London, 1973.

[2] N. N. Bautin and E. A. Leontovich, *Methods and modes of the qualitative study of dynamical systems on the plane*, "Nauka", Moscow, 1976. (Russian)

[3] A. F. Filippov, *Differential equations with discontinuous right-hand side*, "Nauka", Moscow, 1985. (Russian)

[4] E. Roxin and V. Spinadel, *Reachable zones in autonomous differential systems*, Contrib. Differ. Equations **1** (1963), 275–315.

[5] M. M. Baitman, *Domains of controllability in the plane*, Differentsial′nye Uravneniya **14** (1978), no. 4, 579–593; English transl. in Differential Equations **14** (1978).

[6] N. N. Butenina and N. A. Sizova, *Singular intervals of a controllable dynamical system of the second order*, Vestnik NNGU. Mathematical Modeling and Optimal Control, N. Novgorod, 1997, pp. 108–115. (Russian)

[7] N. N. Butenina, Z. G. Pavlyuchonok, and V. P. Savel′ev, *Qualitative methods for the global investigation of controllability regions on a plane*, Differentsial′nye Uravneniya **31** (1995), no. 4, 555–568; English transl., Differential Equations **31** (1995), no. 4, 515–527.

[8] ———, *Controllability of dynamical systems*, N. Novgorod, 1997. (Russian)

DEPARTMENT OF COMPUT. MATH. AND CYBERN., NIZHNY NOVGOROD UNIVERSITY, GAGARIN AVE., 23, NIZHNY NOVGOROD 603600

Translated by O. YA. VIRO

Amer. Math. Soc. Transl.
(2) Vol. **200**, 2000

Orbit Structure in a Neighborhood of a Separatrix Cycle Containing Two Saddle-Foci

V. V. Bykov

ABSTRACT. Codimension two dynamical systems possessing a contour formed with two saddle-foci and two of their heteroclinic orbits are studied. The existence of a countable set of heteroclinic orbits to the saddle-foci is proved. An open region is found on the corresponding bifurcation surface in the space of smooth vector fields where systems with a nonrough heteroclinic orbit to the saddle-foci are dense everywhere. The simplest bifurcations of homoclinic orbits are also considered.

1. Introduction

Consider a two-parameter family of C^r-smooth ($r \geqslant 3$) vector fields $X_\mu(x)$, $x \in \mathbb{R}^3$, $\mu \in \mathbb{R}^2$ having two equilibrium states O_1 and O_2 of saddle type. We assume that the characteristic exponents at O_k are $\{\alpha_k \pm i\omega_k, \gamma_k\}$, where $\alpha_k\gamma_k < 0$, $\alpha_1 < 0$, $\alpha_2 > 0$, $\omega_k \neq 0$ ($k = 1,2$), i.e., O_1 and O_2 are saddle-foci of different topological types. Let Γ_i be a one-dimensional separatrix of O_i, and let $W^s(O_1)$ and $W^u(O_2)$ be the two-dimensional invariant manifolds of the saddle-foci O_1 and O_2, respectively.

The class of systems under consideration is distinguished by the following conditions:

A) the system $X_0(x)$ has a transverse heteroclinic trajectory
$$\Gamma_0 \subseteq W^s(O_1) \cap W^u(O_2);$$

B) for $\mu = 0$, the one-dimensional separatrices Γ_1 and Γ_2 merge and form a (nonrough) heteroclinic trajectory Γ^0;

C) the ratio $\tau = \nu_2\omega_1/(\nu_1\omega_2)$ is not equal to 1 (here $\nu_i \equiv -\alpha_i/\gamma_i$ are the so-called saddle indices).

It follows from A) and B) that the system X_0 has a separatrix contour $\mathcal{L} = O_1 \cup O_2 \cup \Gamma_0 \cup \Gamma^0$ in the phase space (Figure 1). Such systems form a codimension-two bifurcation set H^0 in the space of dynamical systems in the C^3 topology.

It is known [**1, 2**] that if a saddle-focus has a homoclinic trajectory Γ and if the saddle index is less than 1, then the set of trajectories lying entirely in a neighborhood of Γ has a rather nontrivial structure. Moreover, the saddle index is an invariant of topological conjugacy for systems with a separatrix loop [**3**]. In the

2000 *Mathematics Subject Classification.* Primary 37G20.

The author is grateful to the Russian Foundation for Basic Research (grant 97-01-00015) for financial support.

©2000 American Mathematical Society

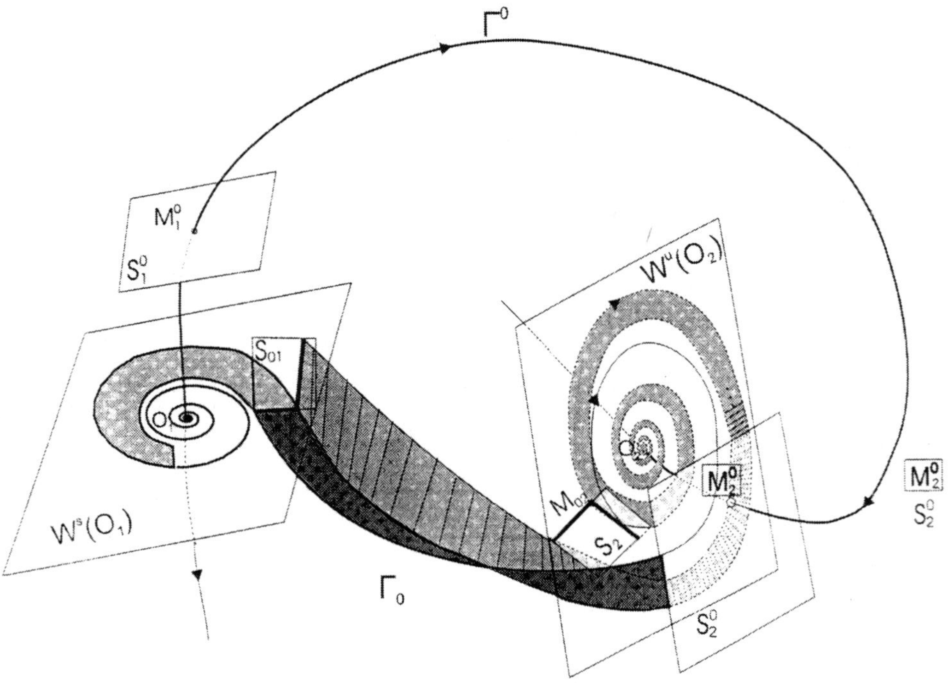

FIGURE 1

case of a separatrix contour containing a saddle and a saddle-focus, the complex behavior of orbits exists independently of the value of the saddle indices [**4, 5, 6**].

In this paper we establish that every system $X_0 \in H^0$ has an infinite set of heteroclinic trajectories Γ_{0i} in a small neighborhood $\mathcal{U}(\mathcal{L})$ of the contour $\mathcal{L}$. We also show that the bifurcation set H^0 is split into two components H_1^0 and H_2^0 (which are distinguished by the following condition: the parameter G given by formula (3.3) is positive in H_1^0 and negative in H_2^0) so that the heteroclinic orbits $\{\Gamma_{0i}\}$ are transverse for the systems in H_1^0, whereas systems with nontransverse heteroclinic trajectory Γ_{0*} along which W_1^s and W_2^u have a tangency are dense everywhere in H_2^0. Notice that the coexistence of the nonrough trajectory Γ^0 and the trajectory of heteroclinic tangency Γ_{0*} means that a degeneration of codimension exceeding 2 occurs, thereby resulting in the impossibility of obtaining a complete description within the framework of the two-parameter family.

Apart from studying heteroclinic trajectories, we consider the simplest bifurcations that appear when the nonrough heteroclinic trajectory Γ^0 is split into one-dimensional separatrices Γ_1 and Γ_2. Here, we find two bifurcation curves l_1 and l_2 corresponding to separatrix loops of saddle-foci O_1 and O_2, respectively. We show that these curves have infinitely many intersection points μ^k.

This fact and the results of [**2, 7, 8**] imply that if $1/2 < \nu_i < 1$ or $0 < \nu_i < 1/2$ simultaneously for both saddle indices ($i = 1, 2$), then in a neighborhood of each point μ^k there exist open sets in which the values of the parameters μ are dense wherever the system $X_\mu(x)$ has coexisting attractor and repeller: the former contains infinitely many stable periodic orbits, whereas the latter contains infinitely many completely unstable periodic orbits.

2. The Poincaré map

By local straightening of the invariant manifolds and by rescaling time, the system can be written in the following form in a neighborhood of an equilibrium state of saddle-focus type:

$$\begin{aligned} \dot{x} &= -\nu x - \omega y + f_1(x,y,z)x + f_2(x,y,z)y, \\ \dot{y} &= \omega x - \nu y + g_1(x,y,z)x + g_2(x,y,z)y, \\ \dot{z} &= z, \end{aligned} \tag{2.1}$$

where $f_i, g_i \in C^{r-1}$ and $f_i(0,0,0) = g_i(0,0,0) = 0$, $i = 1, 2$.

LEMMA 2.1 ([**2**]). *The coordinates in a neighborhood of a saddle-focus can be chosen in such a way that the following identities hold:*

$$f_i(x,y,0) \equiv 0, \quad g_i(x,y,0) \equiv 0, \quad f_i(0,0,z) \equiv 0, \quad g_i(0,0,z) \equiv 0. \tag{2.2}$$

In the cylindrical coordinates

$$x = \rho\cos\varphi, \quad y = \rho\sin\varphi, \quad z = z$$

system (2.1) takes the form

$$\begin{aligned} \dot{\rho} &= -\nu\rho + F_1(\rho,\varphi,z)\rho, \\ \dot{\varphi} &= \omega + F_2(\rho,\varphi,z), \\ \dot{z} &= z, \end{aligned}$$

where, by virtue of (2.2), $F_1, F_2 \in C^{r-1}$ and $F_i(0,\varphi,z) \equiv 0$, $F_i(\rho,\varphi,0) \equiv 0$.

Since $\omega \neq 0$, in a sufficiently small neighborhood of O one can transform the time variable $d\tau = (\omega + F_2)\,dt$, obtaining the system

$$\begin{aligned} \dot{\rho} &= -\nu\rho/\omega + R(\rho,\varphi,z)\rho, \\ \dot{z} &= \omega^{-1}z + \Phi(\rho,\varphi,z)z, \\ \dot{\varphi} &= 1, \end{aligned} \tag{2.3}$$

where

$$R(0,\varphi,z) \equiv 0, \quad \Phi(0,\varphi,z) \equiv 0, \quad R(\rho,\varphi,0) \equiv 0, \quad \Phi(\rho,\varphi,0) \equiv= 0. \tag{2.4}$$

As in [**9**], we consider the following system of integral equations

$$\begin{aligned} \rho(t) &= \rho_0 e^{-\nu/\omega t} + \int_0^t e^{-\nu(t-\tau)/\omega} R(\rho(\tau),\varphi(\tau),z(\tau))\rho(\tau)\,d\tau, \\ z(t) &= z_1 e^{(t-t_0)/\omega} + \int_{t_0}^t e^{(t-t_0)/\omega}\Phi(\rho(\tau),\varphi(\tau),z(\tau))z(\tau)\,d\tau, \end{aligned} \tag{2.5}$$

where $t_0 > 0$ and $\varphi(t) = \varphi_0 + t$. It can be shown [**9**] that for any given $t_0 \geqslant 0$ and all sufficiently small ρ_0 and z_1, system (2.5) has a unique solution for $t \in [0, t_0]$:

$$(\rho(t; t_0, \rho_0, \varphi_0, z_1), z(t; t_0, \rho_0, \varphi_0, z_1)),$$

which is exactly the solution of the system (2.3) hitting the cross-section ($\rho = \rho_0$, $\varphi = \varphi_0$) at $t = 0$ and the cross-section $z = z_1$ at $t = t_0$.

In the same way as in [**2**], applying the method of successive approximations and using identities (2.4), one can prove the following lemma.

LEMMA 2.2. *The solution of the boundary problem* (2.5) *can be represented in the form*

$$\rho(t) = \rho_0 e^{-\nu t/\omega} + \tilde{\rho}(t, t_0, \rho_0, \varphi_0, z_1)\rho_0,$$
$$z(t) = z_1 e^{(t-t_0)/\omega} + \tilde{z}(t, t_0, \rho_0, \varphi_0, z_1), \tag{2.6}$$

where the following estimates hold at $t = t_0$:

$$\|\tilde{\rho}\| + \left\|\frac{\partial\tilde{\rho}}{\partial\rho_0}\right\| + \left\|\frac{\partial\tilde{\rho}}{\partial t}\right\| + \left\|\frac{\partial\tilde{\rho}}{\partial t_0}\right\| \leqslant M e^{-(\nu/\omega+\eta-\epsilon)t_0}, \tag{2.7}$$

whereas at $t = 0$ *we have*

$$\|\tilde{z}\| + \left\|\frac{\partial\tilde{z}}{\partial\rho_0}\right\| + \left\|\frac{\partial\tilde{z}}{\partial t}\right\| + \left\|\frac{\partial\tilde{z}}{\partial t_0}\right\| \leqslant M e^{-(1/\omega+\eta-\epsilon)t_0}, \tag{2.8}$$

with $\eta = \min(\nu/\omega, 1/\omega)$ *and* $\epsilon > 0$ *sufficiently small.*

It is seen from (2.3) and (2.4) that one can find sufficiently small δ such that for all sufficiently small ρ^* and d the rectangles

$$S_0 = \{(\rho, z, \varphi) : \varphi = 0,\ |\rho - \rho^*| < \delta,\ |z| < \delta\},$$
$$S^0 = \{(\rho, z, \varphi) : |\rho| < \delta,\ z = d\} \tag{2.9}$$

are local cross-sections to trajectories of system (2.3). Therefore, substituting $\varphi_0 = 0$, $z_1 = d$ into (2.6) setting $t = t_0$ in the first equation and $t = 0$ in the second equation of (2.6), and taking into account the fact that $\varphi(t_0) = t_0$, we find the following parametric form for the map $T_0 \colon S_0 \to S^0$:

$$\rho = \rho_0 e^{-\nu\varphi/\omega}(1 + \chi_1(\rho_0, \varphi)),$$
$$z_0 = d e^{-\varphi/\omega}(1 + \chi_2(\rho_0, \varphi)), \tag{2.10}$$

where it follows from the estimates (2.7), (2.8) that

$$\|\chi_i\| + \|\partial\chi_i/\partial\varphi\| + \|\partial\chi_i/\partial\rho_0\| \leqslant M e^{(\eta-\epsilon)\varphi}, \qquad i = 1, 2. \tag{2.11}$$

Let us now return to the study of the behaviour near a heteroclinic cycle. Let S_i^0 and S_{0i} denote cross-sections defined by (2.9) in the neighborhood of the saddle-focus O_i ($i = 1, 2$): S_i^0 is assumed to intersect Γ_i transversely at some point M_i^0 (this determines the sign of d in (2.9)) and S_{0i} is transverse to the heteroclinic orbit Γ_0, so the constant ρ^* in (2.9) is the ρ-coordinate of the point $M_{0i} = \Gamma_0 \cap S_{0i}$ (of course, we have different values ρ_1^* and ρ_2^* for S_{01} and S_{02}).

We denote the polar coordinates in S_i^0 by (ρ_i, φ_i) and the (Cartesian) coordinates in S_{0i} by (ξ_i, u_i) (precisely, $\xi = z_0$ and $u = \rho_0 - \rho^*$).

In a neighborhood of each saddle-focus O_i, $i = 1, 2$, and in a neighborhood of segments of the trajectories Γ^0 and Γ_0 bounded by the points M_1^0, M_2^0 and M_{02}, M_{01}, respectively, the trajectories of the system define the correspondence maps: $T_{0i} \colon S_{0i} \to S_i^0$, $T_1 \colon S_1^0 \to S_2^0$, and $T_2 \colon S_{02} \to S_{01}$. It follows from general theorems in the theory of differential equations that the map T_2 can be written as follows:

$$T_2\colon \begin{cases} u_1 = F_1(u_2, \xi_2, \mu), \\ \xi_1 = F_2(u_2, \xi_2, \mu), \end{cases} \tag{2.12}$$

where

$$\left.\frac{\partial(F_1, F_2)}{\partial(u_2, \xi_2)}\right|_{u_2=\xi_2=0} \neq 0.$$

Condition A) means that $\partial F_2(0,0,0)/\partial u_2 \neq 0$. Therefore, the second equation (2.12) can be resolved with respect to u_2 and we find

$$u_1 = a_1(\xi_1, \xi_2, \mu), \qquad u_2 = a_2(\xi_1, \xi_2, \mu), \tag{2.13}$$

where $a_i(\xi_1, \xi_2, \mu)$ are sufficiently smooth functions and $\partial a_i(0,0,\mu)/\partial \xi_j \neq 0$, $i = 1, 2$, $j = 3 - i$. The latter inequality allows us to take (ξ_1, ξ_2) as the coordinates on the cross-sections S_{02} and S_{01}. Thus, formula (2.13) defines the relation between these coordinates and u, and the map T_2 becomes the identity.

In order to obtain formulas determining the map T_{0i}, we replace ρ_0 and z_0 in (2.10) by $\rho^*_{0i} + a_i(\xi_1, \xi_2)$ and ξ_i, respectively (recall that ρ^*_{0i} is the distance from O_i to the point M_{0i} at which the heteroclinic trajectory Γ_0 intersects the surface $\varphi_0 = 0$). This gives

$$\begin{aligned} \rho_i &= (\rho^*_{0i} + a_i(\xi_1, \xi_2, \mu))e^{-\nu_i\varphi_i/\omega_i}(1 + \chi_{1i}(\xi_1, \xi_2, \varphi_i, \mu)), \\ \xi_i &= d_i e^{-\varphi_i/\omega_i}(1 + \chi_{2i}(\xi_j, \varphi_i, \mu)), \qquad j = 3 - i,\ i = 1, 2, \end{aligned} \tag{2.14}$$

where instead of (2.11) the following estimates hold:

$$\begin{aligned} \|\chi_{1i}\| + \left\|\frac{\partial \chi_{1i}}{\partial \xi_1}\right\| + \left\|\frac{\partial \chi_{1i}}{\partial \xi_2}\right\| + \left\|\frac{\partial \chi_{1i}}{\partial \varphi_i}\right\| &\leqslant Me^{-(\eta_i - \epsilon)\varphi_i}, \\ \|\chi_{2i}\| + \left\|\frac{\partial \chi_{2i}}{\partial \xi_j}\right\| + \left\|\frac{\partial \chi_{2i}}{\partial \varphi_i}\right\| &\leqslant Me^{-(\eta_i - \epsilon)\varphi_i}, \qquad i = 1, 2,\ j = 3 - i. \end{aligned} \tag{2.15}$$

The domain of the map T_{0i} is $\sigma_0 = I_1 \times I_2$, where $I_i = (0 < \xi_i \leqslant \delta)$, $i = 1, 2$. Let $\sigma_{1i} = T_{0i}\sigma_0 \subset S^0_i$ be the range of this map. Define the map $T_{21} = T_{01} \circ T_2 \circ T^{-1}_{02} : S^0_2 \to S^0_1$ which takes σ_{12} onto σ_{11}.

Formulas (2.14) allow one to easily obtain the following geometric picture for the action of the map T_{21}. Note first that the region σ_{1i} is bounded by two spirals winding towards the point M_{0i} (one of these spirals on σ_{11} (on σ_{12}) is the connected component $w_1(O_2)$ of $W^u(O_2) \cap S^0_1$ (respectively, $w_2(O_1)$ of $W^s(O_1) \cap S^0_2$)). The map T_{0j} transforms an arbitrary segment

$$I_{\xi_i}(\xi_j) = \{(\xi_1, \xi_2) : \xi_j \in I_j,\ \xi_i = \text{const}\}, \qquad i = 1, 2,\ j = 3 - i,$$

into a spiral on S^0_j, while T_{0i} transforms that segment into a segment whose ends lie on the boundaries of σ_{1i} (Figure 2). Thus, one can view σ_{1i} as stratified into the segments $\sigma_{1i} = \bigcup_{0 \leqslant \xi_i \leqslant \delta} T_{0i} I_{\xi_i}(\xi_j)$ or into the spirals $\sigma_{1i} = \bigcup_{0 \leqslant \xi_j \leqslant \delta} T_{0i} I_{\xi_j}(\xi_i)$ in such a way that every spiral on S^0_2 is taken by the map T_{21} into the corresponding segment on S^0_1, and vice versa.

To obtain the required formulas for the map $T_1 : S^0_1 \to S^0_2$, we first use the Cartesian coordinates on S^0_i: $(x_i = \rho_i \cos\varphi_i,\ y_i = \rho_i \sin\varphi_i)$. The map T_1 is written as follows

$$\begin{pmatrix} x_2 \\ y_2 \end{pmatrix} = E(\mu) + D(\mu)\begin{pmatrix} x_1 \\ y_1 \end{pmatrix} + O(\|x_1, y_1\|^2), \tag{2.16}$$

where $\det(D) > 0$.

Note that formulas (2.14) are invariant with respect to rotations and uniform rescaling of the coordinates on S^0_i (with the values of some coefficients multiplied by irrelevant positive factors). In formula (2.16) these transformations result in the multiplication of the matrix D by some rotation matrices from the left and right and, also, in the multiplication by some positive constant. One can see that the

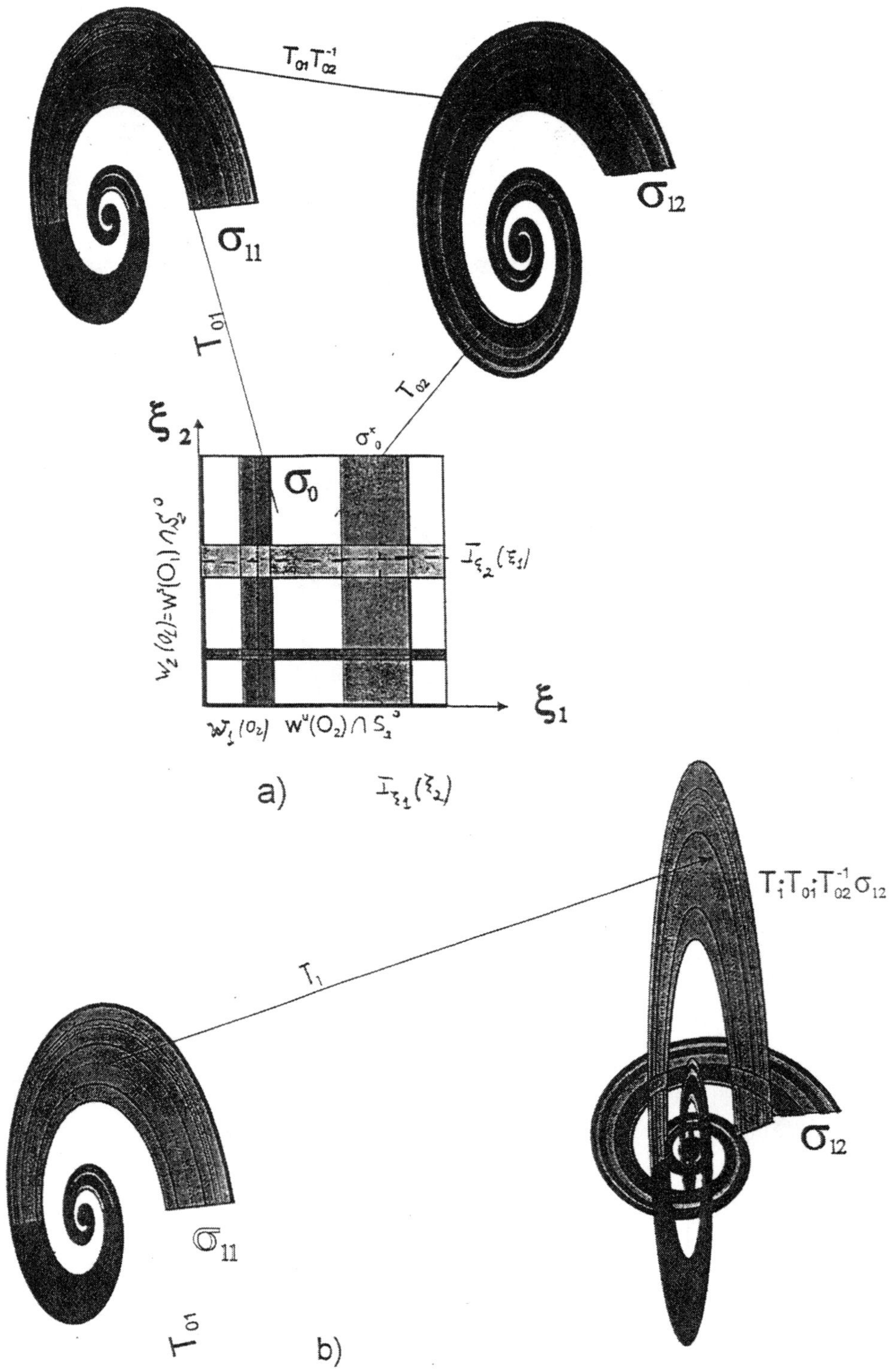

T01T02-1
σ11
σ12
T01
T02
ξ2
ξ1
σ0
a)
T1
T1T01T02-1σ12
σ11
σ12
T01
b)

FIGURE 2

rotations and uniform rescaling of the coordinates on S_i^0 can always be chosen so that the matrix D takes the form

$$D = \begin{pmatrix} \lambda^{-1} & 0 \\ 0 & \lambda \end{pmatrix} \tag{2.17}$$

for some $\lambda > 0$.

According to condition B), the function $E(\mu)$ vanishes at $\mu = 0$. Therefore, generically, we may assume that

$$E(\mu) = (\mu_1, \mu_2). \tag{2.18}$$

Let us transform the coordinates on S_2^0 by the formula:

$$\begin{pmatrix} x_2' \\ y_2' \end{pmatrix} = E(\mu) + D(\mu)T_1^{-1}\begin{pmatrix} x_2 \\ y_2 \end{pmatrix}.$$

By definition, in the new coordinates the map T_1 becomes linear:

$$\begin{pmatrix} x_2 \\ y_2 \end{pmatrix} = E(\mu) + D(\mu)\begin{pmatrix} x_1 \\ y_1 \end{pmatrix} \tag{2.19}$$

and one can easily see that formulas (2.14) for the map T_{02} will retain their form (with the functions a_2 and χ_{12}, χ_{22} slightly modified).

In polar coordinates, formula (2.16) is rewritten as

$$\rho_2 = \rho_1\sqrt{\lambda^{-2}\cos^2\varphi_1 + \lambda^2\sin^2\varphi_1}, \tag{2.20}$$

$$\varphi_2 = 2\pi m + \begin{cases} \arctan(\lambda^2\tan\varphi_1) & \text{if } \cos\varphi_1 \geqslant 0, \\ \pi + \arctan(\lambda^2\tan\varphi_1) & \text{if } \cos\varphi_1 < 0, \end{cases} \tag{2.21}$$

where m can be an arbitrary integer.

3. Periodic and heteroclinic orbits

In this section we will study orbits that lie entirely in a small neighborhood $U(\mathcal{L})$.

THEOREM 3.1. *In any neighborhood of the contour $\mathcal{L}$ there exists a countable set of single-round*[1] *periodic orbits.*

PROOF. The single-round periodic orbits correspond to the fixed points of the map $T_1 \circ T_{21} = T_1 \circ T_{01} \circ T_2 \circ T_{02}^{-1}$. By (2.14) and (2.20), these fixed points are found by solving the following system:

$$\begin{aligned} &(\rho_{02}^* + a_2(\xi_1, \xi_2, 0))e^{-\nu_2\varphi_2/\omega_2}(1 + \chi_{12}(\xi_1, \xi_2, \varphi_2, 0)) \\ &\quad = (\rho_{01}^* + a_1(\xi_1, \xi_2, 0))e^{-\nu_1\varphi_1/\omega_1}(1 + \chi_{11}(\xi_1, \xi_2, 0)) \\ &\qquad \times \sqrt{\lambda^{-2}\cos^2\varphi_1 + \lambda^2\sin^2\varphi_1}, \\ &\xi_1 = d_1 e^{-\varphi_1/\omega_1}(1 + \chi_{21}(\xi_2, \varphi_1, 0)), \qquad \xi_2 = d_2 e^{-\varphi_2/\omega_2}(1 + \chi_{22}(\xi_1, \varphi_1, 0)), \end{aligned} \tag{3.1}$$

where φ_1 and φ_2 are related by (2.21).

[1]Such orbits have exactly one intersection point with S_{01} and S_{02}.

Solving the last two equations of (3.1) with respect to ξ_1 and ξ_2, plugging the obtained expressions into the first equation of (3.1) and taking the logarithm of both sides of the first equation, one obtains

$$\tau\varphi_2 = \varphi_1 + \frac{\omega_1}{\nu_1}\ln\frac{\rho_{02}^*}{\rho_{01}^*} - \frac{\omega_1}{2\nu_1}\ln(\lambda^{-2}\cos^2\varphi_1 + \lambda^2\sin^2\varphi_1) + \dots, \tag{3.2}$$

where the dots stand for smooth terms that tend to zero along with their derivatives as $\varphi_{1,2} \to +\infty$.

Every solution of the system composed of equations (3.2) and (2.21) corresponds to a single-round periodic orbit. Note that (2.21) defines an infinite set of smooth curves on (φ_1, φ_2)-plane (one curve for each integer m) which tend to infinity with the same asymptotic direction $\lim \varphi_1/\varphi_2 = 1$. Obviously, each of these curves has at least one intersection with the curve (3.2): the latter has the asymptotic direction $\lim \varphi_1/\varphi_2 = \tau$, and $\tau \neq 1$ by assumption C). □

The next theorem deals with the existence of heteroclinic trajectories, different from Γ_0, that belong to the intersection of the two-dimensional manifolds $W^s(O_1)$ and $W^u(O_2)$. We restrict our consideration to the trajectories that have only one intersection point with S_{01} and S_{02} (we call them single-round).

Let us introduce the quantity

$$G \equiv \nu_1^2/\omega_1^2 + \nu_2^2/\omega_2^2 + 2\nu_1\nu_2/(\omega_1\omega_2)\mathcal{E} + 1 - \mathcal{E}^2, \tag{3.3}$$

where

$$\mathcal{E} = \frac{1}{2}\left(\lambda^2 + \frac{1}{\lambda^2}\right) > 1.$$

We denote the component of the bifurcation surface H^0 corresponding to $G > 0$ by H_1^0 and those corresponding to $G < 0$ by H_2^0.

THEOREM 3.2. *The system X_0 has an infinite sequence of single-round heteroclinic trajectories $\Gamma_{0i} \in W^s(O_1) \cap W^u(O_2)$, $i = 1, 2, \dots$. If $G > 0$, then the manifolds $W^s(O_1)$ and $W^u(O_2)$ intersect transversely at the points of each of these orbits. On the contrary, if $G < 0$, then systems having a nontransverse single-round heteroclinic trajectory Γ_{0*} are dense everywhere in H_2^0.*

PROOF. The intersection of the locally unstable manifold of O_2 with the cross-section S_{02} is the segment $\xi_2 = 0$ and the intersection of the locally stable manifold of O_1 with the cross-section S_{01} is the segment $\xi_1 = 0$. Therefore, $\xi_2 = 0$ and $\xi_1 = 0$ also give the equations of the images of these segments on the cross-sections S_{01} and S_{02} by the maps T_2 and T_2^{-1}, respectively. Further applying the maps T_{01} and T_{02}, respectively, to these images we obtain two curves, $w_1 \subset S_1^0$ and $w_2 \subset S_2^0$. By (2.14), they are spirals defined by the following equations:

$$w_1: \begin{cases} \rho_1 &= (\rho_{01}^* + a_1(\xi_1, 0, 0))e^{-\nu_1\varphi_1/\omega_1}(1 + \chi_{11}(\xi_1, 0, \varphi_1, 0)), \\ \xi_1 &= d_1 e^{-\varphi_1/\omega_1}(1 + \chi_{21}(0, \varphi_1, 0)), \end{cases} \tag{3.4}$$

$$w_2: \begin{cases} \rho_2 &= (\rho_{02}^* + a_2(\xi_2, 0, 0))e^{-\nu_2\varphi_2/\omega_2}(1 + \chi_{12}(\xi_2, 0, \varphi_2, 0)), \\ \xi_2 &= d_2 e^{-\varphi_2/\omega_2}(1 + \chi_{22}(0, \varphi_2, 0)). \end{cases} \tag{3.5}$$

The curves w_1 and w_2 are the pieces of the intersections of, respectively, the unstable manifold of O_2 and the stable manifold of O_1 with the cross-sections S_1^0 and S_2^0. By construction, the points of intersection of w_2 and $T_1 w_1$ correspond to single-round heteroclinic trajectories $\Gamma_{0i} \in W^s(O_1) \cap W^u(O_2)$; namely, they are exactly the intersection points of the heteroclinic trajectories with S_2^0. Therefore, the single-round heteroclinic orbits are in one-to-one correspondence with the solutions of the system of equations (2.20), (2.21), (3.4), (3.5). In the same way as in Theorem 3.1, this system is reduced to the system of equations (2.21) and (3.2), where the dots in (3.2) stand for terms tending to zero as $\varphi_{1,2} \to +\infty$ along with their derivatives (now these terms are, of course, different from those in Theorem 3.1, but they are of no importance).

Repeating the arguments of Theorem 3.1, we obtain the existence of infinitely many solutions of system (2.21), (3.2), which yields the existence of infinitely many single-round heteroclinic orbits, as claimed.

The transverse intersections of the curves given by (2.21) and (3.2) on the (φ_1, φ_2)-plane cannot be removed by small perturbations of the system X_0 on H^0. Hence, they correspond to transverse heteroclinic orbits. Similarly, the tangency of the two curves given by (2.21) and (3.2) corresponds to a nontransverse heteroclinic orbit.

The condition of tangency at some $\varphi_1 = \theta + \pi n$ (where $\theta \in [-\pi/2, \pi/2]$) is written as

$$
\begin{aligned}
k\tau - n &= \frac{\theta}{\pi} + C - \frac{\tau}{\pi}\arctan(\lambda^2\tan\theta) - \frac{\omega_1}{2\pi\nu_1}\ln(\lambda^{-2}\cos^2\theta + \lambda^2\sin^2\theta) \\
&\quad + \cdots + \frac{\lambda^2 - \lambda^{-2}}{2}\left[\frac{\omega_1}{\nu_1}\sin 2\theta + \cos 2\theta\right] \\
&= \tau + \frac{\lambda^2 - \lambda^{-2}}{2} + \ldots,
\end{aligned} \tag{3.6}
$$

where $k = 2m$ if n is even, and $k = 2m + 1$ if n is odd, i.e., k and n may be arbitrarily large integers of the same parity, the dots here stand for terms that tend to zero along with their derivatives as $k, n \to +\infty$, and

$$C = \frac{\omega_1}{\pi\nu_1}\ln\frac{\rho_{02}^*}{\rho_{01}^*}.$$

It is easy to see that the second equation here does not have solutions for $G > 0$, which proves the transversality of all single-round heteroclinics in this case.

On the contrary, if $G < 0$, then the second equation has two roots θ_1^0 and θ_2^0. By an arbitrarily small perturbation of the system X_0, one can make the value of τ irrational. In this case the left-hand side of the first equation in (3.6) will run over a dense set of values in $\mathbb{R}^1$ when k and n take sufficiently large integer values. Hence, by an arbitrarily small additional perturbation of the system that keeps the irrational value τ unchanged, but changes the value of C, one can make, say, θ_1^0 a solution of (3.6) for some sufficiently large integers k and n of the same parity. This solution will correspond to the nontransverse heteroclinic trajectory. □

We see that the existence of the heteroclinic contour $\mathcal{L}$ with the transverse heteroclinic trajectory Γ_0 implies the simultaneous existence of more complicated heteroclinic trajectories Γ_{0i}. When at least one of these orbits is transverse (and this is the generic case), we may apply Theorem 3.2 once again to the contour $\mathcal{L}_i = O_1 \cup O_2 \cup \Gamma^0 \cup \Gamma_{0i}$ and deduce the existence of infinitely many *double-round* heteroclinic orbits, etc. Thus, using inductive arguments, we obtain, in the generic situation, the existence of infinitely many k-round heteroclinic orbits in the set $W^s(O_1) \cap W^u(O_2)$ for every k.

4. Bifurcation of homoclinic trajectories

Here we consider the birth of homoclinic loops when the separatrix connection Γ^0 is split. We assume that the family X_μ intersects the bifurcational surface H^0 for $\mu = 0$ transversely, which means that (2.18) is satisfied for the map T_1.

Recall that we assumed $X_\mu \in C^r$ with $r \geqslant 3$. When transforming the local vector field near a saddle-focus to the form (2.1), (2.2), the smoothness of the system with respect to μ becomes C^{r-2}; however, this coordinate transformation is C^{r-1} with respect to the phase variables and its first derivatives with respect to the phase variables are C^{r-2}-functions with respect to all variables and parameters (see [**10**]). Therefore, the function $D(\mu)$ in (2.16) is also C^{r-2} and we do not lose smoothness when bringing the map T_1 to the form (2.19). So, all functions introduced in Section 1 are at least C^1 in μ.

LEMMA 4.1. *For the family X_μ with $X_0 \in H^0$, on the plane of parameters $\mu = (\mu_1, \mu_2)$ there exist bifurcation curves l_1 and l_2 such that for $\mu \in l_1$ ($\mu \in l_2$) the system has a single-round separatrix loop Γ_1 (respectively, Γ_2) to the saddle-focus O_1 (respectively, O_2). Each of these curves has the form of a spiral winding towards the origin.*

PROOF. The single-round homoclinic loop to the saddle-focus O_1 exists when the curve $T_1^{-1}w_2 = T_1^{-1} \circ T_{02} \circ T_2^{-1}(W^s_{\text{loc}}(O_1) \cap S_{01})$ contains the origin (i.e., the intersection point with W^u_{loc}) on the cross-section S_1^0. Thus, the curve l_1 is defined by the condition

$$T_1(0,0) \in w_2.$$

According to (2.19), the point $T_1(0,0)$ on S_2^0 has (Cartesian) coordinates (μ_1, μ_2). The equation of the curve w_2 is given by the first equation (2.14) for $\xi_1 = 0$, where ξ_2 is replaced by its expression given by the second equation of (2.14). Denoting $\mu_1 = r\cos\psi$, $\mu_2 = r\sin\psi$ (i.e., $r = \sqrt{\mu_1^2 + \mu_2^2}$), we obtain the following equation for the bifurcational curve l_1:

$$l_1: \quad r = \rho_{02}^* e^{-\nu_2\psi/\omega_2}(1 + R_2(\psi)), \tag{4.1}$$

where

$$|R_2(\psi)| + |dR_2/d\psi| \leqslant Me^{-\eta_2\psi} \tag{4.2}$$

for some $\eta_2 > 0$.

Similarly, the equation for the bifurcation curve l_2 is determined from the condition

$$T_1^{-1}(0,0) \in w_1 = T_{01} \circ T_2(W^u_{\text{loc}}(O_2) \cap S_{02})$$

and it is written as

$$l_2: \quad r = \rho_{01}^* e^{-\nu_1(\pi+\psi)/\omega_1}(1 + R_1(\psi))\sqrt{\lambda^2\cos^2\psi + \lambda^{-2}\sin^2\psi}, \tag{4.3}$$

where

$$|R_1(\psi)| + |dR_1/d\psi| \leqslant Me^{-\eta_1\psi} \tag{4.4}$$

for some $\eta_1 > 0$. □

Looking at equations (4.3) and (4.4), one notices that the intersection points of the curves l_1 and l_2 are found by the same procedure that was used in Theorems 3.1 and 3.2. To avoid further repetition, we just formulate the result.

THEOREM 4.1. *For each two-parameter family $X_\mu(x)$ transverse to H^0 at $\mu = 0$, on the plane of parameters μ there is an infinite sequence of points $\mu^k \in l_1 \cap l_2$ at each of which the system has both a homoclinic loop to the saddle-focus O_1 and a homoclinic loop to the saddle-focus O_2. If $X_0(x) \in H_1^0$ (i.e., $G > 0$), then at every point μ^k the curves l_1 and l_2 intersect transversely. On the contrary, for $G < 0$, by an arbitrary small smooth perturbation of the family $X_\mu(x)$, one can obtain a point of tangency of l_1 and l_2 arbitrarily close to $(\mu_1, \mu_2) = (0, 0)$.*

Acknowledgments. The author is grateful to S. Gonchenko and D. Turaev for formulating the problem and many useful discussions, and to L. Lerman and D. Turaev for tremendous help in the desperate fight with English grammar.

References

[1] L. P. Shilnikov, *A contribution to the problem of the structure of an extended neighborhood of a rough equilibrium state of saddle-focus type*, Mat. Sb. (N.S.) **81 (123)** (1970), 92–103; English transl., Math. USSR-Sb. **10** (1970), 91–102.
[2] I. M. Ovsyannikov and L. P. Shilnikov, *On systems with a saddle-focus homoclinic curve*, Mat. Sb. **130 (172)** (1986), no. 4, 552–570; English transl., Math. USSR-Sb. **58** (1987), 557–574.
[3] V. I. Arnold, V. S. Afraimovich, Yu. S. Il′yashenko, and L. P. Shilnikov, *Dynamical systems* V, Encyclopaedia of Mathematical Sciences, vol. 5, Springer-Verlag, 1993.
[4] V. V. Bykov, *On the structure of the neighborhood of a separatrix contour containing a saddle-focus*, Methods of the Qualitative Theory of Differential Equations (E. A. Leontovich-Andronova, ed.), Gor′kov. Gos. Univ., Gor′kiĭ, 1978, pp. 3–32. (Russian)
[5] ———, *On the structure of bifurcation sets of dynamical systems that are systems with a separatrix contour containing a saddle-focus*, Methods of the Qualitative Theory of Differential Equations (E. A. Leontovich-Andronova, ed.), Gor′kov. Gos. Univ., Gor′kiĭ, 1980, pp. 44–72. (Russian)
[6] ———, *The bifurcation of separatrix contour and chaos*, Physica D **62** (1993), 290–299.
[7] N. K. Gavrilov and L. P. Shilnikov, *On 3-dimensional systems close to systems with a structurally unstable homoclinic curve*, II, Mat. Sb. **90 (132)** (1973), 139–156, 167; English transl., Math. USSR-Sb. **19** (1973), 139–156.
[8] S. Newhouse, *The abundance of wild hyperbolic sets and nonsmooth stable sets for diffeomorphisms*, Publ. Math. IHES **50** (1979), 101–151.
[9] L. P. Shilnikov, *On the Poincaré–Birkhoff problem*, Mat. Sb. (N.S.) **74 (116)** (1967), 378–397; English transl., Math. USSR-Sb. **3** (1967), 91–102.
[10] L. Shilnikov, A. Shilnikov, D. Turaev, and L. Chua, *Methods of qualitative theory in nonlinear dynamics. Part* I, World Scientific, Singapore, 1998.

DEPARTMENT OF DIFFERENTIAL EQUATIONS, INSTITUTE FOR APPLIED MATHEMATICS AND CYBERNETICS, 10 UL′YANOVA ST., NIZHNY NOVGOROD, 603005, RUSSIA

Translated by A. A. KIRILLOV

Amer. Math. Soc. Transl.
(2) Vol. **200**, 2000

Example of a Blue Sky Catastrophe

Nikolaĭ Gavrilov and Andrey Shilnikov

To the memory of Professor E. A. Leontovich-Andronova

ABSTRACT. We present a low-order system of ODEs exhibiting the blue sky catastrophe—a new codimension one bifurcation of periodic orbits.

1. Introduction

The known boundaries of stability (or existence) regions of periodic orbits in systems of differential equations can be classified by means of the following formal criterion: whether a bifurcating periodic orbit on the boundary exists or not. The codimension-one bifurcations listed below comprise the first group:

- A saddle-node (fold) bifurcation: two periodic orbits, one stable and one unstable, coalesce on the stability boundary and annihilate beyond it;
- Period-doubling (flip) bifurcation: one multiplier of the bifurcating periodic orbit equals -1;
- A bifurcation from a periodic orbit to a two-dimensional invariant torus; by rephrasing A. Andronov: "a cycle loses its skin".

All three cases above reduce to the analysis of stability of the corresponding fixed point of the Poincaré map on a cross-section transverse to the periodic orbit.

The codimension-one bifurcations of the second group have much in common with the situations in which a periodic orbit merges with an equilibrium state. The enumeration begins with the Andronov-Hopf bifurcation in which a periodic orbit shrinks into an equilibrium state with a pair of characteristic exponents $\pm i\omega$, $\omega > 0$. It should be noted that the period T of such bifurcating orbit can be estimated as $T \sim 2\pi/\omega$. In the other two cases the periodic orbit adheres to a homoclinic loop to an unstable equilibrium state which can be either a saddle with characteristic exponents in both open left and right half-planes, or a simple saddle-node with one zero real exponent. Since the vector field vanishes at the equilibrium state, the period of the bifurcating orbit tends to infinity as it approaches the homoclinic loop, while its perimeter remains of finite length.

2000 *Mathematics Subject Classification.* Primary 37J20.

This work was supported in part by the Russian Foundation for Basic Research (grants 96-01-01135 and 99-01-00231), INTAS (grant No. 93-05-70 ext), and the program "Universities of Russia" (project No. 1905).

©2000 American Mathematical Society

One may ask if the list of bifurcations and the associated stability boundaries is complete or not. This appeared to be the case until Shilnikov and Turaev [**1**] suggested a new scenario for the saddle-node bifurcation of periodic orbits, which may result in the appearance of a stable periodic orbit of an infinitely long period and infinitely large perimeter. Due to this feature, a virtual bifurcation of this kind was called, following R. Abraham, "a blue sky catastrophe". Moreover, this boundary may, under certain conditions, separate Morse–Smale systems from systems with hyperbolic attractors in the parameter space [**2**].

Here we present an example of a three-dimensional system in which the blue sky catastrophe develops in accordance with the phenomenological mechanism devised by Shilnikov and Turaev. Let us first discuss its geometrically comprehensive construction. The existence of the paper by Shilnikov and Turaev in this volume, to which the reader is referred to for a deeper insight, allows us to give only a brief explanation of the bifurcation setup.

It is assumed that there exists a saddle-node periodic orbit L^* whose unstable manifold $W^u_{L^*}$ returns to L^* as $t \to +\infty$, as shown in Figure 1; hence, the closure of the unstable manifold is not a Hausdorff manifold.

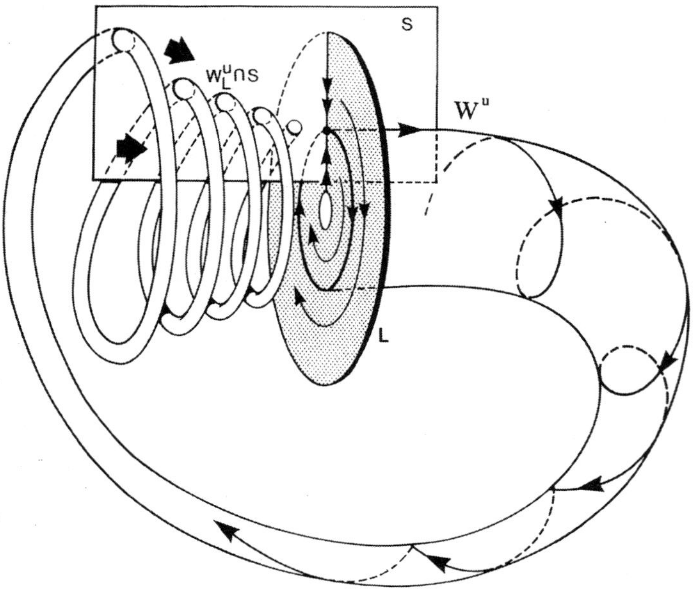

FIGURE 1. Geometrical interpretation of a blue sky bifurcation

The second assumption concerns the global property of the vector field; namely, the contraction of the phase volume in the direction transverse to the global part of the unstable manifold is so strong that the disappearance of L^* is smoothly followed by the birth of a stable large-amplitude periodic orbit upon crossing the stability boundary.

2. The model

Example. Consider the family of systems

$$(1)\qquad \begin{cases} \dot{x} = x(2+\mu - B(x^2+y^2)) + z^2 + y^2 + 2y \equiv P, \\ \dot{y} = -z^3 - (1+y)(z^2+y^2+2y) - 4x + \mu y \equiv Q, \\ \dot{z} = (1+y)z^2 + x^2 - \varepsilon \equiv R, \end{cases}$$

where μ, ε, and B are parameters (we let $B = 10$). The system 1 for $\mu = \varepsilon = 0$ has a closed integral curve given by ($x = 0$, $z^2 + (y+1)^2 = 1$) in the phase space. There are two equilibrium states on this curve: the low one $O'(0,-2,0)$ is a saddle-node with one zero $\lambda_1 = 0$ and two negative characteristic exponents found from the equation $\lambda^2 + 40\lambda + 68 = 0$; the upper equilibrium state $O(0,0,0)$ has one zero exponent $\lambda_1 = 0$ and a pair of purely imaginary characteristic exponents $\lambda_{2,3} = \pm 2i$ (Figure 3a). Therefore, the point O' is of codimension one because $R_{zz} \neq 0$, whereas the point O is of codimension three because the two-dimensional divergence $\sigma(z) = P'_x + Q'_y = -z^2 + \cdots$ at O starts with a quadratic term [4]. This means that a double (semistable) cycle can be generated from O in the (x,y)-plane because "an embryo" of the first Lyapunov value vanishes here.

The unfolding of the bifurcation diagram is shown in Figure 2. It includes three bifurcations, all of codimension one.

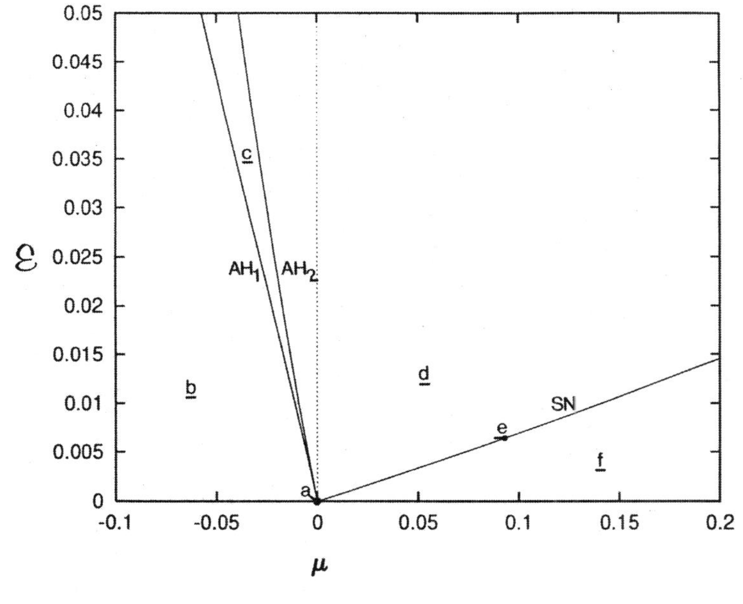

FIGURE 2. The (μ,ε)-bifurcation diagram for $B = 10$.

Let us describe the bifurcation occurring while moving in the clock-wise direction around the origin on (μ,ε)-parameter plane. When ε becomes positive, the saddle-node O' disappears, while the equilibrium state O is decomposed into two equilibria O_1 and O_2, where $z_{O_{1,2}} = \mp\sqrt{\varepsilon} + \ldots$. In the region $\underline{\text{b}}$, the point O_1 is stable and O_2 is a saddle-focus of type $(2,1)$ whose one-dimensional separatrices converge to O_1 as $t \to +\infty$ (Figure 3b). Upon entering the region $\underline{\text{c}}$, the point O_1 loses its stability through a super-critical Andronov–Hopf bifurcation (this is guaranteed by the choice of large $B > 0$) on the curve AH_1 and becomes a saddle-focus (1,2). Here, both unstable separatrices of the saddle-focus O_2 tend to a new stable periodic L_1, see Figure 3c. The equilibrium state O_2 undergoes a similar Andronov–Hopf bifurcation on the curve AH_2 and becomes a totally repelling point in the region d. The unstable manifold of the just born saddle periodic orbit L_2 continues to tend to L_1, as shown in Figure 3d. On the bifurcation curve SN, both periodic orbits coalesce, thereby composing a saddle-node cycle L^* whose unstable manifold, due to continuity, is bi-asymptotic to L^* as $t \to \pm\infty$, as shown in Figure 3e. The cycle L^* vanishes in the region f, but the local stability of the system is inherited by a new, unique, stable large-amplitude periodic orbit L_{bs}, which is not homotopic to either of the former cycles, see Figure 3f.

If one attains the curve SN from the region f, the number of scrolls of the orbit L_{bs} increases uncountably near the "phantom" of the saddle-node orbit L^* in the phase space. Since both equilibrium states O_1 and O_2 are located at a finite distance from the bifurcating periodic orbit, the infinite perimeter of the latter implies its infinite length on the boundary SN, and vice versa; indeed, the flight time of any trajectory passing close by the helix-like density of orbits, from which the saddle-node limit cycle appears, tends to infinity as one gets closer to SN. Thus, we have a mechanism for the blue sky catastrophe bifurcation of the desired codimension one in the two-parameter family. It is worth noticing that the corresponding stability boundary is classified as safe in the sense that the representing phase point does not desert towards any other attractor after the bifurcation, which is apparently invertible.

The stability analysis of this system in a small neighborhood $\mu = \varepsilon = 0$ is carried out analytically. The local consideration does not involve numeric computations unless we need to show that the one-dimensional stable manifold $W^s_{O_1}$ of the saddle-focus O_1 is not enclosed in the unstable manifold of the periodic orbit L_{bs}, similarly to Figure 3d, but for parameter values on the curve SN. Otherwise, this would lead to the appearance of a two-dimensional torus instead of the stable periodic orbit. To verify this condition, one should continue $W^s_{O_1}$ backward in time in order to find its intersection point with some transverse cross-section drawn nearby on the left from O_1, and to check that this point is not surrounded by the trace of the intersection of $W^u_{L_{bs}}$ with the cross-section. This turns out to be true indeed, because our construction follows the phenomenological scenario in [**3**]; namely the presence of saddle-node point O' on the integral curve at the initial stage guarantees the further contraction in this region, so that by that time the manifold $W^u_{L_{bs}}$ passes by the hollow and reaches the cross-section, its trace finally shrinks almost to a point distant from the corresponding intersection point of $W^s_{O_1}$.

If we step back from a vicinity of the origin ($\mu = \varepsilon = 0$), the unstable manifold $W^u_{O_2}$ of the saddle-focus O_2 may no longer follow along the prescribed global pathway. Of special interest is the situation where $W^u_{O_2}$ becomes homoclinic to O_2. Provided that Shilnikov's condition holds here, i.e., the saddle value is positive at the saddle-focus near such a homoclinic loop, one should expect the onset of dynamical chaos, as depicted in Figure 4.

Moreover, we should emphasize that the bifurcation of an equilibrium state with eigenvalues $(0, \pm i\omega)$ may, under some matching condition, generate "miniature" chaos induced by heteroclinic connections between both saddle-foci, as well as by homoclinic loops to either one. The associated chaotic limit set may be a strange attractor or a strange repeller (an attractor as $t \to -\infty$), depending locally on

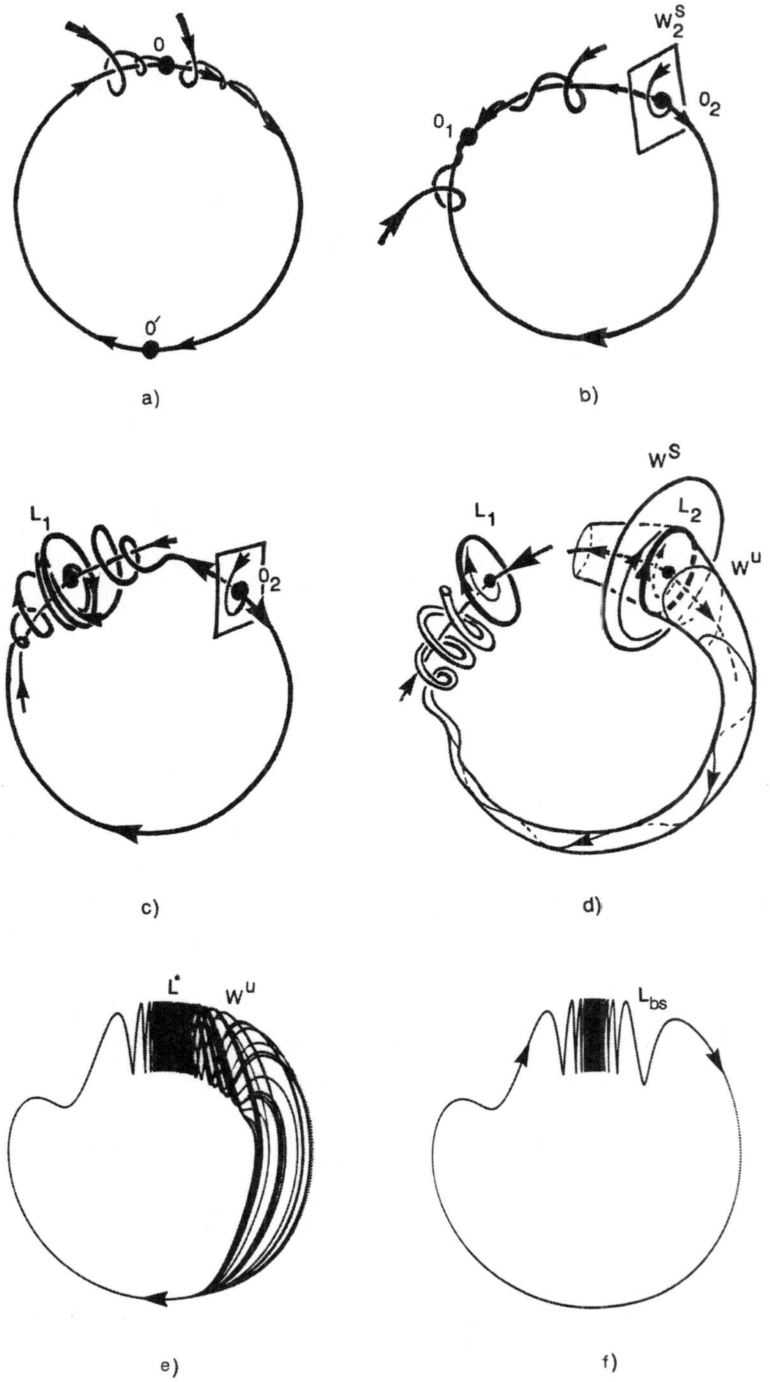

FIGURE 3. Mechanizm of the blue-sky catastrophe. (f) A stable periodic orbit at ($\mu = 0.3, \varepsilon = 0.021$)

the sign of divergence of the vector field. There is a 3D-exotic example of this bifurcation resulting in the co-appearance of the chaotic spiral repeller and attractor

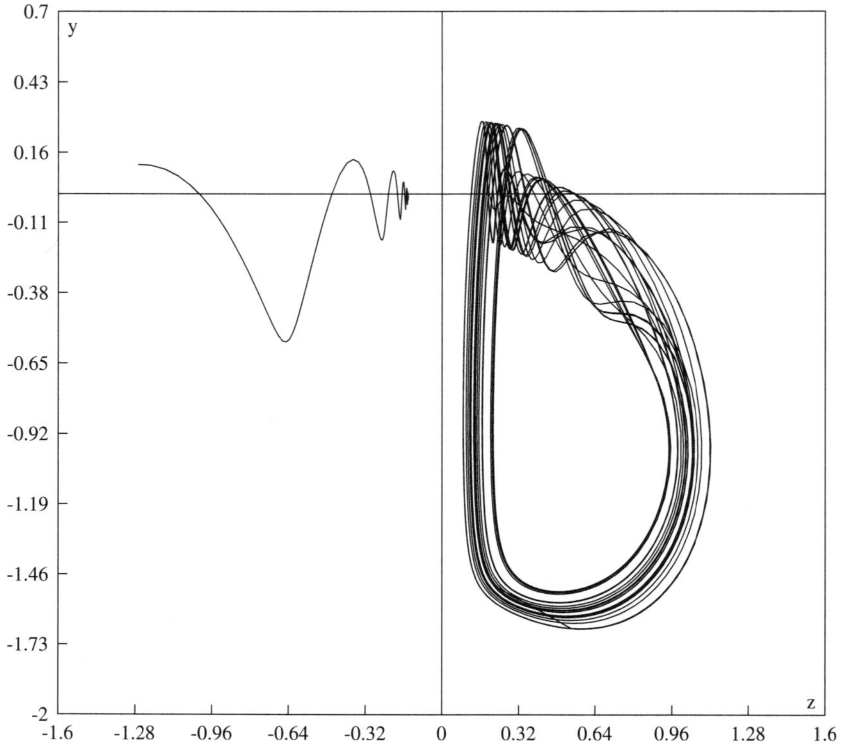

FIGURE 4. Spiral chaos in region b̲ for ($\mu = -0.173$, $\varepsilon = 0.01$).

which are the α- and ω-limit sets, respectively, for trajectories close in the phase space to such a twice-degenerate equilibrium state [**5**].

3. Conclusion

To conclude, we note that the blue-sky catastrophe in action may be helpful for a qualitative explanation of the frequently observed transition from low-amplitude oscillations (spikes) to large-amplitude burstings in models for neuron activity, or to flow surges in models of jet engines.

References

[1] L. P. Shilnikov and D. V. Turaev, *On a blue sky catastrophe*, Dokl. Akad. Nauk SSSR **342** (1995), no. 5, 596–599; English transl. in Russian Math. Dokl.

[2] ———, *Simple bifurcations leading to hyperbolic attractors*, Computers Math. Applic. **34** (1997), no. 2–4, 173–180.

[3] ———, *A new simple bifurcation of a periodic orbit of "blue sky catastrope" type.* This volume.

[4] N. K. Gavrilov, *On some bifurcations of an equilibrium state with one zero and a pair of purely imaginary roots*, Methods of Qualitative Theory of Differential Equations, Gorky State University, Gorky, 1978.

[5] A. L. Shilnikov, C. Nicolis and G. Nicolis, *Bifurcation and predicability analysis of a low-order atmospheric circulation model*, Int. J. Bifurcation and Chaos **5** (1995), no. 6, 1701-1711.

INSTITUTE FOR APPLIED MATHEMATICS & CYBERNETICS, 10 UL'YANOV ST., NIZHNY NOVGOROD 603106, RUSSIA

UNIVERSITY OF CALIFORNIA AT BERKELEY, 253 CORY HALL, BERKELEY, CA 94720
E-mail address: andrey@fred.eecs.berkeley.edu

Translated by ANDREY SHILNIKOV

Amer. Math. Soc. Transl.
(2) Vol. **200**, 2000

Dynamics and Moduli of Ω-conjugacy of 4D-diffeomorphisms with a Structurally Unstable Homoclinic Orbit to a Saddle-focus Fixed Point

S. V. Gonchenko

To the memory of Professor E. A. Leontovich-Andronova

ABSTRACT. We consider a four-dimensional diffeomorphism having a structurally unstable homoclinic orbit to a saddle-focus fixed point with eigenvalues $\lambda e^{\pm i\varphi}$, $\gamma e^{\pm i\psi}$, where $0 < \lambda < 1 < \gamma$. Let N be the set of orbits entirely lying in a small neighborhood of the homoclinic orbit. We specify nontrivial hyperbolic subsets in N and prove that the quantities φ, ψ, and $\theta = -\ln\lambda/\ln\gamma$ are moduli of Ω-conjugacy (i.e., continuous invariants of topological conjugacy on nonwandering sets) of the diffeomorphisms under consideration.

Introduction

The subject of the qualitative theory of differential equations is the description of the structure of dynamical systems up to topological equivalence. In this connection, the fundamental problem is the topological classification of dynamical systems and the determination of topological invariants. This problem has been completely solved only for some classes of systems. The greatest progress achieved here is in the case of two-dimensional flows. For flows on the plane (2-sphere) that have a finite number of singular orbits (equilibria, limit cycles, separatrices), a total topological invariant (the so-called *scheme of the dynamical system*) was found by E. A. Leontovich and A. G. Maier [**1**] (see [**2, 3**] for details). Peixoto [**4**] has given a complete topological classification of structurally stable flows on closed two-dimensional surfaces. His topological invariant (the *distinguished graph of the flow*) is, in some sense, analogous to the scheme of the dynamical system.

The above fundamental problem is wide open in the multidimensional case (i.e., if the dimension of the phase space is greater than 3 for flows and greater than 2 for maps). Here only some particular classes of systems have been classified. For instance, in the class of Morse–Smale systems, complete topological invariants are

---

2000 *Mathematics Subject Classification.* Primary 37C05, 37C15.

This research was supported in part by Russian Foundation for Basic Research (grant 99-01-00231), INTAS-grant 97-804, INCAS-grant 98-1-01 (N. Novgorod), and grant 97-0-1.8-59 of the Education Ministry of Russia.

©2000 American Mathematical Society

known for three-dimensional flows with a finite number of heteroclinic orbits [**5, 6**], and also for diffeomorphisms of 2-manifolds that are gradient-like [**7**] or have orientable heteroclinic sets [**8**]. In the class of structurally stable systems with nontrivial hyperbolic sets, the complete topological classification is known, for instance, for some two-dimensional diffeomorphisms [**9**] and also for Anosov diffeomorphisms [**10, 11**].

In the case of structurally unstable multidimensional systems, the situation is significantly more complicated, since, in the first place, such systems can form open regions in the space of dynamical systems, and, in the second place, there may exist *moduli* of both topological and Ω-equivalence (conjugacy).

Recall that moduli are continuous invariants whose values divide systems into continuum of equivalence classes (see the exact definition below in Section 3). Among the different types of equivalences of dynamical systems, the best known are topological and Ω-equivalence (equivalence on nonwandering sets), but there are others as well [**12**].

The existence of moduli is, in some sense, a characteristic property of multidimensional systems whose structural instability is associated with nontransversal intersections of invariant manifolds of periodic orbits. Palis [**13**] was the first to call attention to this fact. He established that moduli of topological conjugacy occur even in the simplest structurally unstable situation, namely, for two-dimensional diffeomorphisms with structurally unstable heteroclinic orbits [**13, 14**]. In the case of systems with structurally unstable homoclinic orbits, as it was shown in [**15, 16**], Ω-moduli may already exist. The principal Ω-moduli for a homoclinic situation were found in [**15**] for the case of two-dimensional diffeomorphisms and in [**16**] for multidimensional systems provided that the weakest stable and unstable multipliers[1] of saddle periodic orbits (whose manifolds have a tangency) are real and simple. In the present paper, we find principal Ω-moduli for four-dimensional diffeomorphisms having a homoclinic tangency of invariant manifolds of a saddle-focus fixed point (namely, in the case where the fixed point has both stable and unstable complex conjugate eigenvalues). Note that the main results of this article were announced earlier in [**17**].

We consider a four-dimensional C^r-smooth ($r \geqslant 4$) diffeomorphism T satisfying the following conditions.

A. T has a saddle fixed point O with eigenvalues $\lambda_{1,2} = \lambda e^{\pm i\varphi}$, $\gamma_{1,2} = \gamma e^{\pm i\psi}$, where $0 < \lambda < 1 < \gamma$.
B. The saddle quantity $\sigma = \lambda\gamma$ is not equal to 1. We consider the case $\sigma < 1$, since the case $\sigma > 1$ can be reduced to $\sigma < 1$ by means of considering T^{-1} instead of T.
C. The stable W^s and unstable W^u invariant manifolds of O have the simplest tangency (quasi-transversal intersection) at the points of the homoclinic orbit Γ.

The last condition can be made more precise. Namely, denote by $E_M W$ the tangent space to W at a point $M \in W$. Let M^* be some point of Γ. Then C is equivalent to the following conditions.

C.1. $\dim(E_{M^*}W^s \cap E_{M^*}W^u) = 1$.
C.2. The tangency of W^s and W^u at the point M^* is quadratic.

[1] I.e., multipliers which, in absolute value, are maximal (respectively, minimal) among those lying inside (respectively, outside) the unit circle.

In turn, one can give a coordinate formulation of Condition C. Namely, quasi-transversal intersection means that in some local C^2-coordinates (x_1, x_2, y_1, y_2) near the intersection point M^* one has [**14**]

$$W^s = \{y_1 = 0,\ y_2 = 0\} \quad \text{and} \quad W^u = \{x_2 = 0,\ y_1 = x_1^2\}. \tag{0.1}$$

Let U be a sufficiently small neighborhood of the set $O \cup \Gamma$. The neighborhood U can be represented as the union of a small four-dimensional cube U_0 containing the point O and of a finite number of small four-dimensional neighborhoods of those points of Γ that do not lie in U_0. Denote by N the set of orbits of T which lie entirely in U.

The main goal of the paper is to study the structure of N and to find the principal moduli of Ω-conjugacy. The main results are, in the first place, Theorems 1–3 related to the structure of the nontrivial nonuniformly hyperbolic subsets of N, and, in the second place, Theorem 4, in which it is proved that the quantities φ, ψ, and $\theta = -\ln\lambda/\ln\gamma$ are moduli of Ω-conjugacy of diffeomorphisms satisfying A–C. The latter means that if such diffeomorphisms T and T' are Ω-conjugate, then $\varphi(T) = \varphi(T')$, $\psi(T) = \psi(T')'$ and $\theta(T) = \theta(T')$ (the necessity of the last equality is proved provided that $\varphi \neq \psi$).

The contents of the paper are as follows. In Section 1 we describe some analytical properties of T. In Section 2 we study the structure of nontrivial hyperbolic subsets (Theorems 1–3). In Section 3 we study questions about Ω-moduli of diffeomorphisms satisfying conditions A–C and establish (Theorem 4) the existence of the three principal Ω-moduli φ, ψ, and θ.

1. Some analytical properties of diffeomorphism T

In this section we study some analytical properties of T. First of all, we consider a local map $T_0 \equiv T|_{U_0}$, which we reduce to a special form (Lemma 1), we also show that iterations of the map T_0 (in that form) are asymptotically close to iterations of the corresponding linear map (Lemma 2). Next, we study a global map T_1 defined along orbits close to a "global piece" of Γ.

1.1. The maps T_0 and T_1. The map $T_0 \equiv T|_{U_0}$ is a saddle map having the fixed point O with eigenvalues $\lambda_{1,2} = \lambda e^{\pm i\varphi}$, $\gamma_{1,2} = \gamma e^{\pm i\psi}$, where $0 < \lambda < 1 < \gamma$. Naturally, the best form of this map is the linear form:

$$\begin{pmatrix}\bar x_1\\ \bar x_2\end{pmatrix} = \lambda\begin{pmatrix}\cos\varphi & -\sin\varphi\\ \sin\varphi & \cos\varphi\end{pmatrix}\begin{pmatrix}x_1\\ x_2\end{pmatrix}, \qquad \begin{pmatrix}\bar y_1\\ \bar y_2\end{pmatrix} = \gamma\begin{pmatrix}\cos\psi & -\sin\psi\\ \sin\psi & \cos\psi\end{pmatrix}\begin{pmatrix}y_1\\ y_2\end{pmatrix}. \tag{1.1}$$

However, smooth linearizing changes of coordinates do not necessarily exist (especially in the general multidimensional case). But this is not a serious problem, because we prove the following result

LEMMA 1. *By a C^{r-1}-smooth coordinate change, the map T_0 can be brought to the form*

$$\begin{aligned}\begin{pmatrix}\bar x_1\\ \bar x_2\end{pmatrix} &= \lambda\begin{pmatrix}\cos\varphi & -\sin\varphi\\ \sin\varphi & \cos\varphi\end{pmatrix}\begin{pmatrix}x_1\\ x_2\end{pmatrix} + O(\|x\|^2\|y\|),\\ \begin{pmatrix}\bar y_1\\ \bar y_2\end{pmatrix} &= \gamma\begin{pmatrix}\cos\psi & -\sin\psi\\ \sin\psi & \cos\psi\end{pmatrix}\begin{pmatrix}y_1\\ y_2\end{pmatrix} + O(\|x\|\|y\|^2).\end{aligned} \tag{1.2}$$

These coordinates are very convenient, since iterations of the map T_0 in the form (1.2) are asymptotically close to iterations of the linear map (1.1). Namely, the following result is valid.

LEMMA 2. *The map $T_0^k\colon (x_0, y_0) \to (x_k, y_k)$ can be represented in the following form:*

$$(1.3)\qquad \begin{aligned} \begin{pmatrix} x_{k1} \\ x_{k2} \end{pmatrix} &= \lambda^k \begin{pmatrix} \cos k\varphi & -\sin k\varphi \\ \sin k\varphi & \cos k\varphi \end{pmatrix} \begin{pmatrix} x_{01} \\ x_{02} \end{pmatrix} + \lambda^k \gamma^{-k} \rho_k(x_0, y_k) x_0^2 y_k, \\ \begin{pmatrix} y_{01} \\ y_{02} \end{pmatrix} &= \gamma^{-k} \begin{pmatrix} \cos k\psi & -\sin k\psi \\ \sin k\psi & \cos k\psi \end{pmatrix} \begin{pmatrix} y_{k1} \\ y_{k2} \end{pmatrix} + \gamma^{-2k} h_k(x_0, y_k) x_0 y_k^2, \end{aligned}$$

where the functions in the right-hand side of (1.3) are C^{r-1}-smooth and uniformly bounded in k with all derivatives up to the order $r-1$.

These lemmas will be proved in Subsection 1.2.

In U_0 we shall use the coordinates of Lemma 1. By (1.2), the equations of $W^s_{\mathrm{loc}}(O)$ and $W^u_{\mathrm{loc}}(O)$ in U_0 are $y_1 = y_2 = 0$ and $x_1 = x_2 = 0$, respectively. In U_0 we choose a pair of homoclinic points of the orbit Γ. Let these points be $M^+(x^+, 0) \in W^s_{\mathrm{loc}}(O)$ and $M^-(0, y^-) \in W^u_{\mathrm{loc}}(O)$, where $x^+ = (x_1^+, x_2^+)$, $y^- = (y_1^-, y_2^-)$ and ${x_1^+}^2 + {x_2^+}^2 \neq 0$, ${y_1^-}^2 + {y_2^-}^2 \neq 0$. Evidently, there exists an integer n such that $T^n(M^-) = M^+$. Let Π^+ and Π^- be small cubic neighborhoods of M^+ and M^-. We denote the coordinates in Π^+ and Π^- by $x_0 \equiv (x_{01}, x_{02})$ and $y_0 \equiv (y_{01}, y_{02})$ and by $x_1 \equiv (x_{11}, x_{12})$ and $y_1 \equiv (y_{11}, y_{12})$, respectively. Then the map $T_1 \equiv T^n\colon \Pi^- \to \Pi^+$ can be written as

$$(1.4)\qquad \bar{x}_0 - x^+ = F(x_1, y_1 - y^-), \qquad \bar{y}_0 = G(x_1, y_1 - y^-),$$

where $F(0) = 0$, $G(0) = 0$.

We shall find a form of functions F and G using condition C. First, consider the map $\widehat{T}_1$

$$(1.5)\qquad \bar{x}_0 - x^+ = \hat{a} x_1 + \hat{b}(y_1 - y^-), \qquad \bar{y}_0 = \hat{c} x_1 + \hat{d}(y_1 - y^-)$$

corresponding to the linear part of T_1, where $\hat{a}$, $\hat{b}$, $\hat{c}$, and $\hat{d}$ are (2×2)-matrices. Condition C.1 implies that the plane

$$(1.6)\qquad \bar{x}_0 - x^+ = \hat{b}(y_1 - y^-), \qquad \bar{y}_0 = \hat{d}(y_1 - y^-)$$

intersects with the plane $\bar{y}_0 = 0$ along a straight line. This implies that

$$\det\{\hat{d}\} = 0 \quad \text{and} \quad \operatorname{rank}\{\hat{d}\} = 1.$$

By an orthogonal coordinate transformation in the (y_1, y_2)-plane,[2] we bring $\widehat{T}_1$ to the form (1.5), where

$$\hat{d} = \begin{pmatrix} 0 & 0 \\ d_{21} & d_{22} \end{pmatrix} \quad \text{and} \quad d_{21}^2 + d_{22}^2 \neq 0.$$

We assume that $d_{22} \neq 0$. If $d_{22} = 0$ (and consequently $d_{21} \neq 0$), then we consider homoclinic points M^+ and $T_0^{-1} M^-$ instead of M^+ and M^-. In this case, the role of T_1 is played by the map $T_1' = T_1 T_0$ and the matrix $\hat{d}$ is transformed to $\hat{d}'$ where

$$\hat{d}' = \gamma \cdot \begin{pmatrix} 0 & 0 \\ d_{21} & 0 \end{pmatrix} \cdot \begin{pmatrix} \cos\psi & -\sin\psi \\ \sin\psi & \cos\psi \end{pmatrix} = \gamma \cdot \begin{pmatrix} 0 & 0 \\ d_{21}\cos\psi & -d_{21}\sin\psi \end{pmatrix}.$$

---

[2] Orthogonality is necessary in order to preserve the form (1.2) of the local map T_0.

Since $\sin\psi \neq 0$ (because $\psi \neq 0, \pi$), it follows that $d'_{22} = -d_{21}\sin\psi \neq 0$.

Thus, for the map T_1, its Jacobi matrix Δ calculated in M^- has the form

$$\Delta = \begin{pmatrix} a_{11} & a_{12} & b_{11} & b_{12} \\ a_{21} & a_{22} & b_{21} & b_{22} \\ c_{11} & c_{12} & 0 & 0 \\ c_{21} & c_{22} & d_{21} & d_{22} \end{pmatrix},$$

where $d_{22} \neq 0$. Since T_1 is a diffeomorphism, $\det(\Delta) \neq 0$. This implies that

1) $c_{11}^2 + c_{12}^2 \neq 0$;

2) at least one of the minors

$$\begin{pmatrix} b_{11} & b_{12} \\ d_{21} & d_{22} \end{pmatrix} \quad \text{or} \quad \begin{pmatrix} b_{21} & b_{22} \\ d_{21} & d_{22} \end{pmatrix}$$

is not equal to zero.

Now T_1 can be written in the form

$$\begin{aligned} \bar{x}_{01} - x_1^+ &= a_{11}x_{11} + a_{12}x_{12} + b_{11}(y_{11} - y_1^-) + b_{12}(y_{12} - y_2^-) + \dots, \\ \bar{x}_{02} - x_2^+ &= a_{21}x_{11} + a_{22}x_{12} + b_{21}(y_{11} - y_1^-) + b_{22}(y_{12} - y_2^-) + \dots, \\ \bar{y}_{01} &= c_{11}x_{11} + c_{12}x_{12} + d_1(y_{11} - y_1^-)^2 \\ &\quad + d_2(y_{11} - y_1^-)(y_{12} - y_2^-) + d_3(y_{12} - y_2^-)^2 + \dots, \\ \bar{y}_{02} &= c_{21}x_{11} + c_{22}x_{12} + d_{21}(y_{11} - y_1^-) + d_{22}(y_{12} - y_2^-) + \dots . \end{aligned} \tag{1.7}$$

where the dots denote higher order terms. Since $d_{22} \neq 0$, from the last equation in (1.7) we obtain

$$y_{12} - y_2^- = \frac{\bar{y}_{02}}{d_{22}} - \frac{c_{21}}{d_{22}}x_{11} - \frac{c_{22}}{d_{22}}x_{12} - \frac{d_{21}}{d_{22}}(y_{11} - y_1^-) + \dots .$$

Thus, (1.7) can be rewritten as follows

$$\begin{aligned} \bar{x}_{01} - x_1^+ &= \tilde{a}_{11}x_{11} + \tilde{a}_{12}x_{12} + \left(b_{11} - b_{12}\frac{d_{21}}{d_{22}}\right)(y_{11} - y_1^-) + \frac{b_{12}}{d_{22}}\bar{y}_{02} + \dots, \\ \bar{x}_{02} - x_2^+ &= \tilde{a}_{21}x_{11} + \tilde{a}_{22}x_{12} + \left(b_{21} - b_{22}\frac{d_{21}}{d_{22}}\right)(y_{11} - y_1^-) + \frac{b_{22}}{d_{22}}\bar{y}_{02} + \dots, \\ \bar{y}_{01} &= c_{11}x_{11} + c_{12}x_{12} + (y_{11} - y_1^-)^2\left[d_1 - d_2\frac{d_{21}}{d_{22}} + d_3\frac{d_{21}^2}{d_{22}^2}\right] \\ &\quad + A(\bar{y}_{02})^2 + B\bar{y}_{02}(y_{11} - y_1^-) + \dots, \\ y_{12} - y_2^- &= \frac{\bar{y}_2}{d_{22}} - \frac{c_{21}}{d_{22}}x_{11} - \frac{c_{22}}{d_{22}}x_{12} - \frac{d_{21}}{d_{22}}(y_{11} - y_1^-) + \dots . \end{aligned} \tag{1.8}$$

Denote

$$b_0 = b_{11} - b_{12}\frac{d_{21}}{d_{22}}, \quad b_1 = b_{21} - b_{22}\frac{d_{21}}{d_{22}}, \quad \tilde{d} = \frac{d_{21}}{d_{22}},$$
$$d_0 = \frac{1}{d_{22}}, \quad d \equiv \left[d_1 - d_2\frac{d_{21}}{d_{22}} + d_3\frac{d_{21}^2}{d_{22}^2}\right].$$

Let us make one more coordinate transformation to simplify T_1. Namely, by means of the following orthogonal transformation in the (x_1, x_2)-plane

$$x_1 = \cos\vartheta\xi_1 + \sin\vartheta\xi_2, \qquad x_2 = -\sin\vartheta\xi_1 + \cos\vartheta\xi_2,$$

where

$$\sin\vartheta = -\frac{b_1}{\sqrt{b_0^2+b_1^2}}, \qquad \cos\vartheta = \frac{b_0}{\sqrt{b_0^2+b_1^2}},$$

we bring T_1 to the final form

$$
\begin{aligned}
\bar{x}_{01} - x_1^+ &= a_{11}x_{11} + a_{12}x_{12} + b(y_{11} - y_1^-) + \tilde{b}_1\bar{y}_{02} + \dots, \\
\bar{x}_{02} - x_2^+ &= a_{21}x_{11} + a_{22}x_{12} + \tilde{b}_2\bar{y}_{02} + \dots, \\
\bar{y}_{01} &= c_{11}x_{11} + c_{12}x_{12} + d(y_{11} - y_1^-)^2 \\
&\quad + A_1(\bar{y}_{02})^2 + B_1\bar{y}_{02}(y_{11} - y_1^-) + \dots, \\
y_{12} - y_2^- &= c_{21}x_{11} + c_{22}x_{12} - \tilde{d}(y_{11} - y_1^-) + d_0\bar{y}_{02} + \dots,
\end{aligned}
\tag{1.9}
$$

where $b = \sqrt{b_0^2 + b_1^2}$. Here the coefficients x_1^+, x_2^+ and at "x"-coordinates in the right-hand side of (1.9) may differ from the corresponding coefficients in (1.7) and (1.8), but they are denoted by the same letters for economy of symbols. Note, that by condition C.1 one has $b \neq 0$, $d_0 \neq 0$, and $c_{11}^2 + c_{12}^2 \neq 0$.

By (1.9), the equations of $T_1(W^u_{\text{loc}}(O)) \cap \Pi^+$ in parametric form are

$$
\begin{aligned}
\bar{x}_{01} - x_1^+ = b(y_{11} - y_1^-) + \tilde{b}_1\bar{y}_{02} + \dots, &\qquad \bar{x}_{02} - x_2^+ = \tilde{b}_2\bar{y}_{02} + \dots, \\
\bar{y}_{01} = d(y_{11} - y_1^-)^2 + A_1(\bar{y}_{02})^2 &+ B_1\bar{y}_{02}(y_{11} - y_1^-) + \dots, \\
y_{12} - y_2^- = d_0\bar{y}_{02} &- \tilde{d}(y_{11} - y_1^-) + \dots,
\end{aligned}
\tag{1.10}
$$

where the parameters are $u \equiv y_1 - y_1^-$ and $v \equiv y_2 - y_2^-$. The vector $l_u = (u_0, -\tilde{d}u_0)$ in the parameter plane $W^u_{\text{loc}}(O) \cap \Pi^-$ is mapped under DT_1 to the vector $l_s = (bu_0, 0, 0, 0)$ which is tangent, at the point M^+, to both W^u and W^s (just as for (0.1)). Thus, x_{01}-axis is the tangent to W^u at M^+. It follows from (1.10) that the curve $\{x_{02} = 0,\ y_{01} = db^{-2}x_1^2 + \dots,\ y_{02} = 0\}$ in W^u is projected to the x_{01}-axis (with respect to the projection onto $W^s_{\text{loc}}(O) \cap \Pi^+$). Since $d \neq 0$, it means that the tangency of W^u and W^s, in M^+, along x_{01}-direction is quadratic. Since $d_0 \neq 0$, other directions are not tangent. Thus, $T_1(W^u_{\text{loc}}(O)) \cap \Pi^+$ has the shape of a "parabolic cylinder" (as in (0.1)). This cylinder is folded in the direction of y_{01}-axis and prolate in the direction of y_{02}-axis (Figure 1). Besides, its fold-line[3] is close to the line $\{x_{02} - x_2^+ = \tilde{b}_2y_{02},\ x_{01} = x_1^+,\ y_{01} = 0\}$.

We shall assume from now on that the coefficient d is positive (otherwise, we can turn the coordinate axes y_1 and y_2 by 180°).

1.2. Proof of Lemmas 1 and 2.

PROOF OF LEMMA 1. We shall prove Lemma 1 under weaker assumptions. Namely, let the map T_0 have the form

$$\bar{x} = Ax + f(x,y), \qquad \bar{y} = By + g(x,y), \tag{1.11}$$

where $x = (x_1, x_2)$, $y = (y_1, y_2)$; the matrices A and B have the eigenvalues λ_1, λ_2 and γ_1, γ_2, respectively, where $|\lambda_1| = |\lambda_2| = \lambda < 1$ and $|\gamma_1| = |\gamma_2| = \gamma > 1$; further $f(0,y) \equiv 0$, $g(x,0) \equiv 0$. The last equalities correspond to straightened stable and unstable manifolds. (If $\lambda_{1,2} = \lambda e^{\pm i\varphi}$ and $\gamma_{1,2} = \gamma e^{\pm i\psi}$, then one has the situation of Lemma 1.)

---

[3]Such a line for W^u from (0.1) is the y_2-axis.

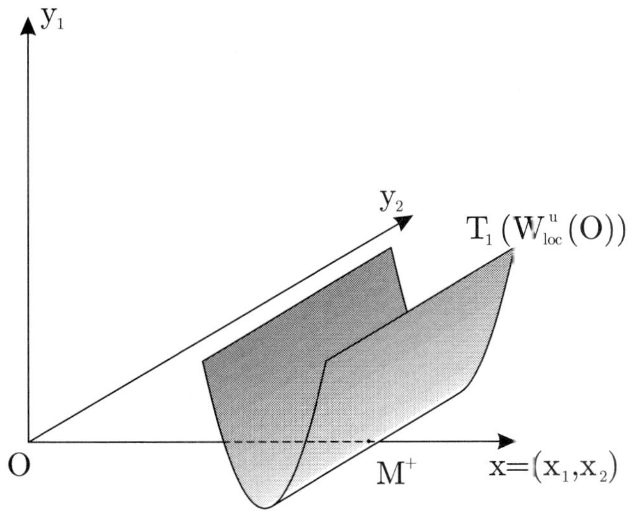

FIGURE 1. The piece $T_1(W^u_{\text{loc}}(O)) \cap \Pi^+$ has the form of a two-dimensional "parabolic cylinder", which is folded in the direction of the y_1-axis and prolate in the direction of y_2-axis.

Let us rewrite the map (1.11) in the following extended form:

$$\begin{aligned} \bar{x} &= Ax + R(x)x + H(y)x + O(x^2|y|), \\ \bar{y} &= By + P(y)y + S(x)y + O(|x|y^2), \end{aligned} \tag{1.12}$$

where $R(0) = 0$, $P(0) = 0$, $H(0) = 0$, $S(0) = 0$.

Let us consecutively perform the following coordinate changes:

1°. $\xi = x + h(y)x$, $\eta = y$, where $h(0) = 0$;
2°. $\xi = x$, $\eta = y + s(x)y$, where $s(0) = 0$;
3°. $\xi = x + r(x)x$, $\eta = y$, where $r(0) = 0$;
4°. $\xi = x$, $\eta = y + p(y)y$, where $p(0) = 0$.

The changes 1°, 2°, 3°, and 4° annihilate the terms H, S, R, and P in (1.12), respectively.

1) After the change 1°, by (1.12), one has

$$\begin{aligned} \bar{\xi} &= \bar{x} + h(\bar{y})\bar{x} \\ &= A(\xi - h(\eta)x) + R(x)x + H(\eta)x + h(B\eta + P(\eta)\eta)Ax + O(\|\xi^2\|\|\eta\|)) \\ &= A\xi + R(x)x + [H(\eta)x - Ah(\eta)x + h(B\eta + P(\eta)\eta)Ax] + O(\|\xi^2\|\|\eta\|)), \\ \bar{\eta} &= B\eta + P(\eta)\eta + S(\xi)\eta + O(\|\xi\|\|\eta^2\|). \end{aligned} \tag{1.13}$$

Assume that $h(y)$ is a function such that the expression in the square brackets in (1.13) is equal to zero identically, i.e.,

$$h(B\eta + P(\eta)\eta) \cdot A = A \cdot h(\eta) - H(\eta). \tag{1.14}$$

Consider the map K_1 given by the formula

$$\bar{h} = A \cdot h \cdot A^{-1} - H(\eta) \cdot A^{-1}, \qquad \bar{\eta} = B\eta + P(\eta)\eta. \tag{1.15}$$

where $h \in \mathbb{R}^4$, $\eta \in \mathbb{R}^2$. The map K_1 has a fixed point $O_1(0,0)$ whose eigenvalues constitute the spectrum of the linear operator DK_1:

$$\bar{h} = A \cdot h \cdot A^{-1} - \left(\frac{DH_1}{D\eta}\right)\eta \cdot A^{-1}, \qquad \bar{\eta} = B\eta.$$

The matrix of DK_1 is of block-triangular form, therefore, its spectrum is the union of the spectra of the linear operators $\bar{h} = A \cdot h \cdot A^{-1}$ and $\bar{\eta} = B\eta$. We now use the following well-known fact from the theory of matrices (see, for instance, [**18**]): the spectrum of a linear operator $L \circ K$ consists of all possible products of elements from spec(L) by elements from spec(K). Then the fixed point O_1 has four eigenvalues of absolute value 1 (which are λ_i/λ_j, $i,j = 1,2$) and two eigenvalues (γ_1 and γ_2) of absolute value greater than 1. Thus [**19**], K_1 has an invariant two-dimensional strong unstable manifold whose equation is $h = h(\eta)$, where $h(0) = 0$. Moreover, $h(\eta)$ satisfies (1.14) and is C^{r-1}-smooth since the functions in the right-hand side of (1.15) are C^{r-1}-smooth. Returning to the old coordinates, we see that after the change 1°, T_0 takes the form (1.12), where $H(y) \equiv 0$.

2) Let us perform the change 2°. We obtain

$$\begin{aligned} &\bar{\xi} = A\xi + R(\xi) + O(\|\xi\|^2\|\eta\|), \\ (1.16)\quad &\bar{\eta} = \bar{y} + s(\bar{\xi})\bar{y} = B(\eta - s(\xi)y) + P(y)y + S(\xi)y + s(\bar{\xi})(By + S(\xi)y) + \dots \\ &\quad = B\eta + P(\eta) + [-Bs(\xi) + S_1(\xi) + s(A\xi + R(\xi))(B + S(\xi))]y + \dots, \end{aligned}$$

where the dots stand for terms of order $O(\|\xi\|\|\eta^2\|)$. Assume now that the expression in brackets in (1.16) is equal to zero identically, i.e.,

$$(1.17) \qquad s(A\xi + R(\xi)\xi) = (Bs(\xi) - S(\xi))(B + S(\xi))^{-1}.$$

Consider the map K_2:

$$\bar{s} = (Bs - S(\xi))(B + S(\xi))^{-1}, \qquad \bar{\xi} = A\xi + R(\xi)\xi,$$

where $s \in \mathbb{R}^4$, $\xi \in \mathbb{R}^2$. This map has the fixed point $O_2(0,0)$ whose four eigenvalues ($\{\gamma_i/\gamma_j\}$, where $i,j = 1,2$) are of absolute value 1 and the two others (λ_1 and λ_2) are are of absolute value less than 1. Thus, K_2 has an invariant two-dimensional strong stable manifold whose equation is $s = s(\xi)$, where $s(0) = 0$. Moreover, $s(\xi)$ satisfies (1.17) and is C^{r-1}-smooth. Returning to old coordinates, we see that T_0, after the changes 1°, 2°, is of the form (1.12), where $H(y) \equiv 0$ and $S(x) \equiv 0$.

3) After the change 3°, we obtain

$$\begin{aligned} &\bar{\xi} = \bar{x} + r(\bar{x})\bar{x} \\ (1.18)\quad &\quad = A(\xi - r(x)x) + R(x)x + r(Ax + R(x)x)(Ax + R(x)x) + \dots \\ &\quad = A\xi + x[r(Ax + R(x)x)(A + R(x)) + R(x) - Ar(x)] + \dots, \end{aligned}$$

where the dots stand for terms of order $O(\|\xi\|^2\|\eta\|)$. Assume that the expression in brackets in (1.18) is equal to zero identically, i.e.,

$$r(Ax + R(x)x) = (Ar(x) - R(x))(A + R(x))^{-1}.$$

Consider the map K_3:

$$\bar{r} = (Ar - R(x))(A + R(x))^{-1}, \qquad \bar{x} = Ax + R(x)x,$$

where $r \in \mathbb{R}^4$, $x \in \mathbb{R}^2$. This map has the fixed point $O_3(0,0)$ with six eigenvalues, four of which are of absolute value 1 and the two others are of absolute value less than 1. Thus, K_3 has an invariant two-dimensional C^{r-1}-smooth strong stable manifold whose equation is $r = r(x)$, where $r(0) = 0$. Returning to the old coordinates, we see that after the changes 1°–3°, T_0 takes the form (1.12), where $H(y) \equiv 0$, $S(x) \equiv 0$, and $R(x) \equiv 0$.

4) Finally, let us perform the change 4°. We obtain

$$\begin{aligned}\bar\eta = \bar y + p(\bar y)\bar y &= B\eta - Bp(y)y + P(y)y + p(By + P(y)y)(By + P(y)y) + \dots \\ &= B\eta + y[p(By + P(y)y)(B + P(y)) + P(y)y - Bp(y)] + \dots,\end{aligned}$$

where dots stand for terms of order $O(\|\xi\|\|\eta\|^2)$. Assume that the expression in the square brackets is equal to zero identically, i.e.,

$$p(By + P(y)y)(B + P(y)) = Bp(y) - P(y).$$

Consider the map K_4:

$$\bar p = (Bp - P(y))(B + P(y))^{-1}, \qquad \bar y = By + P(y)y,$$

where $p \in \mathbb{R}^4$, $y \in \mathbb{R}^2$. This map has a fixed point $O_4(0, 0)$ with six eigenvalues, four of which are of absolute value equal to 1 and the two others are of absolute value greater than 1. Thus, K_4 has an invariant two-dimensional C^{r-1}-smooth strong unstable manifold whose equation is $p = p(y)$, where $p(0) = 0$. Returning to the old coordinates, we see that after the changes 1°–4°, T_0 is written in the required form. □

Proof of Lemma 2. Let the C^{r-1}-smooth ($r \geqslant 4$) map T_0 have the form (1.11), but now $f(x,y) \equiv f_1(x,y)x^2y$ and $g(x,y) \equiv g_1(x,y)xy^2$, in contrast to (1.12).

The proof of the lemma is based on the boundary value problem method [**20, 21**]. We consider the following operator Φ:

$$(1.19)\quad \bar x_j = A^j x_0 + \sum_{s=0}^{j-1} A^{j-s-1} f(x_s, y_s), \qquad \bar y_j = B^{j-k} y_k - \sum_{s=j}^{k-1} B^{j-s-1} g(x_s, y_s)$$

(where $j = 0, 1, \dots, k$) defined on the set

$$R(\delta) = \{z = [(x_j, y_j)]_{j=0}^k, \ \|z\| \leqslant \delta\};$$

here δ is a small positive number and $\|z\|$ is the maximum of the absolute value of the components x_j, y_j of vector z. If $z_0 = [(x_j^0, y_j^0)]_{j=0}^k$ is a fixed point of Φ, then we have the following diagram:

$$(x_0^0, y_0^0) \xrightarrow{T_0} (x_1^0, y_1^0) \xrightarrow{T_0} \cdots \xrightarrow{T_0} (x_k^0, y_k^0).$$

It was proved in [**22**] that, for any sufficiently small $\delta = \delta_0$ and for $\|x_0\| \leqslant \delta_0/2$, $\|y_k\| \leqslant \delta_0/2$, the operator Φ maps the set $R(\delta_0)$ into itself and is a contraction. Thus, system (1.19) has a unique fixed point $z_0 = [x_j^0(x_0, y_k), y_j^0(x_0, y_k)]_{j=0}^k$. In order to estimate the norms of x_j^0 and y_j^0, we proceed as follows.

Consider a set $\Pi_{x_0 y_k} \in R(\delta_0)$ such that x_0 and y_k are fixed and x_j, y_j satisfy the inequalities

$$(1.20)\qquad \|x_j - A^j x_0\| \leqslant \delta_0 \lambda^j \gamma^{j-k}, \qquad \|y_j - B^{j-k} y_k\| \leqslant \delta_0 \gamma^{j-2k}.$$

Note that $\Phi(\Pi_{x_0y_k})$ belongs to $\Pi_{x_0y_k}$ if δ_0 is sufficiently small. Indeed, according to Lemma 1 we obtain from (1.19) the inequalities

$$\begin{aligned}\|\bar{x}_j - A\lambda^j x_0\| &\leqslant \sum_{s=0}^{j-1} \|A\|^{j-s-1}\|f_1\|\|x_s\|^2\|y_s\| \\ &\leqslant L\sum_{s=0}^{j-1} \lambda^{j-s-1}\lambda^{2s}\gamma^{s-k}\delta_0^3 \leqslant L\delta_0^3\lambda^j\gamma^{-k}\frac{\lambda^{-1}}{1-\lambda\gamma}, \\ \|y_j - B^{j-k}y_k\| &\leqslant \sum_{s=j}^{k-1} \|B^{-1}\|^{s+1-j}\|g_1\|\|x_s\|\|y_s\|^2 \\ &\leqslant L\sum_{s=j}^{k-1} \gamma^{j-s-1}\lambda^s\gamma^{2(s-k)}\delta_0^3 \leqslant L\delta_0^3\gamma^{j-2k}\frac{\gamma^{-1}}{1-\lambda\gamma},\end{aligned}$$

where $L = \max\{\|f_1\|, \|g_1\|\}$. Thus, $z_0 \in \Pi_{x_0y_k}$ for sufficiently small δ_0. Since Φ is a contraction operator, estimates (1.20) are valid for x_j^0 and y_j^0.

Estimates for the derivatives of x_j^0 and y_j^0 with respect to x_0 and y_k can be obtained in exactly the same way as in [**16**] (see the proof of Lemma 1.2 there). □

2. Nontrivial hyperbolic subsets of T

In this section we describe the structure of nontrivial hyperbolic subsets of the diffeomorphism T (Theorems 2 and 3). To do this, we construct a special neighborhood (Lemma 3) and determine the character of the intersections of horseshoes and strips (Theorem 1).

2.1. A special neighborhood of the structurally unstable homoclinic orbit. Given a small enough neighborhood $U = U(O \cup \Gamma)$, it is convenient to take a smaller, so-called *special neighborhood* (as it was done in [**23, 24**] for the case of structurally unstable homoclinic orbits of a saddle fixed point with real weakest eigenvalues). We assume that the neighborhood U_0 is chosen so that the iterations $T(M^-)$ and $T^{-1}(M^+)$ of the points M^- and M^+ lie outside U_0. Moreover, we take small neighborhoods $\Pi^+ \subset U_0$ and $\Pi^- \subset U_0$ of M^+ and M^-, respectively, such that U can be represented as

$$U = U_0 \cup (T(\Pi^-) \cap T^{-(n-1)}(\Pi^+)) \cup \cdots \cup (T^{n-1}(\Pi^-) \cap T^{-1}(\Pi^+)). \tag{2.1}$$

This means that the orbits lying in U enter U_0 through Π^+ and leave U_0 through Π^-.

Let N be the set of all orbits which lie in U entirely. Choose a sufficiently large integer $\bar{k}$. A neighborhood $U_{\bar{k}}$ is called *special* if it contains those and only those orbits of N which, each time they enter Π^+, need at least $\bar{k}$ iterations to reach Π^- (we denote the set of these orbits by $N_{\bar{k}}$). In fact, if $\bar{k}$ is large enough, the special neighborhood $U_{\bar{k}}$ does not depend on the original neighborhood U. Namely, the following lemma holds.

LEMMA 3. *Without loss of generality, Π^+ and Π^- can be chosen as follows:*

$$\begin{aligned}\Pi^+ &= \{(x_0, y_0) : \|x_0 - x^+\| \leqslant \rho_{\bar{k}},\ \|y_0\| \leqslant \gamma^{-\bar{k}}(|y_1^-| + |y_2^-| + \rho_{\bar{k}})\}, \\ \Pi^- &= \{(x_1, y_1) : \|\|x_1\| \leqslant \lambda^{\bar{k}}(|x_1^+| + |x_2^+| + \rho_{\bar{k}}),\ \|y_1 - y^-\| \leqslant \rho_{\bar{k}}\},\end{aligned} \tag{2.2}$$

where $\rho_{\bar{k}} = L_1\gamma^{-\bar{k}/2}$ and L_1 is a positive constant independent of $\bar{k}$.

PROOF. Indeed, no orbit of N will be lost if the initial small neighborhoods Π^+ and Π^- are enlarged to $2\varepsilon_0$-cubes centered at the points M^+ and M^- respectively (where ε_0 is a sufficiently small positive number). Since the special neighborhood does not contain points which, going from Π^+ into Π^-, use less than $\bar{k}$ iterations of T, from (1.3) it follows that for sufficiently large $\bar{k}$ the cubes can be shrunk along coordinates y and x so as to have

$$\|y_0\| \leqslant \gamma^{-\bar{k}}(|y_1^-| + |y_2^-| + 3\varepsilon_0), \qquad \|x_1\| \leqslant \lambda^{\bar{k}}(|x_1^+| + |x_2^+| + 3\varepsilon_0). \tag{2.3}$$

Note that if ε_0 is small enough, then the orbits staying in U_0 for less than $\bar{k}$ iterations of T do not enter these new neighborhoods: by virtue of (1.3) we would have

$$\begin{aligned}\|y_0\| &= \sqrt{|y_0^1|^2 + |y_0^2|^2} \geqslant \gamma^{1-\bar{k}}(|y_1^-| + |y_2^-| - 3\varepsilon_0)\\ &= \gamma^{-\bar{k}}(|y_1^-| + |y_2^-| + (\gamma - 1)|y_1^-| + |y_2^-| - 3\gamma\varepsilon_0) \geqslant \gamma^{-\bar{k}}(|y_1^-| + |y_2^-| + 3\varepsilon_0)\end{aligned}$$

for such an orbit (for $k \leqslant \bar{k} - 1$), which contradicts (2.3).

Further, one can also exclude from Π^- the part that does not intersect $T^{-1}(\Pi^+)$. According to (1.9), $\bar{y}_{01} = c_{11}x_{11} + c_{12}x_{12} + d(y_{11} - y_1^-)^2 + \dots$ and, since for a point (x_1, y_1) in $T^{-1}(\Pi^+)$ the value of $\bar{y}_{01}$ must satisfy (2.3) and the saddle quantity $\lambda\gamma$ is less than 1, the neighborhood Π^- can again be decreased so that the coordinates y_{11} and y_{12} satisfy the inequalities

$$|y_{11} - y_1^-| \leqslant C_1\gamma^{-\bar{k}/2}, \qquad |y_{12} - y_2^-| \leqslant C_2\gamma^{-\bar{k}/2}, \tag{2.4}$$

where C_1 and C_2 are some positive constants. Using (1.9) and (2.3), (2.4), one can diminish the neighborhood Π^+ so that the coordinates x_{01} and x_{02} will satisfy the inequalities

$$|\bar{x}_{01} - x_1^+| \leqslant C_3\gamma^{-\bar{k}/2}, \qquad |\bar{x}_{02} - x_2^+| \leqslant C_4\gamma^{-\bar{k}/2}. \tag{2.5}$$

Repeating the same procedure with the obtained neighborhoods we come to formulas (2.2). □

The sizes of the new neighborhoods $\Pi^\pm$ tend to zero as $\bar{k} \to +\infty$, so they lie inside the initial neighborhoods. By construction, the new neighborhood, being composed of iterations of $\Pi^\pm$ by the rule (2.1), contains no orbits of N except those of $N_{\bar{k}}$. Thus, relations (2.2) indeed determine a special neighborhood, and we can assume that these relations are fulfilled throughout the rest of the paper.

2.2. "Strips and horseshoes" and their intersections. By construction, all orbits of $N_{\bar{k}}$ (except for O) intersect the neighborhoods Π^+ and Π^-. Substituting the bounds (2.2) into the right-hand side of (1.3), one can immediately see that the set of points of Π^+ whose iterations reach Π^- consists of the infinite number of four-dimensional "strips"

$$\begin{aligned}\sigma_k^0 \equiv \Pi^+ \cap T_0^{-k}\Pi^- = \{(x_0, y_0) : \|x_0 - x^+\| &\leqslant \rho_{\bar{k}},\\ |y_{01} - \gamma^{-k}(\cos k\psi \cdot y_1^- - \sin k\psi \cdot y_2^-)| &\leqslant \gamma^{-k}\rho_{\bar{k}},\\ |y_{02} - \gamma^{-k}(\cos k\psi \cdot y_2^- + \sin k\psi \cdot y_1^-)| &\leqslant \gamma^{-k}\rho_{\bar{k}}\},\end{aligned} \tag{2.6}$$

where $k = \bar{k}, \bar{k}+1, \dots$. As $k \to +\infty$, the strips σ_k^0 accumulate to $W^s_{\text{loc}}(O)$. These strips lie inside a "roll" R_s which winds around the small two-dimensional area $W^s_{\text{loc}}(O) \cap \Pi^+$ (Figure 2).

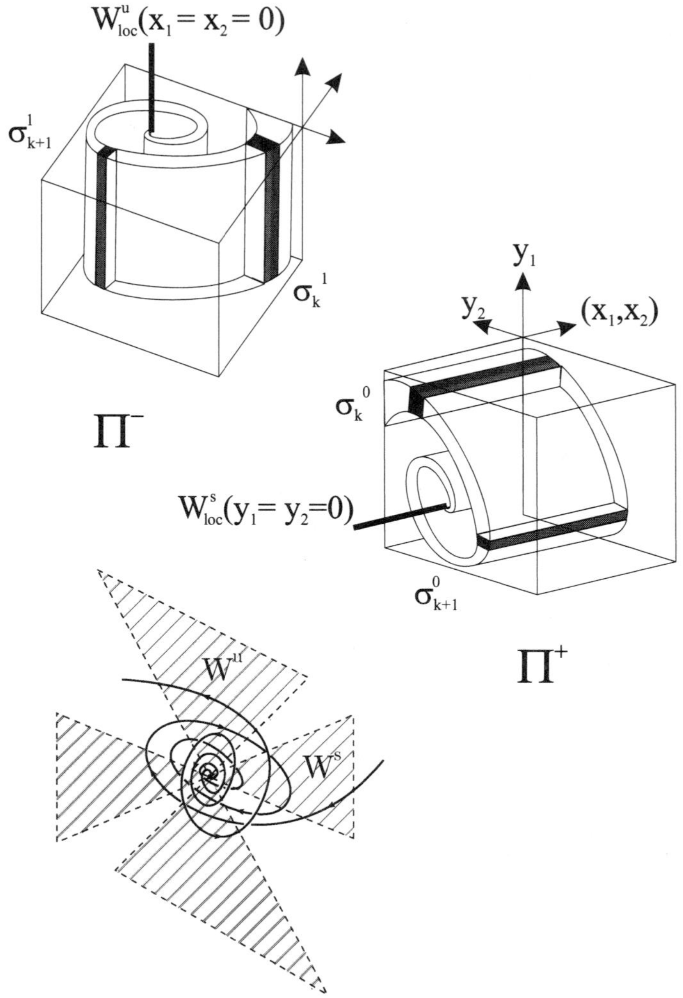

FIGURE 2. The four-dimensional strips σ_k^0 lie inside a "roll" R_s which winds around the small two-dimensional area $W^s_{\text{loc}}(O) \cap \Pi^+$. The strips σ_k^0 accumulate on this area as $k \to +\infty$. The four-dimensional strips σ_k^1 lie inside a "roll" R_u, which winds around the small two-dimensional area $W^u_{\text{loc}}(O) \cap \Pi^-$. The strips σ_k^1 accumulate on this area as $k \to +\infty$. For the sake of a visualization, the coordinate planes (y_{11}, y_{12}) in Π^- and (x_{01}, x_{02}) in Π^+ are shown as straight lines.

Similarly, the image of the strip σ_k^0 under the map T_0^k is the following strip in Π^-:

$$
\begin{aligned}
\sigma_k^1 &\equiv \Pi^- \cap T_0^k \Pi^+ \\
&= \{(x_1, y_1) : |x_{11} - \lambda^k(\cos k\varphi \cdot x_1^+ + \sin k\varphi \cdot x_2^+)| \leqslant \lambda^k \rho_{\bar{k}}, \\
&\quad |x_{12} - \lambda^k(\cos k\varphi \cdot x_2^+ - \sin k\varphi \cdot x_1^+)| \leqslant \lambda^k \rho_{\bar{k}},\ \|y_1 - y^-\| \leqslant \rho_{\bar{k}}\},
\end{aligned}
\tag{2.7}
$$

As $k \to +\infty$, the strips σ_k^1 accumulate to $W^u_{\rm loc}(O)$ (Figure 2b). These strips also lie inside a "roll" R_u, which winds around the small two-dimensional area $W^u_{\rm loc}(O) \cap \Pi^-$ (Figure 2).

According to (1.9), the images $T_1(\sigma_j^1)$ of the strips σ_j^1 have the shape of four-dimensional "horseshoes" accumulating to the two-dimensional "parabolic cylinder" $T_1(W^u_{\rm loc}(O)) \cap \Pi^+$.

Evidently, the orbits of the set $N_{\bar{k}}$ must intersect Π^+ at points of intersection of the horseshoes $T_1(\sigma_j^1)$ and the strips σ_i^0 for various $i \geqslant \bar{k}$ and $j \geqslant \bar{k}$. Hence, the structure of $N_{\bar{k}}$ depends strongly on the character of these intersections.

We say that the horseshoe $T_1(\sigma_j^1)$ has a *regular intersection* with the strip σ_i^0 if the following conditions are fulfilled:

1) the set $T_1(\sigma_j^1) \cap \sigma_i^0$ consists of two connected components, Δ_{ij}^1 and Δ_{ij}^2;
2) the map $T_1 T_0^j$ restricted to the preimage $(T_1 T_0^j)^{-1} \Delta_{ij}^\alpha \subset \sigma_j^0$ of the component Δ_{ij}^α, is *a saddle map* in the sense of [**20**] (roughly speaking, this map is expanding with respect to the coordinates $y_0 \equiv (y_{01}, y_{02})$ and contracting with respect to $x_0 \equiv (x_{01}, x_{02})$; see the exact condition in the proof of Theorem 1 below).

Note that by the construction of T_1, the projections of the four-dimensional strips and horseshoes to the plane (x_{01}, y_{01}) in Π^+ have the shape of ordinary two-dimensional strips and horseshoes. Therefore, we can determine the character of the intersections of the horseshoes $T_1(\sigma_j^1)$ and strips σ_i^0 from the intersection of their projections. Different types of intersections of the projection of the horseshoe $T_1(\sigma_j^1)$ with the projections of the strips are shown in Figure 3. According to the figure, one can see that the intersection of $T_1(\sigma_j^1)$ with the strips σ_i^0, σ_k^0, and σ_s^0 is regular, irregular and empty, respectively.

THEOREM 1. *There is a positive constant S_1 and a sufficiently large integer $\bar{k}$ such that for any $i \geqslant \bar{k}$, $j \geqslant \bar{k}$*

1) *the inequality*

$$A\gamma^{-i}\cos(i\psi + \alpha) - B\lambda^j \cos(i\varphi + \beta) < -S_{ij}(\bar{k}) \tag{2.8}$$

implies $T_1(\sigma_j^1) \cap \sigma_i^0 = \varnothing$, where

$$S_{ij} = (\gamma^{-i} + \lambda^j)\varepsilon_{\bar{k}}, \quad \varepsilon_{\bar{k}} = S_1(\lambda^{\bar{k}} + \gamma^{-\bar{k}}),$$

$$A = \sqrt{(y_1^-)^2 + (y_2^-)^2}, \quad B = \sqrt{[(x_1^+)^2 + (x_2^+)^2][c_{11}^2 + c_{12}^2]},$$

$$\sin\alpha = \frac{y_2^-}{A}, \quad \cos\alpha = \frac{y_1^-}{A}, \quad \sin\beta = \frac{c_{12}x_1^+ - c_{11}x_2^+}{B}, \quad \cos\beta = \frac{c_{11}x_1^+ + c_{12}x_2^+}{B}.$$

2) *the inequality*

$$A\gamma^{-i}\cos(i\psi + \alpha) - B\lambda^j \cos(i\varphi + \beta) > S_{ij}(\bar{k}) \tag{2.9}$$

implies that the intersection of the horseshoe $T_1(\sigma_j^1)$ with the strip σ_i^0 is regular.

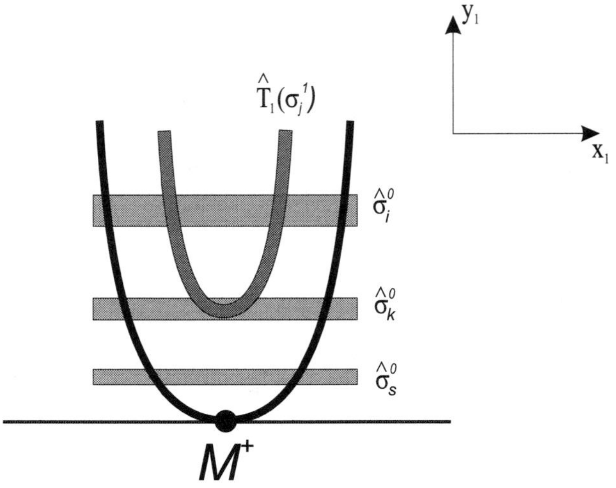

FIGURE 3. One can determine the character of the intersection of the horseshoes and strips according to the intersection of their projections. The figure corresponds to the situation where the horseshoe $T_1(\sigma_j^1)$ intersects the strip σ_i^0 regularly, the strip σ_k^0 irregularly, and does not intersect the strip σ_s^0. Here (and in the figures below) we denote by $\tilde{\sigma}_i^0$ the projection of the strip σ_i^0 to the plane (x_{01}, y_{01}) in Π^+. Respectively, $T_1(\tilde{\sigma}_j^1)$ is the projection of the horseshoe $T_1(\sigma_j^1)$ to the same plane.

PROOF. The intersection points of $T_1(\sigma_j^1)$ with σ_i^0 can be found from (1.9) provided that

$$
\begin{aligned}
x_{11} &= \lambda^j(\cos j\varphi \cdot x_{01} + \sin j\varphi \cdot x_{02} + \dots),\\
x_{12} &= \lambda^j(\cos j\varphi \cdot x_{02} - \sin j\varphi \cdot x_{01} + \dots),\\
\bar{y}_{01} &= \gamma^{-i}(\cos i\psi \cdot \bar{y}_{11} - \sin i\psi \cdot \bar{y}_{12} + \dots),\\
\bar{y}_{02} &= \gamma^{-i}(\cos i\psi \cdot \bar{y}_{12} + \sin i\psi \cdot \bar{y}_{11} + \dots)
\end{aligned}
\tag{2.10}
$$

where conditions (2.2) hold for x_{01}, x_{02}, y_{11}, and y_{12}.

A sufficient condition for the intersection of $T_1(\sigma_j^1)$ with σ_i^0 to be empty is that the right-hand side and left-hand side of the third equation in (1.9) have opposite signs. This and (2.10) imply the inequality

$$
\begin{aligned}
&d\{\gamma^{-i}(\cos i\psi \cdot \bar{y}_{11} - \sin i\psi \cdot \bar{y}_{12} + \dots) - c_{11}\lambda^j(\cos j\varphi \cdot x_{01} + \sin j\varphi \cdot x_{02} + \dots)\\
&- c_{12}\lambda^j(\cos j\varphi \cdot x_{02} - \sin j\varphi \cdot x_{01} + \dots) - A_1(\bar{y}_{02})^2 - B_1\bar{y}_{02}(y_1 - y_1^-) + \dots\} < 0.
\end{aligned}
$$

Since $d > 0$ (by the construction of T_1), we can use (2.2) to rewrite the last inequality as (2.8).

Now let us prove the second part of the theorem. In order that the intersection of $T_1(\sigma_j^1)$ with σ_i^0 consist of two connected components, it is sufficient that the inequality

(2.11)
$$d\{\gamma^{-i}(\cos i\psi\cdot\bar y_{11}-\sin i\psi\cdot\bar y_{12}+\dots)-c_{11}\lambda^j(\cos j\varphi\cdot x_{01}+\sin j\varphi\cdot x_{02}+\dots)$$
$$-c_{12}\lambda^j(\cos j\varphi\cdot x_{02}-\sin j\varphi\cdot x_{01}+\dots)-A_1(\bar y_2)^2-B_1\bar y_2(y_1-y_1^-)+\dots\}>0$$

hold for every x_{01}, x_{02}, y_{11}, and y_{12} satisfying (2.2). In this case inequality (2.11) can be rewritten as follows (since $d>0$):

$$A\gamma^{-i}\cos(i\psi+\alpha)-B\lambda^j\cos(j\varphi+\beta)>C_5(\lambda^j+\gamma^{-i})\gamma^{-\bar k/2}, \tag{2.12}$$

where C_5 is a positive constant independent of i and j. This inequality is similar to (2.9), but it is necessary to show also that the maps $T_1T_0^j$ defined on the preimages $(T_1T_0^j)^{-1}\Delta_{ij}^\alpha\subset\sigma_j^0$, $\alpha=1,2$, of the connected components Δ_{ij}^α are *saddle maps.* Note that if inequality (2.12) is fulfilled, then the third equation in (1.9) for $(x_1,y_1)\in\sigma_j^1$, $(\bar x_0,\bar y_0)\in\sigma_i^0$ can be represented as follows:

$$\begin{aligned}y_{11}-y_1^- &=(-1)^\alpha\big[d^{-1}(\gamma^{-i}(\cos i\psi\cdot\bar y_{11}-\sin i\psi\cdot\bar y_{12}+\dots)\\ &\qquad -c_{11}\lambda^j(\cos j\varphi\cdot x_{01}+\sin j\varphi\cdot x_{02})\\ &\qquad -c_{12}\lambda^j(\cos j\varphi\cdot x_{02}-\sin j\varphi\cdot x_{01}+\dots))\big]^{1/2}\\ &\equiv\Phi_{ij}^\alpha(x_{01},x_{02},\bar y_{11},\bar y_{12}).\end{aligned}\tag{2.13}$$

By this and (1.9), the map $T_{ij}^\alpha\equiv T_1T_0^j\colon\sigma_j^0\to\Delta_{ij}^\alpha\in\sigma_i^0$ can be written as follows:

(2.14)
$$\begin{aligned}\bar x_{01}-x_1^+ &=a_{11}\lambda^j(\cos j\varphi\cdot x_{01}+\sin j\varphi\cdot x_{02})+a_{12}\lambda^j(\cos j\varphi\cdot x_{02}-\sin j\varphi\cdot x_{01})\\ &\qquad+b\Phi_{ij}^\alpha(x_{01},x_{02},\bar y_{11},\bar y_{12})+\tilde b_1\gamma^{-i}(\cos i\psi\cdot\bar y_{12}+\sin i\psi\cdot\bar y_{11})+\dots,\\ \bar x_{01}-x_2^+ &=a_{21}\lambda^j(\cos j\varphi\cdot x_{01}+\sin j\varphi\cdot x_{02})+a_{22}\lambda^j(\cos j\varphi\cdot x_{02}-\sin j\varphi\cdot x_{01})\\ &\qquad+\tilde b_2\gamma^{-i}(\cos i\psi\cdot\bar y_{12}+\sin i\psi\cdot\bar y_{11})+\dots,\\ y_{11}-y_1^- &=\Phi_{ij}^\alpha(x_{01},x_{02},\bar y_{11},\bar y_{12}),\\ y_2-y_2^- &=c_{21}\lambda^j(\cos j\varphi\cdot x_{01}+\sin j\varphi\cdot x_{02})+c_{22}\lambda^j(\cos j\varphi\cdot x_{02}-\sin j\varphi\cdot x_{01})\\ &\qquad+d_0\gamma^{-i}(\cos i\psi\cdot\bar y_{12}+\sin i\psi\cdot\bar y_{11})-\tilde d\Phi_{ij}^\alpha(x_{01},x_{02},\bar y_{11},\bar y_{12}).\end{aligned}$$

Consider the following two-dimensional areas:

$$I_0=\{(x_{01},x_{02}):\|x_0-x^+\|\leqslant\rho_{\bar k}\},\qquad J_1=\{(y_{11},y_{12}):\|y_1-y^-\|\leqslant\rho_{\bar k}. \tag{2.15}$$

Relations (2.14) define a *cross-map* $(T_{ij}^\alpha)^\dagger\colon I_0\times J_1\to I_0\times J_1$. Recall that the connection between a map f and its cross-map $(f)^\dagger$ is as follows:

$$(\bar x,y)=(f)^\dagger(x,\bar y)\text{ is equivalent to }(\bar x,\bar y)=f(x,y).$$

According to [**20**], the map T_{ij}^α is saddle if the following inequalities are fulfilled for $(T_{ij}^\alpha)^\dagger$:

$$\left\|\frac{\partial\bar x_0}{\partial x_0}\right\|+\left\|\frac{\partial\bar x_0}{\partial\bar y_1}\right\|<\frac q2,\qquad\left\|\frac{\partial y_1}{\partial x_0}\right\|+\left\|\frac{\partial y_1}{\partial\bar y_1}\right\|<\frac q2, \tag{2.16}$$

where $0 < q < 1$.[4] It follows from (2.13) and (2.14) that inequalities (2.16) are fulfilled if

$$\|\Phi^{\alpha}_{ij}\| > \frac{Lq}{d}(\gamma^{-i} + \lambda^{j}), \tag{2.17}$$

where $L = \max\{|b|A, |b|B, A, B, |\tilde{d}|A, |\tilde{d}|B\}$. By virtue of (2.13) and (2.2), inequality (2.17) can be written as (2.9). This finishes the proof of the theorem. □

2.3. Coding nonwandering orbits and nontrivial hyperbolic subsets. A convenient way of describing the structure of the set $N_{\bar{k}}$ involves coding its orbits. We denote by σ_0 and σ_1 the sets consisting of the strips σ^0_k in Π^+ and σ^1_k in Π^- for $k \geqslant \bar{k}$. Let an orbit Λ belong to $N_{\bar{k}}$ and suppose that Λ is not asymptotic to O. Then Λ intersects Π^+ and Π^- at points belonging to the strips from the sets σ_0 and σ_1 respectively. Let $\{M^s_p\}^{+\infty}_{s=-\infty}$, $p \in \{0, 1\}$, be the consecutive points of intersection of Λ with the strips from the sets σ_p. Evidently, the following relations hold:

(2.18)

$$M^s_0 = T_1(M^{s-1}_1) \equiv T^n(M^{s-1}_1) \in \sigma^0_{k_s} \subset \Pi^+, \quad M^s_1 = T^{k_s}_0(M^s_0) \in \sigma^1_{k_s} \subset \Pi^-,$$
$$M^{s+1}_0 = T_1(M^s_1) \equiv T^n(M^s_1) \in \sigma^0_{k_{s+1}} \subset \Pi^+, \qquad s = 0, \pm 1, \dots,$$

where $\{k_s\}$ is some integer sequence with $k_s \geqslant \bar{k}$ for every s. According to (2.18), the bi-infinite sequence

$$(\dots, k_{-s} + n, \dots, k_0 + n, k_1 + n, \dots, k_s + n, \dots) \tag{2.19}$$

can be regarded as a *natural coding* of the orbit Λ. Relations (2.18) can also be coded by sequences consisting of two symbols $\{0, 1\}$, so, instead (2.19), we can consider the sequence

$$(\dots, 0, 0, 1, \overbrace{0, 0, \dots, 0, 1}^{k_s+n}, \overbrace{0, 0, \dots, 0, 1}^{k_{s+1}+n}, 0, 0, \dots), \tag{2.20}$$

Such codings can also be extended to orbits of $N_{\bar{k}}$ that are asymptotic to O. So, the following codes

$$(\overbrace{\dots, 0, \dots, 0}^{\infty}, 1, \overbrace{0, 0, \dots, 0, 1}^{k_0+n}, \overbrace{0, 0, \dots, 0, 1}^{k_1+n}, 0, \dots), \tag{2.21a}$$

$$(\dots, 0, 1, \overbrace{0, \dots, 0, 1}^{k_{-1}+n}, \overbrace{0, 0, \dots, 0, 1}^{k_0+n}, \overbrace{0, \dots, 0, \dots}^{\infty}), \tag{2.21b}$$

$$(\overbrace{\dots, 0, \dots, 0}^{\infty}, 1, \overbrace{0, 0, \dots, 0, 1}^{k_0+n}, 0, \dots, 0, 1, \overbrace{0, 0, \dots, 0, 1}^{k_l+n}, \overbrace{0, \dots, 0, \dots}^{\infty}), \tag{2.21c}$$

correspond to orbits which are α-limits, ω-limits, and homoclinic to O, respectively. Also, the orbits O and Γ have the codes

$$(\dots, 0, \dots, 0, \dots) \quad \text{and} \quad (\dots, 0, \dots, 0, 1, 0, \dots, 0, \dots).$$

If Γ were a structurally stable homoclinic orbit, then the correspondence between the set of orbits from $N_{\bar{k}}$ and the set of natural codings would be bijective [**20**]. This is not the case if Γ is structurally unstable (see also [**25**]). For instance, it is clear that $N_{\bar{k}}$ cannot contain orbits with codes in which the numbers k_s and k_{s+1} satisfy inequality (2.8) with $j = k_s$ and $i = k_{s+1}$ for some s. Indeed, in this

[4] The validity of (2.16) means that the cross-map $(T^{\alpha}_{ij})^{\dagger}$ is a contraction map.

case Theorem 1 implies that $T_1(\sigma^1_{k_s})\cap\sigma^0_{k_{s+1}}=\varnothing$, i.e., no orbit in $U_{\bar{k}}$ which starts in $\sigma^0_{k_s}$ goes around a global piece of Γ once and reaches $\sigma^0_{k_{s+1}}$. On the other hand, if for all s the numbers k_s and k_{s+1} satisfy inequality (2.9) (with $j=k_s$ and $i=k_{s+1}$), then there is no bijection. This is a consequence of the fact that the intersection of $T_1(\sigma^1_{k_s})$ with $\sigma^0_{k_{s+1}}$ consists of two connected components for every s and we must specify to which of the components the sth point of intersection of the orbit with Π^+ belongs.

We call a pair (j,i) of integers *admissible* if $i\geqslant\bar{k}$, $j\geqslant\bar{k}$ and i and j satisfy inequality (2.9). Improper pairs (∞,i) or (j,∞) will also be called *admissible* if inequality (2.9) is fulfilled for $\lambda^j=0$ or $\gamma^{-i}=0$, respectively. We call *inadmissible* a pair (j,i) if i and j satisfy (2.8) (also in the case of improper pairs).

THEOREM 2. 1) *In $N_{\bar{k}}$ there is no orbit with a code in which at least one pair $(j=k_s,\ i=k_{s+1})$ is inadmissible.*

2) *For any coding of type* (2.20) (*and also of type* (2.21a) *or* (2.21b)), *where for any s the pair $(j=k_s,\ i=k_{s+1})$ is admissible, there is a continuum of orbits of saddle type in $N_{\bar{k}}$ each having the given code. Finally, for any coding of type* (2.21c), *where for any s the pair $(j=k_s,\ i=k_{s+1})$ is admissible, there are 2^l structurally stable homoclinic orbits in $N_{\bar{k}}$ each having the given code.*

PROOF. First of all, we note that a similar result in some sense was established in **[25]** for the case of two-dimensional diffeomorphisms with a structurally unstable homoclinic orbit. Here the proof is carried out by the same method as in **[25]**.

1) By (2.18), the intersection $T_1(\sigma^1_j)\cap\sigma^0_i$ contains the point M^s_0 of the orbit Λ if $j=k_s$ and $i=k_{s+1}$. Thus, this intersection is nonempty for all s, which proves part 1.

2) According to Theorem 1, inequality (2.9) guarantees that the intersection of $T_1(\sigma^1_j)$ with σ^0_i consists of two connected components, which we have denoted by Δ^1_{ij} and Δ^2_{ij}. If an orbit of $N_{\bar{k}}$ has a code of type (2.20), where the numbers $i=k_{s+1}$ and $j=k_s$ satisfy inequality (2.9) for any s, this does not tell us to which of these components the points of intersection of the orbit with Π^+ belong. Thus, we can consider an improved code

$$(\dots,0,0,\alpha_{s-1},\overbrace{0,0,\dots,0,\alpha_s}^{k_s+n},\overbrace{0,0,\dots,0,\alpha_{s+1}}^{k_{s+1}+n},0,0,\dots),\tag{2.22}$$

now composed by using three symbols $\{0,1,2\}$, where the symbol α_s is either "1" or "2" and shows to which of the components, Δ^1_{ij} or Δ^2_{ij} (here $i=k_{s+1}$, $j=k_s$) the sth point of intersection of the orbit with Π^+ belongs. Now, to prove item 2, it is sufficient to prove that given a sequence of type (2.22), where all the pairs $(j=k_s,\ i=k_{s+1})$ are admissible, there exists a unique orbit (of saddle type) whose consecutive points of intersection with Π^+ and Π^- satisfy the following relations (compare with (2.18)):

$$\begin{aligned} M^s_0&=T_1(M^{s-1}_1)\in\Delta^{\alpha_{s-1}}_{k_sk_{s-1}}\subset\sigma^0_{k_s}\subset\Pi^+,\quad M^s_1=T_0^{k_s}(M^s_0)\in\sigma^1_{k_s}\subset\Pi^-,\\ M^{s+1}_0&=T_1(M^s_1)\in\Delta^{\alpha_s}_{k_{s+1}k_s}\subset\sigma^0_{k_{s+1}}\subset\Pi^+,\qquad s=0,\pm1,\dots.\end{aligned}\tag{2.23}$$

Consider, according to (2.23), the following sequence of the maps

$$\dots\xrightarrow{T_1T_0^{k_{s-1}}}\Delta^{\alpha_{s-1}}_{k_sk_{s-1}}\xrightarrow{T_1T_0^{k_s}}\Delta^{\alpha_s}_{k_{s+1}k_s}\longrightarrow\cdots.\tag{2.24}$$

Let us show that the sequence (2.24) has a unique fixed point (by definition, the fixed point is a sequence of points $\{\dots, M_0^s, M_0^{s+1} \dots\}$ such that $T_1 T_0^{k_s} M_0^s = M_0^{s+1}$, i.e., for such fixed point relations (2.23) are fulfilled). According to the *Lemma on a saddle fixed point in a direct product of spaces* [**20**],[5] it suffices to prove that all maps in the sequence (2.24) are saddle, namely, the maps in the cross-form are defined on compact metric spaces (in our case, the spaces are $I_0 \times J_1$ (see formula (2.15)) and are contraction maps with a constant of contraction less than 1/2. Note that we have shown above that if the assumptions of Theorem 1 hold, then the cross-map $(T_{ij}^{\alpha})^{\dagger}$ is defined on $I_0 \times J_1$, its range belongs to $I_0 \times J_1$ also and it is a contraction map. Thus, this finishes the proof for the case of orbits nonasymptotic to O. For orbits that are asymptotic to O, the proof is quite similar (see also [**25**]). The difference is in the fact that instead of (2.24), we must consider sequences that contain an infinite subsequence composed of maps T_0 and standing from the right or from the left side (or from both sides). Since T_0 is a saddle map (in a sufficiently small neighborhood of U_0), it follows that the saddle fixed point lemma is applicable to such sequences also, which gives us the desired existence and uniqueness result. □

Theorems 1 and 2 imply that in $N_{\bar{k}}$ one can detect a nontrivial nonuniformly hyperbolic subset $\widetilde{N}_{\bar{k}}$, which may be described as follows. Let us consider the space of all bi-infinite sequences of three symbols $\{0, 1, 2\}$; recall that the shift map Σ on this space is called a topological Bernoulli scheme on three symbols. Let us glue two homoclinic trajectories $(\dots, 0, \dots, 0, 1, 0, \dots, 0, \dots)$ and $(\dots, 0, \dots, 0, 2, 0, \dots, 0, \dots)$ in one orbit $\widetilde{\omega}$. Consider the subspace $\mathcal{B}$ of the quotient system obtained containing the trajectories $(\dots, 0, \dots, 0, \dots)$ and $\widetilde{\omega}$ and all symbolic sequences satisfying the following conditions:

3.1 The length of any whole segment consisting of zeros is not less than $\bar{k}+n-1$.

3.2 Let k_s+n-1 and $k_{s+1}+n-1$ be the lengths of consecutive whole segments consisting of zeros and separated by the symbol "1" or "2". Then for any s, the pair $(j = k_s,\ i = k_{s+1})$ is admissible.

THEOREM 3. *Let T be the diffeomorphism under consideration. Then for any sufficiently large $\bar{k}$, there exists a subsystem $\widetilde{N}_{\bar{k}}$ of $N_{\bar{k}}$ such that*

1) *$T|_{\widetilde{N}}$ is topologically conjugate to the symbolic subsystem $\Sigma|_{\widetilde{\mathcal{B}}}$;*

2) *all orbits of $\widetilde{N}_{\bar{k}} \setminus \Gamma$ are of saddle type.*

The proof is based on the observation that the correspondence between an orbit and its improved coding of type (2.22) is a continuous map (when the trajectories $(\dots, 0, \dots, 0, 1, 0, \dots, 0, \dots)$ and $(\dots, 0, \dots, 0, 2, 0, \dots, 0, \dots)$ are identified). It follows from the proof of Theorem 2 that coding is a one-to-one map of $N_{\bar{k}}$ onto $\mathcal{B}$; hence this map is a homeomorphism. By construction of the coding, we immediately see that the map is a topological conjugacy between $T|_{N_{\bar{k}}}$ and the symbolic system $\Sigma|_{\mathcal{B}}$.

3. Moduli of Ω-conjugacy

Note that according to Theorem 3 the closure of saddle orbits from $\widetilde{N}_{\bar{k}}$ contains a homoclinic orbit Γ, which is structurally unstable. Hence, the set $\widetilde{N}_{\bar{k}} \setminus \Gamma$ is

[5]One may call this lemma the principle of contraction maps for bi-infinite sequences of saddle maps represented in cross-form.

nonuniformly hyperbolic and may change under perturbations. This is true even for perturbations that do not destroy the initial homoclinic tangency. This gives us a way to find the Ω-moduli for T that we shall work with below.

Let us state the following definition of a modulus.

DEFINITION 1 ([**26**]). Let χ be a topological space with an equivalence relation R on its points. Suppose that a continuous functional $h\colon \chi \to \mathbb{R}^1$ is locally nonconstant (i.e., in each neighborhood of each point $X \in \chi$ there exists a point $\widetilde{X} \in \chi$ such that $h(X) \neq h(\widetilde{X})$). The functional h is called a *modulus of R-equivalence* if the inequality $h(X_1) \neq h(X_2)$ implies that X_1 and X_2 are nonequivalent. In this situation we say that $X \in \chi$ has a *modulus*. We say that X has (at least) m *moduli* if m independent moduli[6] are defined on χ. Finally, we say that X has *infinitely many moduli* if X has m moduli for any given m.

For χ we take a codimension one bifurcation surface H of C^r-diffeomorphisms close to T in the C^r-topology possessing a structurally unstable homoclinic orbit close to Γ. For the equivalence relation, we consider a local Ω-conjugacy (see Definition 2). Finally, we prove that the functionals

$$h_1(T) = \psi, \quad h_2(T) = \varphi, \quad h_3(T) = \theta \equiv -\ln\lambda/\ln\gamma$$

are moduli of local Ω-conjugacy of diffeomorphisms from H.

Let the diffeomorphisms T and T' belong to H, and let U and U' be some sufficiently small neighborhoods of the homoclinic orbits Γ and Γ'. Denote by $\Omega(T)$ and $\Omega(T')$ the sets of nonwandering orbits entirely lying in U and U', respectively. By definition, $\Omega(T) \subseteq N(T)$, where $N(T)$ is the set of all orbits of T lying in U entirely; of course, $\Omega(T)$ does not necessarily coincide with $N(T)$. At the same time, the nontrivial set $\widetilde{N}_{\bar{k}}(T)$ from Theorem 3 is contained in $\Omega(T)$.

DEFINITION 2. We say that T and T' are *locally Ω-conjugate* if there exists a homeomorphism $h\colon \Omega(T) \to \Omega(T')$ such that $h(O) = O'$, $h(\Gamma) = \Gamma'$ and the following diagram

$$\begin{array}{ccc} \Omega(T) & \xrightarrow{\ T\ } & \Omega(T) \\ \Big\downarrow h & & \Big\downarrow h \\ \Omega(T') & \xrightarrow{\ T'\ } & \Omega(T') \end{array} \tag{3.1}$$

is commutative.

For the diffeomorphisms under consideration let us study some properties preserved by the local Ω-conjugacy. Let T and T' belong to H. Suppose that U and U' are constructed as described in Subsection 2.1 (formula (2.2)), where the homoclinic points M^+, M^- and M'^+, M'^- are chosen so that

$$h(M^+) = M'^+, \qquad h(M^-) = M'^-. \tag{3.2}$$

We call these points *conjugate*. Since diagram (3.1) is commutative and h is continuous, it follows from (3.2) that if U and U' are sufficiently small, then h *preserves codings of type* (2.20) and (2.21).

[6]The independence of a system $h_1, \dots, h_m$ of moduli may be understood, for instance, in the sense that for any $X \in \chi$ and $i \in \{1, \dots, m\}$ in any neighborhood of X there exists $\widetilde{X} \in \chi$ such that $h_l(X) = h_l(\widetilde{X})$ for all $l \neq i$ and $h_i(X) \neq h_i(\widetilde{X})$.

Take $\bar{k}$ sufficiently large, namely, such that the special neighborhoods $U_{\bar{k}}$ and $U'_{\bar{k}}$ lie inside U and U', respectively, and, moreover,

$$h(U_{\bar{k}} \cap \Omega(T)) \subset U', \qquad h^{-1}(U'_{\bar{k}} \cap \Omega(T')) \subset U.$$

Since the coding of orbits is preserved by h, we have

$$h(U_{\bar{k}} \cap \Omega(T)) \subset U'_{\bar{k}}. \tag{3.3}$$

Indeed, by the definition of special neighborhood, the code of any orbit lying in $U_{\bar{k}}$ satisfies the condition: $k_s \geqslant \bar{k}$ for all s. The same condition is valid for the code of orbits from $h(U_{\bar{k}} \cap \Omega(T)) \subset U'$ which immediately implies (3.3), again by definition of special neighborhood. Similarly, $h(U_{\bar{k}} \cap \Omega(T)) \subset U'_{\bar{k}}$, and hence

$$h(U_{\bar{k}} \cap \Omega(T)) = U'_{\bar{k}} \cap \Omega(T'). \tag{3.4}$$

Since the map h preserves codes, we get the following result.

LEMMA 4. *Let T and T' be locally Ω-conjugate and let M^+, M^- and M'^+, M'^- be conjugate homoclinic points. Then, for $\bar{k}$ large enough, for the special neighborhoods $U_{\bar{k}}$ and $U'_{\bar{k}}$ corresponding to the given choice of the homoclinic points, the set $\mathrm{Cod}(\Omega_{\bar{k}})$ of codes of the orbits of $\Omega_{\bar{k}} \equiv \Omega(T) \cap U_{\bar{k}}$ coincides with the set $\mathrm{Cod}(\Omega'_{\bar{k}})$ of codes of the orbits of $\Omega'_{\bar{k}} \equiv \Omega(T') \cap U'_{\bar{k}}$.*

In particular, since $\widetilde{N}_{\bar{k}} \subseteq \Omega_{\bar{k}} \subseteq N_{\bar{k}}$, it follows that if T and T' are locally Ω-conjugate, then

$$\mathrm{Cod}(\widetilde{N}_{\bar{k}}(T)) \subseteq \mathrm{Cod}(N'_{\bar{k}}(T')), \tag{3.5}$$

$$\mathrm{Cod}(\widetilde{N}'_{\bar{k}}(T')) \subseteq \mathrm{Cod}(N_{\bar{k}}(T)), \tag{3.6}$$

Thus, conditions (3.5) and (3.6) are necessary for Ω-conjugacy of the diffeomorphisms under consideration and give us a way of finding moduli of Ω-conjugacy using information on the structure of hyperbolic subsets. Theorem 2 shows that the structure of $N_{\bar{k}}$ is closely linked with the structure of the sets of integer-valued solutions of inequalities (2.8), (2.9). We shall find Ω-moduli by using these inequalities. In order to know some of their properties, we consider, first, some properties of ordinary trigonometric inequalities.

LEMMA 5. *For any fixed $\omega_1, \omega_2 \in [0, 2\pi)$ and $\vartheta_1, \vartheta_2 \in (0, \pi)$ such that $\vartheta_1 \neq \vartheta_2$ each of the inequalities*

$$\cos(i\vartheta_1 + \omega_1) \cdot \cos(i\vartheta_2 + \omega_2) < 0, \tag{3.7a}$$

$$\cos(i\vartheta_1 + \omega_1) \cdot \cos(i\vartheta_2 + \omega_2) > 0 \tag{3.7b}$$

(with respect to i) has infinitely many integral solutions.

PROOF. We prove the lemma for the case of inequality (3.7a) (the other inequality is treated similarly). Suppose the contrary, i.e., that there exist i_0 such that, for all $i \geqslant i_0$, the following inequalities hold

$$\cos(i\vartheta_1 + \omega_1) \cdot \cos(i\vartheta_2 + \omega_2) \geqslant 0. \tag{3.8}$$

Let $J_1 = [-\pi/2; \pi/2]$ and $J_2 = [\pi/2; 3\pi/2]$ be arcs of the unit circle. Our assumption means that for all $i \geqslant i_0$ the angles $i\vartheta_1 + \omega_1$ and $i\vartheta_2 + \omega_2$ belong to the same arc J_{ν_i}, $\nu_i = 1, 2$. Denote

$$\beta^j_k = (i_0\vartheta_j + \omega_j)(\mathrm{mod}\, 2\pi) + k\vartheta_j, \qquad j = 1, 2,\ k = 0, 1, 2, \dots . \tag{3.9}$$

Then one has the following series of inclusions

$$\beta_0^1, \beta_0^2 \in J_{\nu_0}, \quad \beta_1^1, \beta_1^2 \in J_{\nu_1}, \quad \dots, \quad \beta_k^1, \beta_k^2 \in J_{\nu_k}, \quad \dots .$$

Since $0 < \vartheta_j < \pi$, it follows that $|\beta_k^1 - \beta_k^2| \leqslant \pi$ for any k (otherwise, for some k the angles β_k^1 and β_k^2 would belong to different arcs of J_1 and J_2). On the other hand, (3.9) implies that $|\beta_k^1 - \beta_k^2| \geqslant k|\vartheta_1 - \vartheta_2| - 2\pi$. Thus, $\pi \geqslant k|\vartheta_1 - \vartheta_2| - 2\pi$ for any k. This contradicts the condition $\vartheta_1 \neq \vartheta_2$. □

LEMMA 6. *For any fixed $\omega_1, \omega_2 \in [0, 2\pi)$ and $\vartheta_1, \vartheta_2 \in (0, \pi)$ such that $\vartheta_1 \neq \vartheta_2$, there exists a positive $\varepsilon \equiv \varepsilon(\vartheta_1, \vartheta_2)$ such that each of the inequalities*

$$\cos(i\vartheta_1 + \omega_1) > \varepsilon, \qquad \cos(i\vartheta_2 + \omega_2) < -\varepsilon, \tag{3.10a}$$

$$\cos(i\vartheta_1 + \omega_1) < -\varepsilon, \qquad \cos(i\vartheta_2 + \omega_2) > \varepsilon, \tag{3.10b}$$

$$\cos(i\vartheta_1 + \omega_1) > \varepsilon, \qquad \cos(i\vartheta_2 + \omega_2) > \varepsilon, \tag{3.10c}$$

$$\cos(i\vartheta_1 + \omega_1) < -\varepsilon, \qquad \cos(i\vartheta_2 + \omega_2) < -\varepsilon \tag{3.10d}$$

has infinitely many integer-valued solutions.

PROOF. One can use the same method of proof, but with one modification: instead of the arcs J_1 and J_2 as above, we must take the arcs

$$J_1' = [-\pi/2 - \delta/2; \pi/2 + \delta/2], \qquad J_2' = [\pi/2 - \delta/2; 3\pi/2 + \delta/2],$$

where δ is a positive constant such that $\delta < \vartheta_1, \vartheta_2 < \pi - \delta$ (if $0 < \vartheta_j < \pi$ and ϑ_1 and ϑ_2 are fixed, then such a constant exists always). □

THEOREM 4. *If T and T' are locally Ω-conjugate, then the following conditions are fulfilled:*

i) $\psi = \psi'$;
ii) $\varphi = \varphi'$;
iii) $\theta = \theta'$ *except possibly for the cases in which* $\varphi = \psi$.

PROOF. We assume that T and T' are locally Ω-conjugate. Then, there exist conjugate homoclinic points M^+, M^- and M'^+, M'^- in U_0 and U_0' such that the inclusions (3.5) and (3.6) hold. By the construction of T_1, we can assume that d and d' are both positive.

1) Suppose that $\psi \neq \psi'$. Then, by Lemma 6, there is a positive ε such that the inequalities

$$\cos(i\psi + \alpha) > \varepsilon, \qquad \cos(i\psi' + \alpha') < -\varepsilon$$

are fulfilled for a countable set of values of i. Let $i = i_0$ be one of them. Since $\lambda\gamma < 1$, the inequality

$$A\gamma^{-i_0}\cos(i_0\psi + \alpha) - B\lambda^{i_0}\cos(i_0\varphi + \beta) > (\lambda^{i_0} + \gamma^{-i_0})\varepsilon_{\bar{k}}$$

holds for sufficiently large i_0. Thus, the pair (i_0, i_0) is admissible and, by Theorem 2, the diffeomorphism T has a one-round saddle periodic orbit[7] intersecting the strip $\sigma_{i_0}^0$ (Figure 4a). On the other hand, the same pair (i_0, i_0) is inadmissible for T' since inequality (2.8) with $i = i_0$, $j = i_0$, i.e.,

$$A'\gamma'^{-i_0}\cos(i_0\psi' + \alpha') - B'\lambda'^{i_0}\cos(i_0\varphi' + \beta') < -(\lambda'^{i_0} + \gamma'^{-i_0})\varepsilon_{\bar{k}},$$

[7] We call a periodic orbit $Q \in N_{\bar{k}}$ *k-round* if the intersection $Q \cap \Pi^+$ (or $Q \cap \Pi^+$) consists of k points.

is fulfilled. This means that T' has no one-round periodic orbits intersecting the strip $\sigma^0_{i_0}$ (Figure 4b). Thus, inclusion (3.5) fails, which proves the first statement of the theorem.

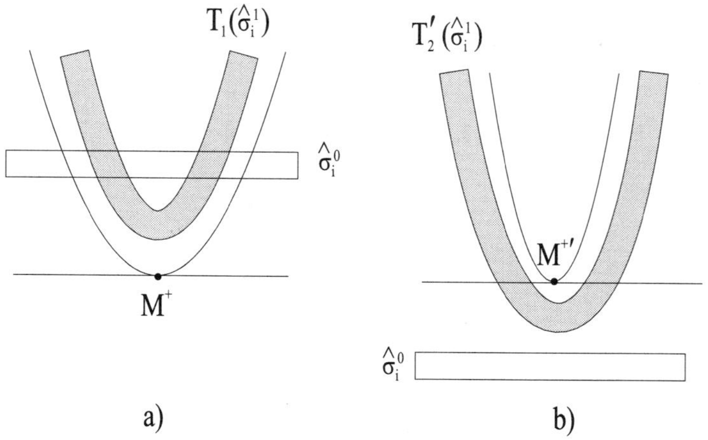

FIGURE 4. The case $\psi(T) \neq \psi(T')$. The diffeomorphism T has a one-round saddle periodic orbit (its intersection point with Π^+ is a fixed point of the map $T_1T_0^{i_0}\colon \sigma^0_{i_0} \to \sigma^0_{i_0}$). The diffeomorphism T' has no corresponding orbits, since $T_1'(\sigma^1_{i_0}) \cap \sigma^0_{i_0} = \varnothing$. The location of the corresponding projections of the strips and horseshoes is shown in Figure 4a for T and in Figure 4b for T'.

2) Now, by part 1), one has $\psi = \psi'$. Suppose that $\varphi \neq \varphi'$. We assume, to be definite, that $\psi \neq \varphi$. It follows from the proof of part 1) that for each sufficiently large i the inequality

$$\cos(i\psi + \alpha) \cdot \cos(i\psi + \alpha') \geqslant 0 \tag{3.11}$$

must be fulfilled. Indeed, if the inequality opposite to (3.11) has an infinite number of integer-valued solutions, then, as in part 1), one can find distinguishing one-round periodic orbits for T and T'.

Now we consider the main case of part 2), i.e., the case in which the inequalities

$$\begin{gathered} \cos(i\psi + \alpha) \cdot \cos(i\psi + \alpha') > \varepsilon_1 > 0, \\ \cos(i\varphi + \beta) < -\varepsilon < 0, \qquad \cos(i\varphi' + \beta') > \varepsilon > 0 \end{gathered} \tag{3.12}$$

are fulfilled for a countable set of integers i. Note that this condition holds for almost all values of ψ (in particular, if $\psi/2\pi$ is irrational). The exclusive cases will be considered later. Let I be a set of integers i satisfying (3.12) and, to be definite, we assume that for all $i \in I$ the following inequalities are fulfilled

$$\cos(i\psi + \alpha) > \varepsilon_1^1 > 0, \qquad \cos(i\psi + \alpha') > \varepsilon_1^2 > 0. \tag{3.13}$$

Let $i_0 \in I$. By (3.12) and (3.13), for the diffeomorphism T the inequality

$$A\gamma^{-k}\cos(k\psi + \alpha) - B\lambda^{i_0}\cos(i_0\varphi + \beta) > (\gamma^{-k} + \lambda^{i_0})\varepsilon_{\bar k}$$

holds for all $k \in I$. By Theorem 1 this means that the horseshoe $T_1(\sigma^1_{i_0})$ regularly intersects all strips σ^0_k with $k \in I$. In other words, any pair (i, j), where $i, j \in I$, is admissible for T. Moreover, since the horseshoes $T_1(\sigma^1_{i_0})$ and $T_1(\sigma^1_k)$ regularly intersect both strips $\sigma^0_{i_0}$ and σ^0_k, by Theorem 2 the diffeomorphism T possesses a two-round saddle periodic orbit that intersects the strips $\sigma^0_{i_0}$ and σ^0_k consecutively (Figure 5a). On the other hand, in the case of the diffeomorphism T', there exists a countable set of inadmissible pairs (k, i_0) with $i_0, k \in I$. Indeed, for such a pair, by Theorem 1, the inequality

$$A'\gamma'^{-k}\cos(k\psi+\alpha') - B'\lambda'^{i_0}\cos(i_0\varphi'+\beta') < -(\gamma^{-k}+\lambda^{i_0})\varepsilon_{\bar{k}} \tag{3.14}$$

must be fulfilled. Since $\cos(i_0\varphi'+\beta') > \varepsilon$, inequality (3.14) may be valid for some $k \in I$. Namely, it is true for any k satisfying the inequality

$$A'\gamma'^{-k}(A'+\varepsilon_{\bar{k}}) < \lambda'^{i_0}(B'\varepsilon+\varepsilon_{\bar{k}}). \tag{3.15}$$

Taking the logarithm of both sides of (3.15), we obtain the inequality

$$k > i_0\theta' - \frac{1}{\ln\gamma'}\ln\frac{B'\varepsilon+\varepsilon_{\bar{k}}}{A'+\varepsilon_{\bar{k}}}$$

which shows (by Theorem 1) that the horseshoe $T_1'(\sigma^1_{i_0})$ does not intersect infinitely many strips σ^0_k with $k \in I$. Thus, in this case inclusion (3.5) fails.

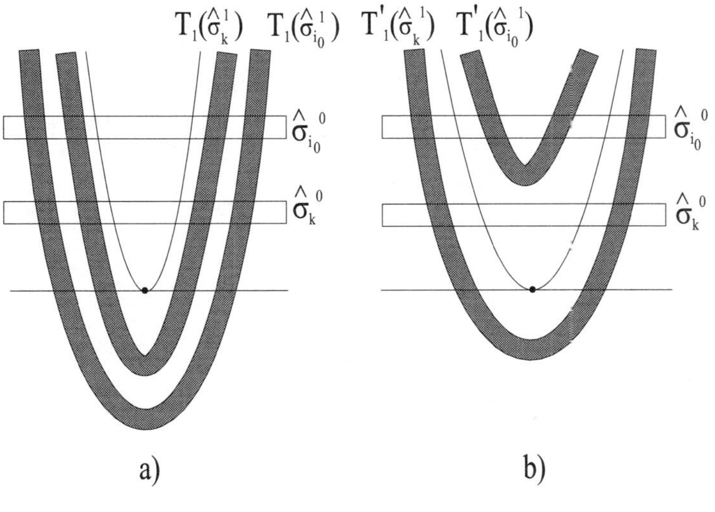

FIGURE 5. The case $\varphi(T) \neq \varphi(T')$. The diffeomorphism T has a one-round saddle periodic orbit (its intersection point with $\sigma^0_{i_0}$ is a fixed point of the map $T_1T_0^kT_1T_0^{i_0}: \sigma^0_{i_0} \to \sigma^0_{i_0}$). For T, the horseshoes $T_1(\sigma^1_{i_0})$ and $T_1(\sigma^1_k)$ intersect regularly with both strips $\sigma^0_{i_0}$ and σ^0_k (Figure 5a). Since $T_1'(\sigma^0_{i_0}) \cap \sigma^0_k = \varnothing$ (Figure 5b), the diffeomorphism T' has no corresponding orbits.

Now we consider the case in which for all sufficiently large integers i the following conditions are valid (remaining ones from part 2))

$$\begin{gathered}\psi \neq \varphi\\ \text{either } \cos(i\psi+\alpha)=0, \text{ and then } |\cos(i\psi+\alpha')| > \varepsilon > 0,\\ \text{or } \cos(i\psi+\alpha')=0, \text{ and then } |\cos(i\psi+\alpha)| > \varepsilon > 0.\end{gathered} \tag{3.16}$$

Note that this case is possible only if $\psi/2\pi$ is rational (and, moreover, only if ψ is equal to $\pi/2$). Also it is clear that if the equalities $\cos(i\psi+\alpha) = \cos(i\psi+\alpha') = 0$

are fulfilled for an infinite number of integers i, then, in turn, there exist infinitely many integers i for which strong inequalities of type (3.12) are valid. Since $\varphi \neq \varphi'$ by assumption, it follows from Lemma 6 that the inequality

$$\cos(j\varphi + \beta) \cdot \cos(j\varphi' + \beta') < -\varepsilon$$

has infinitely many integer solutions. Thus, we can assume that the conditions

$$\begin{gathered} \cos(i\psi + \alpha) = 0, \qquad \cos(i\psi + \alpha') > \varepsilon_1 > 0, \\ \cos(i\varphi + \beta) < -\varepsilon_2 < 0, \qquad \cos(i\varphi' + \beta') > \varepsilon_2 > 0 \end{gathered} \tag{3.17}$$

are fulfilled for a countable set I_1 of integers i. Thus, by (3.17), for T the strips σ_i^0 are "critical" at all $i \in I_1$: their projections to the two-dimensional plane (x_{01}, y_{01}) in Π^+ contain the segment $y_{01} = 0$ (Figure 6a). At the same time, for T' the corresponding strips are "normal" (Figure 6b). Nevertheless, we can specify, for a given $i_0 \in I_1$, a countable set of integers $k \in I_1$ such that a) $T_1'(\sigma_{i_0}^1) \cap \sigma_k^0 = \varnothing$, and b) the intersection of $T_1(\sigma_{i_0}^1)$ with σ_k^0 is regular. Using this, we shall find distinguishing nonwandering (periodic) orbits for T and T'.

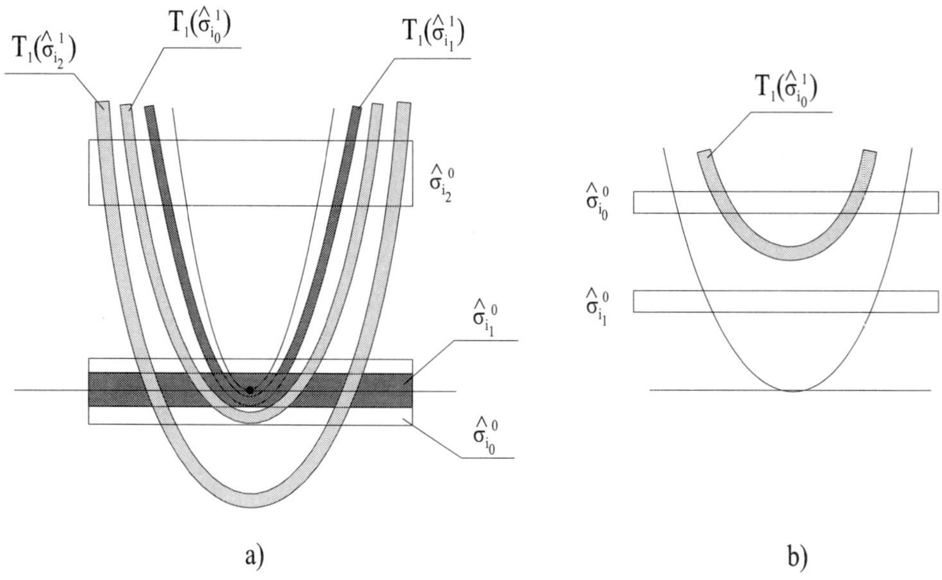

FIGURE 6. Figure 6a) illustrates the case in which the strips $\sigma_{i_0}^0$ and $\sigma_{i_1}^0$ are critical for T (i.e., their projections to the plane (x_{01}, y_{01}) contain the segment $y_{01} = 0$). At the same time, the corresponding strips are "normal" for T' (Figure 6b). Nevertheless, one can find a three-round periodic orbit of T that consecutively intersects the strips $\sigma_{i_0}^0$, $\sigma_{i_1}^0$, and $\sigma_{i_2}^0$. Here, $T_1(\sigma_{i_0}^1)$ and $\sigma_{i_1}^0$, $T_1(\sigma_{i_1}^1)$ and $\sigma_{i_2}^0$, $T_1(\sigma_{i_2}^1)$ and $\sigma_{i_0}^0$ have regular intersections. On the other hand, $T_1'(\sigma_{i_0}^1) \cap \sigma_{i_1}^0 = \varnothing$.

Let $i_0 \in I_1$. Then, by (3.17) and (2.8), $T_1'(\sigma_{i_0}^1) \cap \sigma_k^0 = \varnothing$ for $k \in I_1$ such that

$$A'\gamma'^{-k} - B'\lambda'^{i_0}\varepsilon_2 < -(\gamma'^{-k} + \lambda'^{i_0})\varepsilon_k'. \tag{3.18}$$

Taking the logarithm of both sides of (3.18), we obtain the inequality

$$k > i_0\theta' - \frac{1}{\ln\gamma'}\ln\frac{B'\varepsilon_2 + \varepsilon'_{\bar{k}}}{A' + \varepsilon'_{\bar{k}}}, \tag{3.19}$$

which is valid for an infinite number of values of $k \in I_1$.

At the same time, for T, the inequality

$$-B\lambda^{i_0}\cos(i_0\varphi + \beta) > (\gamma^{-k} + \lambda^{i_0})\varepsilon_{\bar{k}} \tag{3.20}$$

defines, for a given $i_0 \in I_1$, those $k \in I_1$ at which the horseshoe $T_1(\sigma^1_{i_0})$ intersects the critical strip σ^0_k regularly. By (3.17), one can rewrite (3.20) as

$$B\lambda^{i_0}\varepsilon_2 > (\gamma^{-k} + \lambda^{i_0})\varepsilon_{\bar{k}}$$

or as

$$k > i_0\theta - \frac{1}{\ln\gamma}\ln\frac{B\varepsilon_2 - \varepsilon_{\bar{k}}}{\varepsilon_{\bar{k}}}. \tag{3.21}$$

Obviously, for every given i_0 and fixed $\varepsilon_{\bar{k}}$, the last inequality has infinitely many integer solutions $k \in I_1$. Let $k = i_1 \in I_1$ be one of these solutions such that inequalities (3.19) and (3.21) hold simultaneously. Then $T_1(\sigma^1_{i_0})$ intersects $\sigma^0_{i_1}$ regularly and $T'_1(\sigma^1_{i_0}) \cap \sigma^0_{i_1} = \varnothing$ (Figure 6). This allows us to construct distinguishing nonwandering orbits. Note that since the intersection of the horseshoe $T_1(\sigma^1_{i_0})$ with its own (critical) strip $\sigma^0_{i_0}$ may be irregular, we cannot find, as before, a two-round orbit using only the strips $\sigma^0_{i_0}$ and $\sigma^0_{i_1}$. But we can find a three-round periodic orbit in the following way. We introduce a third strip $\sigma^0_{i_2}$ (independently of T') such that $T_1(\sigma^1_{i_2})$ regularly intersects all three strips $\sigma^0_{i_0}$, $\sigma^0_{i_1}$, and $\sigma^0_{i_2}$ while $T_1(\sigma^1_{i_1})$ regularly intersects $\sigma^0_{i_2}$ (Figure 6a). In this case, by Theorem 2, there exists a three-round periodic orbit consecutively intersecting the strips $\sigma^0_{i_0}$, $\sigma^0_{i_1}$, and $\sigma^0_{i_2}$. Now, to complete the proof of part 2) we must find such a number i_2.

For this we use the inequality $\varphi \neq \psi$. Then, by Lemma 6, the inequalities

$$\cos(j\varphi + \alpha_1) < -\varepsilon, \qquad \cos(j\psi + \beta) > \varepsilon \tag{3.22}$$

have infinitely many integer solutions. Consider the strips σ^0_j with such numbers. Then, by (3.22), $T_1(\sigma^1_j)$ regularly intersects σ^0_j. Next, $T_1(\sigma^1_j)$ regularly intersects $\sigma^0_{i_0}$ (and automatically $\sigma^0_{i_1}$) if $B\lambda^j\varepsilon > (\gamma^{-i_0} + \lambda^j)\varepsilon_{\bar{k}}$ or

$$j < \frac{1}{\theta}i_0 + \frac{1}{\theta\ln\gamma}\ln\frac{B\varepsilon - \varepsilon_{\bar{k}}}{\varepsilon_{\bar{k}}}. \tag{3.23}$$

Let $j = i_2$ be some solution of (3.23). Note that since $\theta > 1$, (3.23) implies the inequality $i_2 < i_0$, but we can always take i_2 tend to ∞ as $i_0 \to \infty$. Finally, it follows from (3.17) and (3.22) that the horseshoe $T_1(\sigma^1_{i_1})$ regularly intersects $\sigma^0_{i_2}$ (see Figure 6a). Thus, the required i_0, i_1, and i_2 exist. This finishes the proof of part 2).

3) Now we assume that $\varphi = \varphi'$, $\psi = \psi'$, and under the assumptions of the theorem also $\varphi \neq \psi$. Suppose that $\theta \neq \theta'$, say let $\theta > \theta'$.

Since $\varphi \neq \psi$, it follows from Lemma 6 and parts 1) and 2) that the inequalities

$$\begin{aligned} &\cos(k\psi + \beta) > \varepsilon > 0, \qquad \cos(k\varphi + \alpha) \geqslant 0, \\ &\cos(k\psi + \beta') \geqslant 0, \qquad \cos(k\varphi + \alpha') > \varepsilon > 0 \end{aligned} \tag{3.24}$$

are fulfilled for an infinite number of integer k. Let I_2 be the set of such k.

Now for T we shall find numbers i and j from the set I_2 such that the horseshoes $T_1(\sigma_j^1)$ intersect the strips σ_i^0 regularly. By Theorem 1, such i and j can be found from inequality (2.9). Using (3.24), we can consider the weaker inequality

$$A\gamma^{-i}\varepsilon - B\lambda^j > (\gamma^{-i} + \lambda^j)\varepsilon_{\bar{k}}$$

which is equivalent to the inequality

$$i < j\theta - \frac{1}{\ln\gamma}\ln\frac{B+\varepsilon_{\bar{k}}}{A\varepsilon - \varepsilon_{\bar{k}}}. \tag{3.25}$$

For the diffeomorphism T', on the contrary, we shall find i and j from the set I_2 such that $T_1'(\sigma_j^1)\cap\sigma_i^0 = \varnothing$. By Theorem 1, it is sufficient to consider those i and j for which inequality (2.8) is valid. By (3.24), we can consider the weaker inequality $B'\lambda'^j\varepsilon < (\gamma'^{-i} + \lambda'^j)\varepsilon'_{\bar{k}}$ which is equivalent to

$$i > j\theta' - \frac{1}{\ln\gamma'}\ln\frac{B'\varepsilon - \varepsilon'_{\bar{k}}}{\varepsilon'_{\bar{k}}}. \tag{3.26}$$

Now we consider those i and j for which inequalities (3.25) and (3.26) are fulfilled simultaneously. Since $\theta > \theta'$, the set of such integers is not empty; they satisfy the inequality

$$j\theta' - \frac{1}{\ln\gamma'}\ln\frac{B'\varepsilon - \varepsilon'_{\bar{k}}}{\varepsilon'_{\bar{k}}} < i < j\theta - \frac{1}{\ln\gamma}\ln\frac{B+\varepsilon_{\bar{k}}}{A\varepsilon - \varepsilon_{\bar{k}}},$$

which has infinitely many solutions also satisfying (3.24). Let $(i = i_0,\ j = j_0)$ be one such solution. Then, we have the following situation: for T, the horseshoe $T_1(\sigma_{j_0}^1)$ regularly intersects the strip $\sigma_{i_0}^0$; for T', the horseshoe $T_1'(\sigma_{j_0}^1)$ does not intersect the strip $\sigma_{i_0}^0$. This allows us to construct a distinguishing periodic orbit. Since $\lambda\gamma < 1$ and $i_0, j_0 \in I_2$, the horseshoes $T_1(\sigma_{j_0}^1)$ and $T_1(\sigma_{i_0}^1)$ regularly intersect their own strips $\sigma_{j_0}^0$ and $\sigma_{i_0}^0$, respectively. Next, $T_1(\sigma_{j_0}^1)$ regularly intersects $\sigma_{i_0}^0$ by

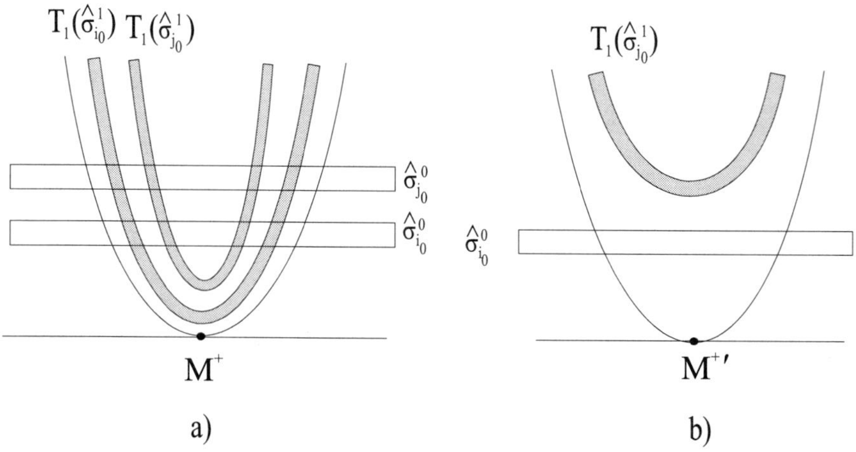

FIGURE 7. The case in which $\theta(T) > \theta(T')$. The diffeomorphism T has a two-round periodic orbit consecutively intersecting the strips $\sigma_{j_0}^0$ and $\sigma_{i_0}^0$ (Figure 7a). The diffeomorphism T' has no such orbit since $T_1'(\sigma_{j_0}^1)\cap\sigma_{i_0}^0 = \varnothing$ (Figure 7b).

construction. Finally, the horseshoe $T_1(\sigma^1_{i_0})$ intersects the strip $\sigma^0_{j_0}$ regularly too, since the inequality

$$A\gamma^{-j_0}\varepsilon - B\lambda^{i_0} > (\gamma^{-j_0} + \lambda^{i_0})\varepsilon_{\bar{k}}$$

is fulfilled because $i_0 > j_0$ by (3.26). Thus, for T, the pairs (j_0, i_0) and (i_0, j_0) are admissible (Figure 7a). By Theorem 2, the diffeomorphism T has a two-round periodic orbit intersecting the strips $\sigma^0_{j_0}$ and $\sigma^0_{i_0}$ consecutively. Again by Theorem 2, the diffeomorphism T' has no such orbit, since the pair $(j = j_0,\ i = i_0)$ is inadmissible for T' by construction (Figure 7b). Thus, inclusion (3.5) fails in this case. This completes the proof of Theorem 3. □

Acknowledgments. I would like to thank L. P. Shilnikov and D. V. Turaev for many useful discussions. I wish to thank B. Fiedler for hospitality during my visit to Frie Universitat, Berlin, at the end of 1996, when the essential part of this paper was written.

References

[1] E. A. Leontovich-Andronova and A. G. Maier, *On a diagram determining the topological structure of a partition into trajectories*, Dokl. Akad. Nauk SSSR **103** (1955), 557–560. (Russian)

[2] A. A. Andronov, E. A. Leontovich, I. I. Gordon, and A. G. Maier, *Theory of bifurcations of dynamical systems on the plane*, "Nauka", Moscow, 1967; English transl., Israel Program of Scientific Translations, Jerusalem and London, 1973.

[3] N. N. Bautin and E. A. Leontovich, *Methods and ways (examples) of the qualitative analysis of dynamical systems in the plane*, "Nauka", Moscow, 1976. (Russian)

[4] M. M. Peixoto, *On the classification of flows on 2-manifolds*, Dynamical Systems (Proc. Sympos., Univ. Bahia, Salvador, Brazil, 1971) (M. M. Peixoto, ed.), Academic Press, New York, 1973, pp. 389–419.

[5] Ya. L. Umanskiĭ, *The diagram of a three-dimensional Morse–Smale system without closed trajectories*, Dokl. Akad. Nauk SSSR **230** (1976), no. 6, 1286–1289; English transl. in Soviet Math. Dokl. **17** (1976).

[6] ———, *Necessary and sufficient conditions for topological equivalence of three-dimensional Morse–Smale dynamical systems with a finite number of singular trajectories*, Mat. Sb. **181** (1990), no. 2, 212–239; English transl. in Math. USSR Sb. **69** (1990).

[7] A. N. Bezdenezhnykh and V. Z. Grines, *Dynamical properties and topological classification of gradient-like diffeomorphisms on two-dimensional manifolds*, Selecta Math. Soviet. **11** (1992), no. 1, 1–17.

[8] ———, *Diffeomorphisms with orientable heteroclinic sets*, in Methods of the Qualitative Theory of Differential Equations (E. A. Leontovich-Andronova, ed.), Gor′kov. Gos. Univ., Gor′kiĭ, 1985, pp. 139–152. (Russian)

[9] V. Z. Grines and H. H. Kalai, *On topological equivalence of diffeomorphisms with nontrivial basic sets on two-dimensional manifolds*, Methods of the Qualitative Theory of Differential Equations (E. A. Leontovich-Andronova, ed.), Gor′kov. Gos. Univ., Gor′kiĭ, 1988, pp. 40–48. (Russian)

[10] Ya. G. Sinai, *Markov partition and U-diffeomorphisms*, Funktsional Anal. i Prilozhen. **2** (1968), no. 1, 64–89; English transl. in Functional Anal. Appl. **2** (1968), no. 1.

[11] A. Manning, *There are no new Anosov diffeomorphisms on tori*, Amer. J. Math. **96** (1974), 422–429.

[12] S. V. Gonchenko, D. V. Turaev, and L. P. Shilnikov, *On models with a structurally unstable homoclinic curve*, Dokl. Akad. Nauk SSSR **320** (1991), no. 2, 269–272; English transl., Soviet Math. Dokl. **44** (1992), no. 2, 422–426.

[13] J. Palis, *A differentiable invariant of topological conjugacies and moduli of stability*, Astérisque **51** (1978), 335–346.

[14] S. Newhouse, J. Palis, and F. Takens, *Bifurcations and stability of families of diffeomorphisms*, IHES Publ. Math. **57** (1983), 5–72.

[15] S. V. Gonchenko and L. P. Shilnikov, *Invariants of Ω-conjugacy of diffeomorphisms with a structurally unstable homoclinic trajectory*, Ukrain. Mat. Zh. **42** (1990), no. 2, 153–159; English transl., Ukrainian Math. J. **42** (1990), no. 2, 134–140.
[16] ———, *On the moduli of systems with a non-rough Poincaré homoclinic curve*, Izv. Ross. Akad. Nauk Ser. Mat. **56** (1992), no. 6, 1165–1197; English transl., Russian Acad. Sci. Izv. Math. **41** (1993), no. 3, 417–445.
[17] S. V. Gonchenko, D. V. Turaev, and L. P. Shilnikov, *Dynamical phenomena in multidimensional systems with a structurally unstable homoclinic Poincaré curve*, Dokl. Akad. Nauk **330** (1993), no. 2, 144–147; English transl., Russian Acad. Sci. Dokl. Math. **47** (1993), no. 3, 410–415.
[18] P. Lancaster, *Theory of matrices*, Academic Press, New York and London, 1969.
[19] M. W. Hirsch, C. C. Pugh, and M. Shub, *Invariant manifolds*, Lecture Notes in Math., vol. 583, Springer-Verlag, Berlin, 1977.
[20] L. P. Shilnikov, *On the Poincaré–Birkhoff problem*, Mat. Sb. (N.S.) **74 (116)** (1967), 378–397; English transl., Math. USSR-Sb. **3** (1967), 91–102.
[21] ———, *A contribution to the problem of the structure of the neighborhood of a homoclinic tube of an invariant torus*, Dokl. Akad. Nauk SSSR **180** (1968), no. 2, 286–289; English transl. in Soviet Math. Dokl. **9** (1968).
[22] V. S. Afraimovich and L. P. Shilnikov, *On critical sets of Morse–Smale systems*, Trudy Moskov. Mat. Obshch. **28** (1973), 181–214; English transl. in Trans. Moscow Math. Soc. **28** (1973).
[23] S. V. Gonchenko, *Nontrivial hyperbolic subsets of systems with a structurally unstable homoclinic curve*, Methods of the Qualitative Theory of Differential Equations (E. A. Leontovich-Andronova, ed.), Gor′kov. Gos. Univ., 1984, pp. 89–102. (Russian)
[24] S. V. Gonchenko and L. P. Shilnikov, *On dynamical systems with structurally unstable homoclinic curves*, Dokl. Akad. Nauk SSSR **286** (1986), no. 5, 1049–1053; English transl., Soviet Math. Dokl. (1986), **33**, no. 1, 234–238.
[25] N. K. Gavrilov and L. P. Shilnikov, *On three-dimensional dynamical systems close to systems with a structurally unstable homoclinic curve*, I, **88 (130)** (1972), 475–492; II, Mat. Sb. (N.S.) **90 (132)** (1973), 139–156; English transl., Math. USSR-Sb. **17** (1972), 467–485; Math. USSR-Sb. **19** (1973), 139–156.
[26] S. V. Gonchenko, *Moduli of systems with structurally unstable homoclinic orbits* (*the cases of diffeomorphisms and vector fields*), Methods in Qualitative Theory and Bifurcation Theory (L. P. Shilnikov, ed.), Gor′kov. Gos. Univ., Gor′kiĭ, 1989, pp. 89–102; English transl. in Selecta Math. Soviet. **11** (1990).

INST. FOR APPL. MATH. AND CYB., 10 UL′YANOV ST., 603005 NIZHNY NOVGOROD, RUSSIA
E-mail address: gonchenko@focus.nnov.ru

Translated by THE AUTHOR

Amer. Math. Soc. Transl.
(2) Vol. **200**, 2000

Topological Classification of Amply Situated Attractors of A-Diffeomorphisms of Surfaces

V. Z. Grines and R. V. Plykin

1. Introduction and statement of results

Consider a diffeomorphism f of a smooth closed 2-dimensional surface M_ν, where $\nu \in \{o, n\}$, M_o is an orientable surface of genus g, $g \geqslant 1$, and M_n is a nonorientable surface of genus g, $g \geqslant 3$. We suppose that f satisfies Smale's axiom A, that is, f is an A-diffeomorphism.

By the Smale theorem on spectral decompositions, the nonwandering set of an A-diffeomorphism f is the finite disjoint union of closed invariant sets (basic sets); each of them contains an everywhere dense trajectory [**29**]. A one-dimensional basic set is either an attractor or a repeller [**25**], and for any point x of this attractor (or repeller) Ω, the unstable (stable) manifold of this point lies in Ω.

Consider A-diffeomorphisms whose nonwandering sets contain one-dimensional basic sets. The problem of topological classification of these diffeomorphisms leads to the following problem of topological conjugacy of restrictions of A-diffeomorphisms to one-dimensional basic sets.

Let Ω and Ω' be one-dimensional basic sets of A-diffeomorphisms f and f'. The problem is to find necessary and sufficient conditions for the existence of a homeomorphism $g\colon M \to M$ such that $g(\Omega) = \Omega'$ and $f'|_\Omega = gfg^{-1}|_{\Omega'}$.

In [**11, 12, 13**] the class of oriented one-dimensional basic sets of A-diffeomorphisms of closed orientable surfaces[1] was introduced. The above-stated problem was solved by the first author in the early seventies during his work in the Laboratory of Differential Equations (directed by E. A. Leontovich-Andronova), Research Institute of Applied Mathematics and Cybernetics. This problem was suggested by L. P. Shilnikov. The principal tool for solving this problem was a new topological invariant of dynamical systems on surfaces, namely, the homotopy winding class,

---

2000 *Mathematics Subject Classification.* Primary 37C15, 37C70.

This research was partially supported by Russian Foundation for Basic Research under grants 96-01-00236, 99-01-00230 and by INTAS under grants Nos. 95-0418, 97-1843.

[1] A nontrivial basic set Ω is said to be *orientable* if for any $x \in \Omega$ and all fixed positive α, β the intersection index of the manifolds $W^s_\alpha(x)$ and $W^u_\beta(x)$ does not depend on the choice of the intersection point; here $W^s_\alpha(x) = \{y \in W^s(x) \mid l(x,y) < \alpha\}$, $W^u_\beta(x) = \{y \in W^u(x) \mid l(x,y) < \beta\}$, and l is a metric on $W^s(x)$ and $W^u(x)$. If the nonwandering set of an A-diffeomorphism of an orientable surface contains an orientable basic set, then the genus of the surface is at least 1 [**11, 12, 13**].

©2000 American Mathematical Society

introduced first in [**3**] for trajectories of flows on surfaces and then in [**11, 12**] for invariant manifolds of points of an orientable basic set. The homotopy winding class was discovered while applying ideas of A. Weil and Anosov (concerning the asymptotic behavior of preimages of non-self-intersecting curves on the universal covering spaces of closed surfaces with nonpositive Euler characteristic) to the study of dynamical systems on these surfaces. For the history of the subject, see the reviews [**1, 4, 5, 6, 7**], which also contain many results and an extensive bibliography.

Further, in [**26**] the class of amply situated one-dimensional basic sets of A-diffeomorphisms was introduced for both orientable and non-orientable closed surfaces,[2] and in [**27**] an idea was presented for generalizing the results of [**11, 12, 13**] to basic sets of this class. The paper [**28**] contains some results announced in [**27**]; however, the complete proofs were not published. The aim of the present paper is to fill this omission.

It was shown in [**20, 21**] that the topological classification of arbitrary one-dimensional basic sets can be reduced to the algebraic classification of generalized hyperbolic automorphisms of the fundamental groups of the supports of the attractors (see the paper [**17**], which contains a detailed exposition of these results; see also [**6, 7, 8**]). The paper [**18**] contains the topological classification up to semi-conjugacy of restrictions of A-diffeomorphisms to one-dimensional basic sets with restrictions of hyperbolic (or generalized hyperbolic) homeomorphisms to the zero-dimensional set which is the intersection of two transversal geodesic laminations.

In the recent series of papers [**30**], new results are obtained on topological classification of attractors. These results are connected with the combinatorial description of automorphisms of the fundamental groups of the supports. In [**31**], a relation is established between one-dimensional basic sets of A-diffeomorphisms of surfaces and generalized pseudo-Anosov homeomorphisms. Further, in [**16**] the topological classification is obtained for structurally stable diffeomorphisms of orientable surfaces for which nonwandering sets contain arbitrary one-dimensional basic sets, and in [**19**] a relationship is established between structural stability and asymptotic behavior of the preimages (on the universal covering surface M_ν) of stable and unstable manifolds of points lying on one-dimensional amply situated attractors and repellers.

Now let us state the results. Below we assume that the nonwandering set of an A-diffeomorphism f contains a one-dimensional amply situated attractor Ω (some obvious changes are necessary in the case of one-dimensional repeller). First, let us consider the case $M_\nu = T^2$ (that is, M_ν is the oriented surface of genus $g = 1$). Represent T^2 as the quotient group of $\mathbb{R}^2$ over the lattice Γ of the Gaussian integers: $T^2 = \mathbb{R}^2/\Gamma$. By π we denote the natural projection of $\mathbb{R}^2$ on the quotient group T^2, by $\bar{f}$ denote the diffeomorphism of $\mathbb{R}^2$ that covers f, that is, $\pi\bar{f} = f\pi$, and by f_* the automorphism of Γ defined by the formula $f_*(\gamma) = \bar{f}(\gamma) - \bar{f}(O)$, where O is the origin in $\mathbb{R}^2$ and $\gamma \in \Gamma$. The results formulated below were obtained in [**13**] for the case of orientable attractors. It can easily be seen that amply situated one-dimensional basic sets on the torus are orientable, and thus the results of [**13**]

[2] A nontrivial basic set Ω is called *amply situated* if there are no loops homotopic to zero formed by a segment of the stable manifold and a segment of the unstable manifold of a point $x \in \Omega$. Orientable basic sets are amply situated. However, the converse is not true. If the nonwandering set of an A-diffeomorphism of M_ν contains an amply situated basic set, then the genus of M_ν is at least 1 in the case $\nu = o$ and at least 3 in the case $\nu = n$ [**26**].

hold for amply situated attractors as well. Theorem 1 in [**13**] ensures that the automorphism f_* is hyperbolic, that is, the unimodular integer matrix determining f_* has no eigenvalues with unit absolute value. Let f_0 be an algebraic automorphism of the torus such that $f_{0*} = f_*$. Then Proposition 2.1 in [**10**] implies that there exists a mapping $h\colon T^2 \to T^2$ such that $hf = f_0 h$ and h is homotopic to the identical mapping. By $B(f,h)$ we denote the subset of T^2 such that the preimage $h^{-1}(b)$ contains more than one point for any $b \in B(f,h)$.

Further, Theorem 2 in [**13**] shows that h takes Ω on the whole torus T^2 and that $B(f,h)$ is formed by a finite number of periodic points $p_1, \dots, p_k$ of f_0, and by the unstable manifolds of these points. In this case each $h^{-1}(p_i) \cap \Omega$, $i = 1, \dots, k$, consists of two boundary periodic points of Ω.[3]

By $P(f,h)$ we denote the union of $p_1, \dots, p_k$.

THEOREM 1. *Let Ω and Ω' be amply situated attractors of A-diffeomorphisms $f, f'\colon T^2 \to T^2$. A homeomorphism $g\colon T^2 \to T^2$ such that $g(\Omega) = \Omega'$ and $f'|_{\Omega'} = gfg^{-1}|_{\Omega'}$ exists if and only if there exists a linear mapping $\psi\colon T^2 \to T^2$ such that $f'_0 = \psi f_0 \psi^{-1}$ and $\psi(P(f,h)) = P'(f',h')$, where f_0 and f'_0 are automorphisms of the torus such that $f_{0*} = f_*$ and $f'_{0*} = f'_*$, while h and h' are mappings of the torus homotopic to identity and satisfying the relations $hf = f_0 h$ and $h'f' = f'_0 h'$.*[4]

Now consider an A-diffeomorphism f of M_ν, where $\nu \in \{o, n\}$ and M_o is an oriented closed surface of genus g, $g > 1$, while M_n is a nonoriented closed surface of genus g, $g \geqslant 3$.

Let us regard M_o as the quotient space $\mathbf{U}/F_o$, where $\mathbf{U}$ is the Lobachevsky plane represented by the Poincaré model in the disk $|z| < 1$ of the complex z-plane with the metric of constant negative curvature, and F_o is a discrete group of isometries of $\mathbf{U}$ isomorphic to the fundamental group of M_o. By π_o denote the natural projection of $\mathbf{U}$ on M_o.

For a nonorientable surface M_n, consider a two-fold covering $\pi\colon \widehat{M} \to M_n$, where $\widehat{M}$ is a closed oriented surface of genus $g-1$. Regard $\widehat{M}$ as $\mathbf{U}/\widehat{F}$, where $\widehat{F}$ is a discrete group of orientation preserving hyperbolic shifts of the Lobachevsky plane isomorphic to the fundamental group of $\widehat{M}$. By $\hat{\pi}$ denote the natural projection of U on $\widehat{M}$, and by $\pi_n = \pi\hat{\pi}$ denote the superposition, which is the projection of the universal covering space $\mathbf{U}$ of M_n. Then there exists a group F_n of hyperbolic shifts of $\mathbf{U}$ such that M_n is the quotient space $\mathbf{U}/F_n$ (see, for example, [**22**]). Note that the hyperbolic shifts that are elements of F_n belong to one of two types: the shifts of the first type preserve the orientation of the Lobachevsky plane, and the shifts of the second type change the orientation. By construction, $\widehat{F}$ is a subgroup of F_n of index 2 formed by all elements of F_n preserving the orientation of the Lobachevsky plane; further, the sets of fixed points for any element (except for the identity) of F_n and for the corresponding element of $\widehat{F}$ coincide.[5]

---

[3]A periodic point $p \in \Omega$ is called a *boundary periodic point* of an attractor Ω if one of the connected components of $W^s(p) \setminus p$ does not intersect Ω. Then the other connected component of $W^s(p) \setminus p$ intersects Ω; we shall denote it by $W^s_\infty(p)$. The set of boundary periodic points in Ω is nonempty and finite [**11, 12, 26**]. The periodic points that are not boundary are said to be *interior*.

[4]A transformation $\psi\colon T^2 \to T^2$ is called *linear* if it can be represented as a composition of an algebraic automorphism and a group shift of the torus.

[5]Recall that any nonidentical element of $\widehat{F}$ or F_n has two fixed points, which lie on the absolute E ($E : |z| = 1$); they are called *rational points*.

Consider a diffeomorphism $\bar{f}_\nu\colon \mathbf{U} \to \mathbf{U}$ covering f (i.e., $\bar{f}_\nu$ satisfies the condition $\pi_\nu \bar{f}_\nu = f\pi_\nu$, where $\nu \in \{o, n\}$). By $\bar{f}_{\nu*}$ denote the automorphism of F_ν defined by the formula $\bar{f}_{\nu*}(\gamma) = \bar{f}_\nu \gamma \bar{f}_\nu^{-1}$, where $\gamma \in F_\nu$. By virtue of the results of [**23**], the automorphism $\bar{f}_{\nu*}$ uniquely induces a homeomorphism $\bar{f}_\nu^*$ of the absolute of the Lobachevsky plane that is an extension of $\bar{f}_\nu$ to E.[6]

For $x \in \Omega$, by $w^u(\bar{x}_\nu)$ denote the preimage of $W^u(x)$ in $\mathbf{U}$ that passes through $\bar{x}_\nu \in \mathbf{U}$ (here $\pi_\nu(\bar{x}_\nu) = x$). It was proved in [**11, 12, 27, 28**] that $w^u(\bar{x}_\nu)$ is a nonclosed smooth curve whose boundary consists of exactly two distinct points μ_1^u and μ_2^u that are irrational points of the absolute E.[7]

Following [**11, 12**], we say that the orbits of μ^{u+} and μ^{u-} under the action of F_ν,

$$\mu_1(W^u(x)) = \bigcup_{\gamma \in F_\nu} \gamma(\mu^{u+}), \qquad \mu_2(W^u(x)) = \bigcup_{\gamma \in F_\nu} \gamma(\mu^{u-}),$$

are the *homotopy winding classes* of $W^u(x)$.

Further, a pair (p, q) of boundary periodic points of an attractor Ω is called a *special pair of boundary periodic points* [**11, 12**] if for any preimage $w^u(\bar{p}_\nu)$ of $W^u(p)$ there exists a preimage $w^u(\bar{q}_\nu)$ of $W^u(q)$ (here $\pi_\nu(\bar{p}_\nu) = p$ and $\pi_\nu(\bar{q}_\nu) = q$) such that the curves $w^u(\bar{p}_\nu)$ and $w^u(\bar{q}_\nu)$ have boundary points in common at the absolute. This definition implies that the union $W^u(p) \cup W^u(q)$ is a boundary attainable from inside an open disk continuously embedded in M_ν. So if (p, q) is a special pair of boundary periodic points, then we have $\mu_i(W^u(p)) = \mu_i(W^u(q))$, $i = 1, 2$ (for a suitable indexing).

Diffeomorphisms $f, f'\colon M_\nu \to M_\nu$ are said to be π_1-*conjugate* if there exist covering diffeomorphisms $\bar{f}_\nu$, $\bar{f}'_\nu$ and an automorphism ψ of F_ν such that $\bar{f}'_{\nu*} = \psi \bar{f}_{\nu*} \psi^{-1}$.

THEOREM 2. *Suppose that $f, f'\colon M_\nu \to M_\nu$ are π_1-conjugate diffeomorphisms (i.e., $\bar{f}'_{\nu*} = \psi \bar{f}_{\nu*} \psi^{-1}$) with nonwandering sets containing amply situated attractors Ω and Ω' that do not contain special pairs of boundary periodic points. Let p be a periodic point of Ω. Then a homeomorphism $g\colon M_\nu \to M_\nu$ such that $\bar{g}_{\nu*} = \psi$ (for some covering homeomorphism $\bar{g}_\nu$), $g(\Omega) = \Omega'$, and $f'|_\Omega = gfg^{-1}|_{\Omega'}$ exists if and only if there exists a periodic point $p' \in \Omega'$ of period equal to that of p such that $\mu_i(W'^u(p')) = \psi^*(\mu_j(W^u(p)))$ for some $i, j \in \{1, 2\}$, where ψ^* is the homeomorphism of the absolute induced by ψ.*

THEOREM 3. *Suppose that $f, f'\colon M_\nu \to M_\nu$ are π_1-conjugate diffeomorphisms (that is, $\bar{f}'_{\nu*} = \psi \bar{f}_{\nu*} \psi^{-1}$); suppose that their nonwandering sets contain amply situated attractors Ω and Ω' with the same number of special pairs of boundary periodic points. Then a homeomorphism $g\colon M_\nu \to M_\nu$ such that $\bar{g}_{\nu*} = \psi$ (for some covering homeomorphism $\bar{g}_\nu$), $g(\Omega) = \Omega'$, and $f'|_\Omega = gfg^{-1}|_{\Omega'}$, exists if and only if for any special pair (p, q) of boundary periodic points in Ω there is a special pair (p', q') of boundary periodic points in Ω' such that $\mu_i(W'^u(p')) = \psi^*(\mu_j(W^u(p)))$ for some $i, j \in \{1, 2\}$.*

---

[6]The paper [**23**] deals with the case of a diffeomorphism f of an orientable surface. However, if f is defined on a nonoriented surface, then it follows from the construction that $\bar{f}_{n*}$ preserves the subgroup $\widehat{F}$. Thus $\bar{f}_n$ is the diffeomorphism covering $\hat{f}$ (which is a diffeomorphism of $\widehat{M}$), as well as the covering mapping for f with respect to the two-fold covering π. Then the restriction of $\bar{f}_{n*}$ to $\widehat{F}$ induces a homeomorphism of the absolute that is just the homeomorphism induced by $\bar{f}_{n*}$ [**23**].

[7]A point of the absolute is called *irrational* if it is not a fixed point for any non-identical element of F_ν.

2. Asymptotic behavior of preimages of stable and unstable points of an attractor on the Lobachevsky plane

It was shown by Anosov and Bowen [**2, 9**] that an attractor Ω can be represented as the finite union $\Omega_1 \cup \dots \cup \Omega_m$, $m \geqslant 1$, of connected closed subsets Ω_i such that $f^m(\Omega_i) = \Omega_i$ and $f(\Omega_i) = \Omega_{i+1}$, where $\Omega_{m+1} = \Omega_1$; these subsets are called *C-dense components of Ω*. The set $W^\sigma(x) \cap \Omega_i$, $\sigma \in \{s, u\}$, is dense in Ω_i for any point x of a C-dense component Ω_i, $i = 1, \dots, m$. Further, any component of $W^\sigma(x) \setminus x$ contains a set dense in Ω_i whenever this component does not contain boundary periodic points [**12**].

Let y be a point of Ω. By $L^\sigma(t)$ (where $t \in [0, +\infty)$ and $\sigma \in \{s, u\}$) we denote the ray in M (moreover, $L^\sigma(t) \subset W^\sigma(y)$) containing a set dense in the C-dense component of Ω that contains y. Consider an arbitrary preimage $l^\sigma_\nu(t)$ of $L^\sigma(t)$ in the universal covering space $\mathbf{U}$, that is, $\pi_\nu(l^\sigma_\nu(t)) = L^\sigma(t)$.

LEMMA 1. *As $t \to +\infty$ the preimage $l^\sigma_\nu(t)$ tends to the unique irrational point $\mu(l^\sigma_\nu(t))$ of E.*

PROOF. By Ω_* we denote a C-dense component of Ω that contains points of $L^\sigma(t)$. For a diffeomorphism f of an orientable surface M_o, put $\Omega_{*o} = \Omega_*$. In the case of a nonorientable surface M_n, by Ω_{*n} denote some connected component of the preimage of Ω_* in $\widehat{M}$ under the projection $\pi\colon \widehat{M} \to M_n$. Note that Ω_{*n} is a C-dense component of some attractor of $\hat{f}$ amply situated in $\widehat{M}$, where $\hat{f}$ is the A-diffeomorphism that covers f under the projection π. Let us define $\widetilde{M}_\nu$, $\widetilde{F}_\nu$, and $\tilde{\pi}_\nu$, where $\nu \in \{o, n\}$, by setting $\widetilde{M}_o = M_o$, $\widetilde{M}_n = \widehat{M}$, $\widetilde{F}_o = F_o$, $\widetilde{F}_n = \widehat{F}$, $\tilde{\pi}_o = \pi_o$, and $\tilde{\pi}_n = \hat{\pi}$.

Let p_ν be an interior periodic point of $\Omega_{*\nu}$, where $\nu \in \{o, n\}$. Since the unstable manifold of p_ν is contained in its closure, there exists a point $x_\nu \in \Omega_{*\nu}$ with the following properties:

1) the curve C_ν formed by the arcs $\lambda^u = [p_\nu, x_\nu]^u \subset W^u(p_\nu)$ and $\lambda^s = [x_\nu, p_\nu]^s \subset W^s(p_\nu)$ is a tame closed curve in $\widetilde{M}_\nu$ nonhomotopic to zero;
2) the arc λ^s, respectively λ^u, is transversal to the unstable, respectively stable, manifolds of points of $\Omega_{*\nu}$; moreover, $\operatorname{int} \lambda^u \cap \Omega_{*\nu} \neq \varnothing$ and $\operatorname{int} \lambda^s \cap \Omega_{*\nu} \neq \varnothing$;
3) the intersection indices of the oriented arcs λ^u and λ^s at the points p_ν and x_ν are equal.

We say that the curve C_ν is a *quasitransversal* for stable and unstable manifolds of points belonging to $\Omega_{*\nu}$.

It follows from the properties of coverings and of the group $\widetilde{F}_\nu$ that the preimage of C_ν in $\mathbf{U}$ under the projection $\tilde{\pi}_\nu$ is the countable union of curves $\{c_\nu\}$ with the following properties:

1) the c_ν are non-self-intersecting curves;
2) different curves c_ν and c'_ν do not intersect one another;
3) for each c_ν, there exists a nonidentical element $\gamma_c \in \widetilde{F}_\nu$ preserving c_ν such that an arc $(\bar{z}_\nu, \gamma_c(\bar{z}_\nu)) \subset c_\nu$ contains no congruent points for any $\bar{z}_\nu \in c_\nu$;
4) each curve c_ν has exactly two boundary points on E (which are the attracting and repelling fixed points of γ_c) and divides $\mathbf{U}$ into two connected components;
5) different curves c_ν and c'_ν have no points in common at the absolute E.

The proof of the following statement is similar to that of Lemma 2 in [**19**]: if z_ν belongs to $\Omega_{*\nu} \cap \operatorname{int} \lambda^{\bar\sigma}$, then the intersection $w^\sigma(\bar z_\nu) \cap c_\nu$ consists of the point $\bar z_\nu$ only; here $\bar\sigma = u$ for $\sigma = s$ and $\bar\sigma = s$ for $\sigma = u$, $\bar z_\nu$ is the preimage of z_ν in $\mathbf{U}$ under the projection $\tilde\pi_\nu$ belonging to $w^\sigma(\bar z_\nu) \cap c_\nu$, and $w^\sigma(\bar z_\nu)$ is the preimage of $W^\sigma(z_\nu)$ passing through $\bar z_\nu$.

The ray L^σ intersects the arc $\operatorname{int} \lambda^{\bar\sigma}$ of the quasitransversal C_ν countably many times. Therefore, there is a countable set of curves $\{c_{\nu k}\}$, $k \in Z^+$, in $\mathbf{U}$ such that

1) their boundary points $c^+_{\nu k}$ and $c^-_{\nu k}$, $k \in Z^+$, belong to E;
2) $\tilde\pi_\nu(c_{\nu k}) = C_\nu$;
3) the curve $l^\sigma(t)$ intersects each of the curves $c_{\nu k}$ exactly once as $t \to +\infty$.

By $c_{\nu *}$ we denote the topological limit of the sequence $\{c_{\nu k}\}$. Let us show that $c_{\nu *}$ is a point of E.

Assume the converse. Without loss of generality, it can also be assumed that the sequence $c^+_{\nu k}$ converges to a point $c^+_{\nu *}$ and the sequence $c^-_{\nu k}$ converges to a point $c^-_{\nu *}$. Then there are two cases:

a) all points of the set $c_{\nu *}$ belong to E;
b) some points of $c_{\nu *}$ lie in $\mathbf{U}$.

Consider case a). Since the topological limit of connected sets is connected and the points $c^+_{\nu *}$, $c^-_{\nu *}$ belong to E, we see that $c_{\nu *}$ is an arc lying in E (the points $c^+_{\nu *}$ and $c^-_{\nu *}$ are different by virtue of our assumption). Further, for any point of the absolute, its $\widetilde F_\nu$-orbit is dense in E. Thus there exists an element $\gamma \in \widetilde F_\nu$ such that $\gamma(c^+_{\nu 1})$ is an interior point of $c_{\nu *}$. Then there is a number $N > 1$ such that $c_{\nu N}$ and $\gamma(c_{\nu 1})$ are different curves intersecting one another. This is impossible.

In case b), consider an arbitrary point $\bar z_\nu$ of $c_{\nu *} \cap \mathbf{U}$. Its ϵ-neighborhood, $\epsilon > 0$, intersects a countable set of different curves $c_{\nu k}$ congruent to each other. Then the intersection of a tame closed curve C_ν with an arbitrary neighborhood of $z_\nu = \tilde\pi(\bar z_\nu)$ consists of countably many pairwise different arcs. This is impossible.

Thus the topological limit $c_{\nu *}$ is a certain point of the absolute; denote this point by $\mu(l^\sigma_\nu(t))$.

Let us show that $\mu(l^\sigma_\nu(t))$ is an irrational point of the absolute. Assume the converse. Then the results of [**23**] (see also [**24**]) imply that there exists an element γ of $\widetilde F_\nu$ with unstable fixed point $\mu(l^\sigma_\nu(t))$ such that any nonidentical element of $\widetilde F_\nu$ preserving $\mu(l^\sigma_\nu(t))$ is a power of γ. Put $l^\sigma_{\nu 1} = \gamma(l^\sigma(t))$ and $l^\sigma_{\nu 2} = \gamma^2(l^\sigma(t))$. Since the curves $l^\sigma(t)$, $l^\sigma_{\nu 1}$, and $l^\sigma_{\nu 2}$ have a common boundary point $\mu(l^\sigma_\nu(t)) \in E$, which is the topological limit of the sequence $\{c_{\nu k}\}$, we conclude that there exists an $\widetilde N$, $\widetilde N > 1$, such that the curve $c_{\nu\widetilde N}$ intersects the curves $l^\sigma(t)$, $l^\sigma_{\nu 1}$, and $l^\sigma_{\nu 2}$. By $\bar z_\nu$, $\bar z_{\nu 1}$, and $\bar z_{\nu 2}$ we denote the points of intersection. Consider the domain $D_1 \subset \mathbf{U}$ bounded by the curves $l^\sigma_\nu(t)$, $l^\sigma_{\nu 1}$ and by the arc $(\bar z_\nu, \bar z_{\nu 1}) \subset c^\nu_{\widetilde N}$, and the domain $D_2 \subset \mathbf{U}$ bounded by the curves $l^\sigma_{\nu 1}$, $l^\sigma_{\nu 2}$ and by the arc $(\bar z_{\nu 1}, \bar z_{\nu 2}) \subset c_{\nu\widetilde N}$; we suppose that D_1 and D_2 do not contain $c_{\nu 1}$. Then the choice of γ ensures that D_1 and D_2 contain no curves congruent to $l^\sigma_\nu(t)$ and having a boundary point $\mu(l^\sigma_\nu(t))$.

Since the image of the ray $l^\sigma_\nu(t)$ in M_ν under $\tilde\pi_\nu$ contains a set dense in $\Omega_{*\nu}$, we see that at least one of the arcs $(\bar z_\nu, \bar z_{\nu 1}) \subset c_{\nu\widetilde N}$ and $(\bar z_{\nu 1}, \bar z_{\nu 2}) \subset c_{\nu\widetilde N}$ contains a point $\bar z_{\nu *}$ such that its image under $\tilde\pi_\nu$ lies at the image of the curve $l^\sigma_\nu(t)$. Then there exists a curve $l^\sigma_{\nu *}$ congruent to $l^\sigma_\nu(t)$ that passes through $\bar z_{\nu *}$. The curves l^σ_* and $c_{\nu\widetilde N}$ have at most one intersection point. Therefore, the boundary point of $l^\sigma_{\nu *}$, which belongs to E, coincides with $\mu(l^\sigma_\nu(t))$. The contradiction proves Lemma 1. □

The statements of the following lemma can be proved similarly to Lemmas 3.3 and 3.4 of [**12**] (for a nonorientable surface M_n, we consider the two-fold covering $\widehat{M}$, as in the proof of Lemma 1).

LEMMA 2. *Let $w^u(\bar{x}_\nu)$ be the preimage of $W^u(x)$ in $\mathbf{U}$, $x \in \Omega$, that passes through $\bar{x}_\nu$ (so $\pi_\nu(\bar{x}_\nu) = x$). Then*

1) the curve $w^u(\bar{x}_\nu)$ has exactly two boundary points μ_1^u, μ_2^u, which are irrational points of E;

2) if $W^s(x)$ does not contain a boundary periodic point of Ω, then $w^s(\bar{x}_\nu)$ has exactly two different boundary points μ_1^s, μ_2^s, which are irrational points of E, and the intersection $w^u(\bar{x}_\nu) \cap w^s(\bar{x}_\nu)$ is precisely the point $\bar{x}_\nu$;

3) if $W^s(x)$ contains a boundary periodic point p, then the preimage $w^s_\infty(\bar{p}_\nu)$ of $W^s_\infty(p)$ (here $p = \pi_\nu(\bar{p}_\nu)$) containing the point $\bar{x}_\nu$ has exactly two boundary points; one of them is $\bar{p}_\nu$, and the other, μ^s_∞, is an irrational point of E; moreover, the intersection $w^u(\bar{x}_\nu) \cap w^s(\bar{p}_\nu)$ is precisely the point $\bar{x}_\nu$;

4) the boundary points of $w^u(\bar{x}_\nu)$ are different from the boundary points of $w^s(\bar{x}_\nu)$ (in case 2) and from the boundary point μ^s_∞ of $w^s_\infty(\bar{p}_\nu)$ (in case 3);

5) if $w^u(\bar{x}_\nu)$ contains no preimages of a boundary periodic point and $\bar{y}_\nu$ is a point such that $\pi_\nu(\bar{y}_\nu) \in \Omega$ and $\bar{y}_\nu \notin w^u(\bar{x}_\nu)$, then the curves $w^u(\bar{y}_\nu)$ and $w^u(\bar{x}_\nu)$ do not intersect one another; if $w^u(\bar{x}_\nu)$ contains a preimage $\bar{p}_\nu$ of a boundary periodic point, then there exists a preimage $\bar{q}_{\nu 1}$ different from $\bar{p}_\nu$, of a boundary periodic point, respectively, a preimage $\bar{q}_{\nu 2}$ different from $\bar{p}_\nu$, such that $w^u(\bar{q}_{\nu 1})$, respectively $w^u(\bar{q}_{\nu 2})$, has a boundary point μ_1^u, respectively μ_2^u; in this case the curve $w^u(\bar{y}_\nu)$ has no boundary points in common with the curves $w^u(\bar{x}_\nu)$, $w^u(\bar{q}_{\nu 1})$, and $w^u(\bar{q}_{\nu 2})$ for any point $\bar{y}_\nu$ satisfying the conditions $\pi_\nu(\bar{y}_\nu) \in \Omega$, $\bar{y}_\nu \notin w^u(\bar{x}_\nu)$, $\bar{y}_\nu \notin w^u(\bar{q}_{\nu 1})$, and $\bar{y}_\nu \notin w^u(\bar{q}_{\nu 2})$;[8]

6) if $w^s_*(\bar{x}_\nu)$ and $w^s_*(\bar{y}_\nu)$ are connected components of $w^s(\bar{x}_\nu)\setminus\bar{x}_\nu$ and $w^s(\bar{y}_\nu)\setminus\bar{y}_\nu$ (here $x, y \in \Omega$) such that $w^s(\bar{x}_\nu) \cap w^s(\bar{y}_\nu) = \varnothing$, then the boundary points of $w^s_*(\bar{x}_\nu)$, $w^s_*(\bar{y}_\nu)$ belonging to E are different.

3. Proof of Theorem 2

The necessity of the assumptions of Theorem 2 follows from Lemma 2. In the proof of their sufficiency, we shall use the following remark.

REMARK. Suppose that the assumptions of Theorem 2 hold. Then there exists a diffeomorphism $h\colon M_\nu \to M_\nu$ such that $\bar{h}_{\nu *} = \psi$ for a certain covering diffeomorphism $\bar{h}_\nu$ [**23, 24**]. Put $f'' = hfh^{-1}$, $\Omega'' = h(\Omega)$, and $p'' = h(p)$. For any preimage $\bar{p}''_\nu$ of a periodic point p'' of f'' there exists a preimage $\bar{p}_\nu$ of a periodic point p of f such that $\bar{h}_\nu(\bar{p}_\nu) = \bar{p}''_\nu$. Then $\bar{h}_\nu(w^u(\bar{p}_\nu)) = w''^u(\bar{p}''_\nu)$, and the homeomorphism $\psi^* = \bar{h}^*_\nu$ of the absolute takes the boundary points of $w^u(\bar{p}_\nu)$ to the boundary points of $w''^u(\bar{p}''_\nu)$.

By the assumptions of Theorem 2, ψ^* takes one of the boundary points of $w^u(\bar{p}_\nu)$ to one of the boundary points of $w'^u(\bar{p}'_\nu)$, where $\bar{p}'_\nu$ is a certain preimage of p'. So the curves $w'^u(\bar{p}'_\nu)$ and $w''^u(\bar{p}''_\nu)$ have a common boundary point on the absolute, and thus for some $i, j \in \{1, 2\}$ we have $\mu_i(W'^u(p')) = \mu_i(W''^u(p''))$.

[8]If an attractor Ω contains no special pairs of boundary periodic points, then the preimages $\bar{q}_{\nu 1}$ and $\bar{q}_{\nu 2}$ are different.

Since $f'' = hfh^{-1}$, for covering diffeomorphisms $\bar{h}_\nu$ and $\bar{f}_\nu$ there exists a covering diffeomorphism $\bar{f}''_\nu$ such that $\bar{f}''_\nu = \bar{h}_\nu \bar{f}_\nu \bar{h}_\nu^{-1}$. Then $\bar{f}''_{\nu *} = \bar{h}_{\nu *} \bar{f}_{\nu *} \bar{h}_{\nu *}^{-1}$, and relations $\bar{f}'_{\nu *} = \psi \bar{f}_{\nu *} \psi^{-1}$ and $\bar{h}_{\nu *} = \psi$ imply that $\bar{f}''_{\nu *} = \bar{f}'_{\nu *}$.

Suppose that we have proved the existence of a homeomorphism $g_1 \colon M_\nu \to M_\nu$ homotopic to the identity that satisfies the conditions $g_1(\Omega) = \Omega'$ and $f'|_{\Omega'} = g_1 f'' g_1^{-1}|_{\Omega'}$. Then $g = g_1 h$ is the required homeomorphism. Thus we can assume without loss of generality that the diffeomorphisms f, f' satisfy the conditions

$$(*) \qquad \bar{f}'_{\nu *} = \bar{f}_{\nu *}, \qquad \mu_i(W'^u(p')) = \mu_j(W^u(p)),$$

where $i, j \in \{1, 2\}$ and $\bar{f}'_\nu$, $\bar{f}_\nu$ are certain covering diffeomorphisms.

The automorphisms $\bar{f}'_{\nu *}$ and $\bar{f}_{\nu *}$ uniquely induce the homeomorphisms $\bar{f}'^*_\nu$, $\bar{f}^*_\nu$ of the absolute such that $\bar{f}'^*_\nu = \bar{f}^*_\nu$ (see [**23, 22**]).

Let Ω, respectively Ω', be an attractor of a diffeomorphism f, respectively f'. By $\overline{\Omega}_\nu$, respectively $\overline{\Omega}'_\nu$, we denote the preimage of Ω, respectively of Ω', in the universal covering space $\mathbf{U}$ under the projection π_ν. The proofs of the statements of the following lemma are similar to those in [**12**, §4].

LEMMA 3. *Suppose that the nonwandering sets of diffeomorphisms f, f' contain amply situated attractors Ω, Ω' without special pairs of boundary periodic points. Also suppose that condition $(*)$ holds for some covering diffeomorphisms $\bar{f}'_\nu$ and $\bar{f}_\nu$, $i, j \in \{1, 2\}$, and some periodic points $p \in \Omega$, $p' \in \Omega'$. Then*

1) *for any point $\bar{x}_\nu \in \overline{\Omega}$ there exists a point $\bar{x}'_\nu \in \overline{\Omega}'$ such that the curves $w^u(\bar{x}_\nu)$ and $w'^u(\bar{x}'_\nu)$ have points in common on the absolute;*

2) *if a point $\bar{q}_\nu \in \overline{\Omega}$ is a preimage of an interior periodic point $q \in \Omega$, then there exists a unique point $\bar{q}'_\nu \in \overline{\Omega}'_\nu$ such that the curve $w^u(\bar{q}_\nu)$, respectively $w^s(\bar{q}_\nu)$, has boundary points in common with the curve $w'^u(\bar{q}'_\nu)$, respectively $w'^s(\bar{q}'_\nu)$, and that the image of $\bar{q}'_\nu$ is an interior periodic point $q' \in \Omega'$ having the same period as q;*

3) *if $\bar{q}_\nu \in \overline{\Omega}_\nu$ is a preimage of a boundary periodic point $q \in \Omega$, then there exists a unique point $\bar{q}'_\nu \in \overline{\Omega}'_\nu$ such that the boundary points of the curve $w^u(\bar{q}_\nu)$, respectively of the curve $w^s_\infty(\bar{q}_\nu)$ (which lies in E), are the boundary points of the curve $w'^u(\bar{q}'_\nu)$, respectively of the curve $w'^s_\infty(\bar{q}'_\nu)$, and the image of $\bar{q}'_\nu$ is a boundary periodic point $q' \in \Omega'$ having the same period as q;*

4) *if $\bar{x}_\nu$ is a point of $\overline{\Omega}_\nu$ such that $w^s(\bar{x}_\nu)$ contains no preimages of a boundary periodic point of Ω, then there exists a point $\bar{x}'_\nu \in \overline{\Omega}'_\nu$ such that the boundary points of the curve $w^u(\bar{x}_\nu)$, respectively $w^s(\bar{x}_\nu)$, are the boundary points of the curve $w'^u(\bar{x}'_\nu)$, respectively of $w'^s(\bar{x}'_\nu)$.*

LEMMA 4 (a proof of the sufficiency of the assumptions of Theorem 2). *Suppose that the assumptions of Lemma 3 hold. Then there exists a homeomorphism $g \colon M_\nu \to M_\nu$ such that $g(\Omega) = \Omega'$, $f'|_{\Omega'} = gfg^{-1}|_{\Omega'}$, and g is homotopic to the identity mapping.*

PROOF. First, let us define a mapping $\bar{h}_\nu$ of $\overline{\Omega}_\nu$ to $\overline{\Omega}'_\nu$. Consider a point $\bar{x}_\nu$ of the set $\overline{\Omega}_\nu$. There are two possible cases:

1) $w^s(\bar{x}_\nu)$ contains no preimages of a boundary periodic point of Ω;
2) $w^s(\bar{x}_\nu)$ contains a preimage of a boundary periodic point p of Ω.

In the first case, Lemma 3 ensures that there exists a point $\bar{x}'_\nu \in \overline{\Omega}'_\nu$ such that the boundary points of $w'^u(\bar{x}'_\nu)$, respectively of $w'^s(\bar{x}'_\nu)$, are the boundary points of $w^u(\bar{x}_\nu)$, respectively of $w^s(\bar{x}_\nu)$.

In the second case, consider the curve $w^s_\infty(\bar{p}_\nu)$ containing the curve $w^s_\infty(\bar{x}_\nu)$. Here $w^s_\infty(\bar{x}_\nu)$ is the preimage of the curve $W^s_\infty(x)$ (where $x = \pi_\nu(\bar{x}_\nu)$), which is a connected component of $W^s(x) \setminus x$ contained in the component $W^s_\infty(p)$. By Lemma 3, the preimage of the boundary periodic point p' of Ω' contains a point $\bar{p}'_\nu \in \overline{\Omega}'_\nu$ such that the boundary points of $w'^s_\infty(\bar{p}'_\nu)$ and $w^s_\infty(\bar{p}_\nu)$ lying at E are the same points. In this case, the boundary points of $w'^u(\bar{p}'_\nu)$ coincide with the boundary points of $w^u(\bar{p}_\nu)$. By Lemma 3, there exists a point $\bar{y}'_\nu \in \overline{\Omega}'_\nu$ such that the boundary points of $w'^u(\bar{y}'_\nu)$ coincide with the boundary points of $w^u(\bar{x}_\nu)$. Then there is an intersection point $\bar{x}'_\nu \in \overline{\Omega}'_\nu$ of the curves $w'^s(\bar{p}'_\nu) \cup \bar{p}'_\nu$ and $w'^u(\bar{y}'_\nu)$.

Thus, taking into account Lemma 3, it is easy to see that for any point $\bar{x}_\nu \in \overline{\Omega}_\nu$ we have determined a unique corresponding point $\bar{x}'_\nu \in \overline{\Omega}'_\nu$. Let us denote the constructed mapping by $\bar{h}_\nu$. Combining the construction of $\bar{h}_\nu$, the proximity of stable and unstable manifolds of points of Ω and Ω' on compact sets, and the properties of coverings, we see that $\bar{h}$ is a homeomorphism of $\overline{\Omega}_\nu$ to $\overline{\Omega}'_\nu$ such that $\bar{h}_\nu(\gamma(\bar{x}_\nu)) = \gamma(\bar{h}_\nu(\bar{x}_\nu))$ for $\bar{x}_\nu \in \overline{\Omega}_\nu$, and $\bar{f}'_\nu \bar{h}_\nu|_{\overline{\Omega}_\nu} = \bar{h}_\nu \bar{f}_\nu|_{\overline{\Omega}_\nu}$.

Then there exists a homeomorphism $h\colon \Omega \to \Omega'$ such that $h(\Omega) = \Omega'$ and $f'|_\Omega = hfh^{-1}|_\Omega$.

The results of [**11, 12, 13, 26, 27**] imply that the boundary of $M \setminus \Omega$ attainable from inside has a unique decomposition into $R(\Omega)$ pencils $C_1, \dots, C_{R(\Omega)}$, where Ω is a one-dimensional attractor or repeller. Each pencil C_r of degree n_r of the attractor (here $n_r \geqslant 1$, $r \in \{1, \dots, R(\Omega)\}$) is a union of n_r unstable manifolds $W^u(p_1) \cup \dots \cup W^u(p_{n_r})$ of boundary periodic points $p_1, \dots, p_{n_r}$ of Ω. Moreover, there is a sequence $x_1, \dots, x_{2n_r}$ such that for any $i \in \{1, \dots, n_r\}$ the following conditions hold:

1) x_{2i-1} and x_{2i} lie in different connected components of $W^u(p_i) \setminus p_i$;
2) $x_{2i+1} \in W^s(x_{2i})$ (we put $x_{2n_r+1} = x_1$);
3) $(x_{2i}, x_{2i+1})^s \cap \Omega = \varnothing$ for $i = 1, \dots, n_r$;
4) the curve $L^u_{2i} \cup (x_{2i}, x_{2i+1})^s \cup L^u_{2i+1}$ is a boundary of D_i attainable from inside. Here D_i is the open unit disk immersed in M continuously, L^u_{2i}, respectively L^u_{2i+1}, is a connected component of $W^u(x_{2i}) \setminus x_{2i}$, respectively of $W^u(x_{2i+1}) \setminus x_{2i+1}$, that does not contain p_i, respectively p_{i+1}.

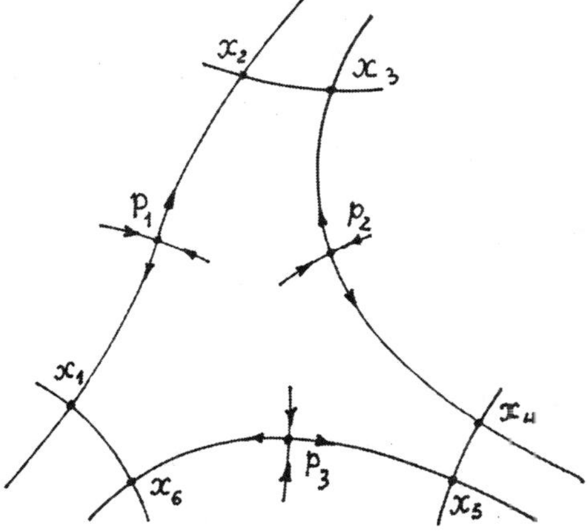

FIGURE 1. A pencil of degree 3.

For each $i \in \{1, \dots, n_r\}$, choose a pair $\tilde{x}_{2i-1}$, $\tilde{x}_{2i}$ and a tame curve l_i with the endpoints $\tilde{x}_{2i-1}$ and $\tilde{x}_{2i}$ so that

1) $\tilde{x}_{2i-1} \in W^s(x_{2i-1})$ and $\tilde{x}_{2i} \in W^s(x_{2i})$;
2) $(\tilde{x}_{2i}, \tilde{x}_{2i+1})^s \subset (x_{2i}, x_{2i+1})^s$ (we put $\tilde{x}_{2n_r+1} = \tilde{x}_1$);
3) l_i intersects the stable manifold of any point of the arc $(x_{2i-1}, x_{2i})^u$ exactly one time, and this intersection is transversal;
4) $L_r = \bigcup_{i=1}^{i=n_r} l_i \cup (\tilde{x}_{2i}, \tilde{x}_{2i+1})^s$ is a tame closed curve;
5) any pencil C_r, as well as any curve L_r, is a boundary of K_r attainable from inside, where K_r is an open annulus immersed in M_ν continuously and consists of wandering points of f.

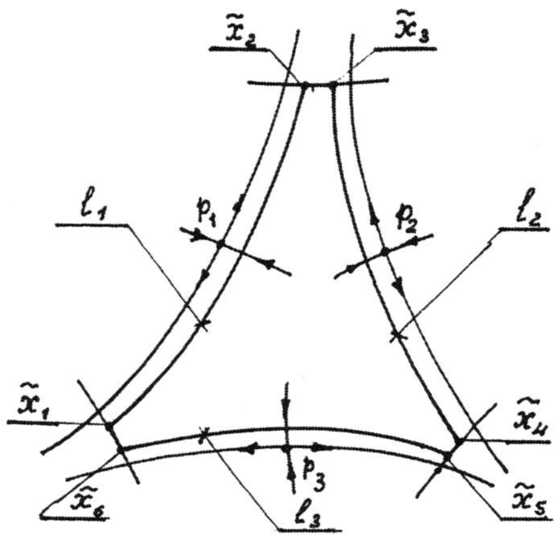

FIGURE 2. The construction of the curve L_r for a pencil of degree 3.

Put

$$K = \bigcup_{r=1}^{r=R(\Omega)} K_r, \qquad L = \bigcup_{r=1}^{r=R(\Omega)} L_r.$$

By $\overline{K}_\nu = \pi_\nu^{-1}(K)$ and $\overline{L}_\nu = \pi_\nu^{-1}(L)$ we denote the preimages of K and L in $\mathbf{U}$. Then $\overline{K}_\nu$ is a countable union of connected components whose images are domains in K. Moreover, there are exactly $R(\Omega)$ pairwise noncongruent components $\bar{K}_1, \dots, \bar{K}_{R(\Omega)}$ such that

$$\overline{K}_\nu = \bigcup_{\gamma \in F_\nu} \gamma(\bar{K}_1 \cup \dots \cup \bar{K}_{R(\Omega)}).$$

The boundary of each component $\bar{K}_r$, $r = 1, \dots, R(\Omega)$, is formed by the closures of the preimages of the unstable manifolds of boundary periodic points of Ω belonging to exactly one of the C_r, and by the closure of a curve $l_{\nu r} \subset \overline{L}_\nu$ such that $\pi_\nu(l_{\nu r}) = L_r$. Divide the set $\{1, \dots, R(\Omega)\}$ of the indices r into two parts R_1 and R_2 as follows: $r \in R_1$ if L_r is homotopic to zero in M_ν, and $r \in R_2$ otherwise.

If $r \in R_1$, then $l_{\nu r}$ and L_r are closed curves homotopic to zero in $\mathbf{U}$.

If $r \in R_2$ (i.e., if L_r is not homotopic to zero in M_ν), then $l_{\nu r}$ has exactly two boundary points on the absolute, which are fixed points for some element $\gamma_r \in F_\nu$ preserving $l_{\nu r}$ such that the arc $\xi_r = (\bar{x}_\nu, \gamma_r(\bar{x}_\nu)) \subset l_{\nu r}$ contains no congruent points for any $\bar{x}_\nu \in l_{\nu r}$.

If $r \in R_1$, then the component $\bar{K}_r$ contains no congruent points and its boundary consists of the curve $l_{\nu r}$ and of exactly n_r pairwise noncongruent curves that are the preimages of the unstable manifolds of boundary periodic points belonging to the pencil C_r. Then put $\bar{K}_r^0 = \bar{K}_r$ (see Figure 3a).

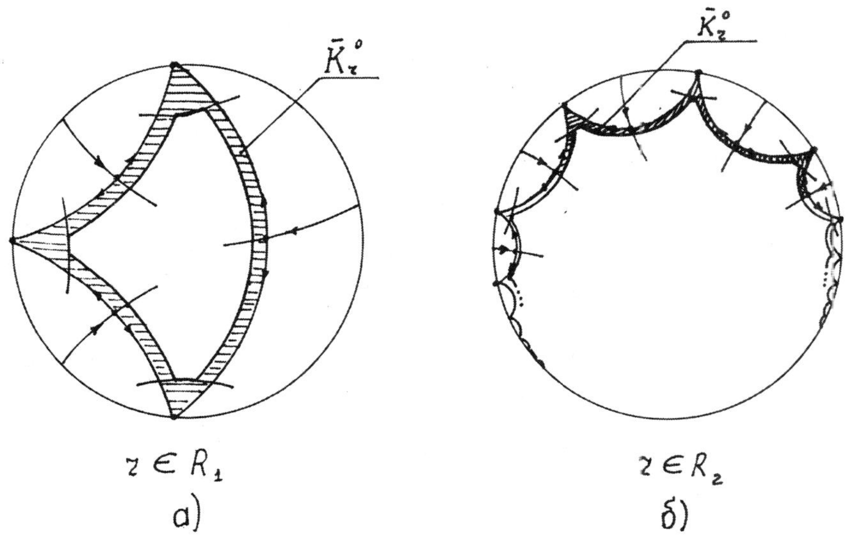

FIGURE 3

If $r \in R_2$, consider a point $\bar{p}_{\nu r}$ on the boundary of the component $\bar{K}_r$ that is a preimage of a boundary periodic point of Ω. Choose a point $\bar{x}_{\nu r} \in l_{\nu r}$ and a tame curve $\lambda_r \in \bar{K}_r$ with endpoints $\bar{p}_{\nu r}$ and $\bar{x}_{\nu r}$ and without congruent points. Put $\tilde{\lambda}_r = \gamma_r(\lambda_r)$. By $\bar{K}_r^0$ we denote the domain in $\bar{K}_r$ bounded by the curves λ_r and $\tilde{\lambda}_r$, by the arc $\xi_r = (\bar{x}_\nu, \gamma_r(\bar{x}_\nu)) \subset l_{\nu r}$, and by the connected component of the boundary of $\bar{K}_r$ containing $\bar{p}_r$ and $\gamma(\bar{p}_r)$. By construction, $\bar{K}_r^0$ contains no congruent points, and $\bar{K}_r = \bigcup_{n \in \mathbb{Z}} \gamma_r^n(\bar{K}_r^0)$ (see Figure 3b).

Since $\Omega' = h(\Omega)$, for any pencil C_r of Ω there is a pencil C_r' such that $C_r' = h(C_r)$. For any C_r', by L_r' denote a curve constructed in a similar way and having properties similar to those of L_r. By K_r' we denote an open annulus continuously immersed in M_ν so that the union of C_r' and L_r' is a boundary of K_r' attainable from inside. Put

$$K' = \bigcup_{r=1}^{r=R(\Omega)} K_r', \qquad L' = \bigcup_{r=1}^{r=R(\Omega)} L_r'.$$

By $\overline{K}_\nu'$ and $\overline{L}_\nu'$ we denote the preimages $\pi_\nu^{-1}(K')$ and $\pi_\nu^{-1}(L')$ in $\mathbf{U}$.

Put $\bar{p}_{\nu r}' = \bar{h}_\nu(\bar{p}_{\nu r})$. By $\bar{K}_r'$ we denote the connected component of $\overline{K}_\nu'$ whose boundary contains $\bar{p}_{\nu r}'$. By construction, $\bar{h}_\nu$ has the following property. Choose different points $\bar{x}_\nu$ and $\bar{y}_\nu$ in $\overline{\Omega}$ so that $\bar{y}_\nu \in w^s(\bar{x}_\nu)$ and $(\bar{x}_\nu, \bar{y}_\nu)^s \cap \overline{\Omega}_\nu = \varnothing$. Then $\bar{h}_\nu(\bar{y}) \in w'^s(\bar{h}_\nu(\bar{x}))$ and $(\bar{h}_\nu(\bar{x}), \bar{h}_\nu(\bar{y}_\nu))^s \cap \overline{\Omega}_\nu' = \varnothing$. Thus the homeomorphism $\bar{h}_\nu$ takes the union of the preimages of the unstable manifolds of the boundary periodic points of Ω belonging to the boundary of $\bar{K}_r$ to the union of the preimages of the unstable manifolds of the boundary periodic points of Ω' belonging to the boundary of $\bar{K}_r'$. By $l_{\nu r}'$ we denote the curve in the boundary of $\bar{K}_{\nu r}'$ such that the image of $l_{\nu r}'$ is L_r'.

If L_r is homotopic to zero in M_ν (that is, if $r \in R_1$ and the boundary of $\bar K_{\nu r}$ contains exactly n_r curves that are preimages of unstable manifolds belonging to C_r), then it follows from Lemma 3 that the boundary of $\bar K'_{\nu r}$ also contains exactly n_r curves that are preimages of unstable manifolds of boundary periodic points belonging to the pencil C'_r of the set Ω'. Thus the component $\bar K'_r$, as well as $\bar K_r$, contains no congruent points, and therefore, $l'_{\nu r}$ is a closed curve homotopic to zero. In this case we put $\bar K'^0_r = \bar K'_r$.

If the curve L_r is not homotopic to zero in M_ν (that is, if $r \in R_2$), then by construction the boundary points of the curve $l'_{\nu r}$ at the absolute are the same as of the curve $l_{\nu r}$; further, $l'_{\nu r}$ is invariant under the element γ'_r, and the arc $\xi'_r = [\bar x'_{\nu r}, \gamma_r(\bar x'_{\nu r})] \in l'_r$ contains no congruent points other than $\bar x'_{\nu r}, \gamma_r(\bar x'_{\nu r})$ (so the curve L'_r also is not homotopic to zero in M_ν and is homotopic to L_r). Let us determine from $\bar p'_{\nu r}$ a point $\bar x'_{\nu r} \in l'_{\nu r}$ and a tame curve $\lambda'_r \in \bar K'_r$ without congruent points and with the endpoints $\bar p'_{\nu r}$ and $\bar x'_{\nu r}$. Put $\tilde\lambda'_r = \gamma_r(\lambda'_r)$. By $\bar K'^0_r$ we denote the domain in $\bar K'_r$ bounded by the curves λ'_r and $\tilde\lambda'_r$, by the arc ξ'_r, and by the connected component of the boundary of $\bar K'_r$ that contains $\bar p'_r$ and $\gamma(\bar p'_r)$. By construction, $\bar K'^0_r$ contains no congruent points, and $\bar K'_r = \bigcup_{n\in\mathbb{Z}} \gamma_r^n(\bar K'^0_r)$.

In the case $r \in R_1$, choose the orientations of the curves $l_{\nu r}$ and $l'_{\nu r}$ so that the domains $\bar K^0_r$ and $\bar K'^0_r$ are to the left of these curves, and fix an orientation-preserving homeomorphism $\phi_{\nu r}\colon l_{\nu r} \to l'_{\nu r}$.

In the case $r \in R_2$, choose a homeomorphism $\phi_{\nu r}\colon \xi_r \cup \lambda_r \to \xi'_r \cup \lambda'_r$ so that

$$\phi_{\nu r}(\xi_r) = \xi'_r, \quad \phi_{\nu r}(\lambda_r) = \lambda'_r, \quad \phi_\nu(\bar p_{\nu r}) = \bar p'_{\nu r},$$
$$\phi_\nu(\bar x_{\nu r}) = \bar x'_{\nu r}, \quad \phi_\nu(\gamma_r(\bar x_{\nu r})) = \gamma_r(\phi_\nu(\bar x_{\nu r})).$$

Let us define a homeomorphism $\varphi_{\nu r}\colon \tilde\lambda_r \to \tilde\lambda'_r$ as follows. For $\bar z_\nu \in \tilde\lambda_r$, put $\varphi_{\nu r}(\bar z_\nu) = \gamma_r\phi_\nu(\gamma_r^{-1}(\bar z))$. By $\overline{\bar K^0_r}$ and $\overline{\bar K'^0_r}$ we denote the closures in $\mathbf{U}$ of the components $\bar K^0_r$ and $\bar K'^0_r$.

By Σ, respectively Σ', we denote the boundary of $\overline{\bar K^0_r}$, respectively of $\overline{\bar K'^0_r}$. Then define a homeomorphism $\delta_{\nu r}\colon \Sigma \to \Sigma'$ as follows.

If $r \in R_1$ and $\bar z_\nu \in \Sigma$, then put

$$\delta_{\nu r}(\bar z_\nu) = \begin{cases} \bar z_\nu & \text{if } \bar z_\nu \in E, \\ \bar h_\nu(\bar z_\nu) & \text{if } \bar z_\nu \in \overline{\Omega}_\nu, \\ \phi_{\nu r}(\bar z_\nu) & \text{if } \bar z_\nu \in l_r. \end{cases}$$

If $r \in R_2$ and $\bar z_\nu \in \Sigma$, then put

$$\delta_{\nu r}(\bar z_\nu) = \begin{cases} \bar z & \text{if } \bar z_\nu \in E, \\ \bar h_\nu(\bar z_\nu) & \text{if } \bar z_\nu \in \overline{\Omega}_\nu, \\ \phi_{\nu r}(\bar z_\nu) & \text{if } \bar z_\nu \in \xi_r \cup \lambda_r, \\ \varphi_{\nu r}(\bar z_\nu) & \text{if } \bar z_\nu \in \tilde\lambda_r. \end{cases}$$

Further, choose a homeomorphism $\Delta_{\nu r}\colon \overline{\bar K^0_r} \to \overline{\bar K'^0_r}$ so that $\Delta_{\nu r}|_\Sigma = \delta_{\nu r}|_\Sigma$. By $\overline{\overline{K}_\nu \cup \overline{L}_\nu}$, respectively by $\overline{\overline{K}'_\nu \cup \overline{L}'_\nu}$, we denote the closure of $\overline{K}_\nu \cup \overline{L}_\nu$, respectively of $\overline{K}' \cup \overline{L}'$. Then define a mapping $\bar\eta\colon \overline{\overline{K}_\nu \cup \overline{L}_\nu} \to \overline{\overline{K}'_\nu \cup \overline{L}'_\nu}$ as follows.

If $\bar z_\nu \in \overline{\overline{K}_\nu \cup \overline{L}_\nu}$, then put $\bar\eta(\bar z_\nu) = \beta_r^{-1}(\Delta_r(\beta_r(\bar z_\nu)))$, where β_r is an element of F_ν such that $\beta_r(\bar z_\nu) \in \overline{\bar K^0_r}$ for some $r \in \{1, \dots, R(\Omega)\}$. It can easily be seen that

1) $\bar\eta|_{\overline{\overline{K}_\nu \cup \overline{L}_\nu} \cap \overline{\Omega}_\nu} = \bar h|_{\overline{\overline{K}_\nu \cup \overline{L}_\nu} \cap \overline{\Omega}_\nu}$;

2) $\bar{\eta}(\gamma(\bar{z})) = \gamma(\bar{\eta}(\bar{z}))$ for $\gamma \in F_\nu$.

Thus there exists a homeomorphism $\eta\colon \Omega \cup K \cup L \to \Omega' \cup K' \cup L'$ such that $\eta\pi_\nu = \pi_\nu\eta$, $\eta(\Omega) = \Omega'$, and $f'|_{\Omega'} = \eta f \eta^{-1}|_{\Omega'}$.

By construction, η takes each closed curve L_r in L to a curve L'_r that is homotopic to L_r and lies in L'. Then it can be proved (similarly to Lemma 6.3 of [**12**]) that there exists a homeomorphism $g\colon M_\nu \to M_\nu$ homotopic to the identity and such that $g|_{\Omega\cup K\cup L} = \eta|_{\Omega\cup K\cup L}$. A homeomorphism with these properties is the required one. This completes the proof of Lemma 4. □

The proof of Theorem 3 is omitted, because it is obtained by a simple modification of the proof of Theorem 2.

References

[1] D. V. Anosov, *Flows on closed surfaces and behavior of trajectories lifted to the universal covering plane*, J. Dynam. Control Systems **1** (1995), no. 1, 125–138.

[2] ———, *On a class of invariant sets of smooth dynamical systems*, Proc. of the fifth international conference on nonlinear oscillations, vol. 2, Quantitative methods, Mathematics Institute of the Academy of Sciences of Ukrainian SSR, (1970), 39–45. (Russian)

[3] S. Kh. Aranson and V. Z. Grines, *On invariants of dynamical systems on two-dimensional manifolds (necessary and sufficient conditions of topological equivalency of transitive dynamical systems)*, Mat. Sb. **90** (1973), no. 3, 372–402; English transl. in Math. USSR-Sb.

[4] ———, *Topological classification of flows on closed two-dimensional surfaces*, Uspekhi Mat. Nauk **41** (1986), no. 1, 149–169; English transl. in Russian Math. Surveys **41** (1986).

[5] ———, *Flows on two-dimensional manifolds*, Encyclopaedia of Mathematical Sciences, vol. 1, Springer-Verlag, Berlin, 1988, pp. 219–233.

[6] ———, *Topological classification of cascades on closed two-dimensional manifolds*, Uspekhi Mat. Nauk **45** (1990), no. 4, 3–32; English transl. in Russian Math. Surveys **45** (1990).

[7] ———, *Cascades on surfaces*, Encyclopaedia of Mathematical Sciences, Vol. 66, Dynamical Systems IX. Springer-Verlag, Berlin and Heidelberg, 1995, pp. 141–175.

[8] ———, *Dynamical systems with minimal entropy on two-dimensional manifolds*, Selecta Math. Soviet **2** (1992), no. 2, 123–158.

[9] R. Bowen, *Periodic points and measures for axiom A diffeomorphisms*, Trans. Amer. Math. Soc. **154**, (1971), 337–397.

[10] J. Franks, *Anosov diffeomorphisms on tori*, Trans. Amer. Math. Soc. **145**, (1969), 117–124.

[11] V. Z. Grines, *On topological conjugacy of diffeomorphisms of a two-dimensional manifold on one-dimensional basic sets*, Uspekhi Mat. Nauk **29 (180)** (1974), no. 6, 163–164. (Russian)

[12] ———, *On topological conjugacy of diffeomorphisms of a two-dimensional manifold on orientable one-dimensional basic sets.* I, Trudy Moskov. Mat. Obshch. **32** (1975), 35–61; English transl. in Trans. Moscow Math. Soc. **32** (1977).

[13] ———, *On topological conjugacy of diffeomorphisms of a two-dimensional manifold on orientable one-dimensional basic sets.* II, Trudy Moskov. Mat. Obshch. **34** (1977), 243–252; English transl. in Trans. Moscow Math. Soc. **1978**, issue 2.

[14] ———, *Topological classification of Morse–Smale diffeomorphisms of surfaces with a finite number of heteroclinic trajectories*, Mat. Zametki **54** (1993), no. 3, 3–17; English transl. in Math. Notes **54** (1993).

[15] ———, *Topological classification of structurally stable diffeomorphisms of surfaces with one-dimensional attractors and repellers*, Mat. Model. **7** (1995), no. 5, 46–47.

[16] ———, *On topological classification of structurally stable diffeomorphisms of surfaces with one-dimensional attractors and repellers*, Mat. Sb. **188** (1997) no. 4, 57–94; English transl. in Russian Acad. Sci. Sb. Math.

[17] ———, *Topological classification of one-dimensional attractors and repellers of A-diffeomorphisms of surfaces by means of automorphisms of the fundamental groups of their supports*, Itogi Nauki i Tekhniki. Sovremennye Problemy Matematiki. Fundamental′nye Napravleniya, VINITI, Moscow, 1998; English transl. in J. Math. Sci. **47** (1998).

[18] ———, *Representation of one-dimensional attractors of A-diffeomorphisms of surfaces by hyperbolic homeomorphisms*, Mat. Zametki **62** (1997), no. 1, 76–87; English transl. in Math. Notes **62** (1997).

[19] ———, *Structural stability and asymptotic behavior of invariant manifolds of A-diffeomorphisms of surfaces*, J. Dynam. Control Systems **3** (1997), no. 1, 91–110.

[20] V. Z. Grines and Kh. Kh. Kalai, *On topological equivalence of diffeomorphisms of two-dimensional manifolds with nontrivial basic sets*, Methods in Qualitative Theory of Differential Equations (E. A. Leontovich-Andronova, ed.), Gor′kov. Gos. Univ., Gor′kiĭ, 1988, pp. 40–49. (Russian)

[21] ———, *Topological classification of basic sets without pairs of conjugate points for A-diffeomorphisms of surfaces*, VINITI, preprint No. 1137-V, Gor′kiĭ, 1988. (Russian)

[22] W. Mangler, *Die Klassen topologischer Abbildungen einer geschlossenen Flache auf sich*, Math. Z. **44** (1939), 541–553.

[23] J. Nielsen, *Untersuhungen zur Topologie der geshlosenen zweiseitigein Flachen.* I, Acta Math. **50** (1927), 189–358; II, Acta Math. **53** (1929), 1–76; III, Acta Math. **58** (1932), 87–167.

[24] ———, *Surface transformation classes of algebraically finite type*, Det. Kgl. Dansk Videnskaternes Selskab. Math.-Phys. Meddelerser. **21** (1944), 1–89.

[25] R. V. Plykin, *On topology of basic sets of the Smale diffeomorphisms*, Mat. Sb. **84** (1971), no. 2, 301–312; English transl. in Math. USSR-Sb. **13** (1971).

[26] ———, *Sources and sinks of A-diffeomorphisms of surfaces*, Mat. Sb. **94** (1974), no. 2, 243–264; English transl. in Math. USSR-Sb. **23** (1974).

[27] ———, *On the geometry of hyperbolic attractors of smooth cascades*, Uspekhi Mat. Nauk **39** (1984), no. 6 (240), 75–113; English transl. in Russian Math. Surveys **39** (1984).

[28] ———, *Geometry and topology of hyperbolic attractors of diffeomorphisms*, PhD Thesis, Moscow, 1984. (Russian).

[29] S. Smale, *Differentiable dynamical systems*, Bull. Amer. Math. Soc. **73** (1967), no. 6, 747–817.

[30] A. Yu. Zhirov, *Hyperbolic attractors of diffeomorphisms of orientable surfaces. 1. Coding, classification, and coverings*, Mat. Sb. **185** (1994), no. 6, 3–50; 2. *Enumeration and applications to pseudo-Anosov diffeomorphisms*, Mat. Sb. **185** (1994), no. 9, 29–80; 3. *Classification algorithm*, Mat. Sb. **186** (1995), no. 9, 59–82; English transl. in Russian Acad. Sci. Sb. Math., **82** (1995); **83** (1995); **85** (1996).

[31] A. Yu. Zhirov and R. V. Plykin, *Correspondence between one-dimensional hyperbolic attractors of diffeomorphisms of surfaces and generalized pseudo-Anosov diffeomorphisms*, Mat. Zametki **58** (1995), no. 1, 149–152; English transl. in Math. Notes **58** (1995).

Nizhny Novgorod State Agricultural Academy, Gagarin Ave., 97, Nizhny Novgorod, Russia

E-mail address: grines@unn.ac.ru

Institute of Atomic Energy, Obninsk, Russia

Translated by S. S. ANISOV

Amer. Math. Soc. Transl.
(2) Vol. **200**, 2000

A Proof of Shilnikov's Theorem for C^1-Smooth Dynamical Systems

Mikhail V. Shashkov and Dmitry V. Turaev

ABSTRACT. Dynamical systems with a homoclinic loop to a saddle equilibrium state are considered. Andronov and Leontovich showed (see [**1, 3**]) that a generic bifurcation of a two-dimensional C^1-smooth dynamical system with a homoclinic loop leads to the appearance of a unique periodic orbit. Shilnikov [**14, 15, 18**] proved that in the case of dynamical systems of sufficiently high smoothness, this result holds true in the multidimensional setting if some additional conditions are satisfied. In the present paper we give a proof of the Shilnikov theorem for dynamical systems in C^1.

1. Main theorem

Let us consider a family of C^1-smooth vector fields X_μ on an $(n+1)$-dimensional manifold. We assume that the vector field X_μ and its first derivatives depend on μ continuously. Let the following hold.

(A) *The system X_μ has a saddle equilibrium state O, and the roots $\lambda_n, \dots, \lambda_1, \gamma$ of the characteristic equation of the linearized system at the point O for $\mu = 0$ satisfy the inequalities* $\operatorname{Re}\lambda_n \leqslant \cdots \leqslant \operatorname{Re}\lambda_1 < 0 < \gamma$.

Thus, we can introduce local coordinates (x, y) ($x \in \mathbb{R}^n$, $y \in \mathbb{R}^1$) in a small neighborhood of O so that the system X_μ takes the following form near O for $\mu = 0$

$$\begin{cases} \dot{x} = \Lambda x + \dots, \\ \dot{y} = \gamma y + \dots . \end{cases} \tag{1.1}$$

Here Λ is an $(n \times n)$-matrix with the eigenvalues $\{\lambda_1, \dots, \lambda_n\}$; the dots stand for nonlinearities.

The unstable manifold W^u of O is one-dimensional (it is tangent to the y-axis at O) and consists of three orbits: the point O itself and two *separatrices* leaving O in opposite directions. The stable manifold W^s is n-dimensional; it divides a small neighborhood of the equilibrium into two parts: U^+ and U^- (see Figure 1). Assume that

(B) *for $\mu = 0$ one of the separatrices Γ is homoclinic to O, i.e.,* $\Gamma \subset (W^s \cap W^u)$.

2000 *Mathematics Subject Classification.* Primary 37C29.

This research was supported by grants RFBR 99-01-00231 and INTAS 97-804.

©2000 American Mathematical Society

Without loss of generality we assume that the separatrix Γ leaves the point O towards the region U^+ (i.e., towards positive y, see Figure 1).

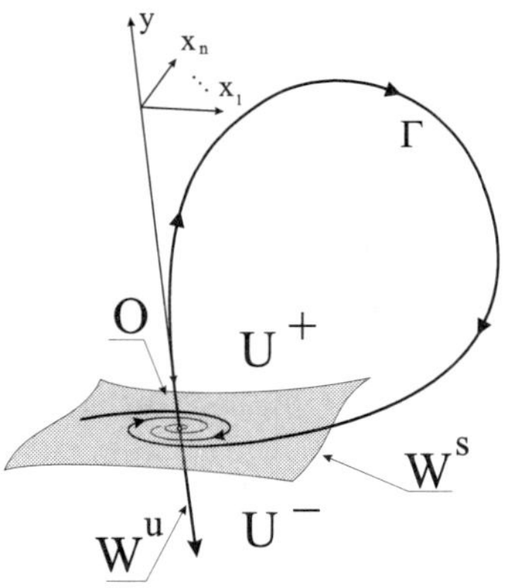

FIGURE 1. The system X_0 has a homoclinic orbit Γ to the saddle equilibrium O. The stable manifold W^s divides a small neighborhood of O into two regions: U^+ and U^-.

We consider the behavior of orbits in a small neighborhood U of the homoclinic loop $\mathcal{L} = O \cup \Gamma$.

For systems on the plane ($n = 1$) this problem was completely solved by Andronov and Leontovich [**1, 2, 3**] (see also [**4**]). In particular, it was shown that if the saddle value $\sigma = \lambda_1 + \gamma$ is nonzero, then bifurcations of the homoclinic loop produce only one periodic orbit. Thus, the bifurcation of such a homoclinic loop was proved to be one of the four main bifurcations of the birth of a limit cycle on a plane.

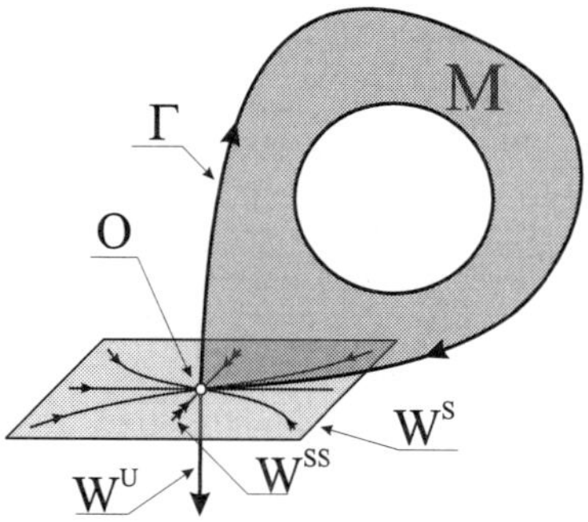

FIGURE 2. A two-dimensional invariant manifold M exists near the homoclinic loop $\mathcal{L} = O \cup \Gamma$ if and only if the leading eigenvalue λ_1 is real and simple, the loop does not lie in the strong stable manifold W^{ss} and some additional transversality conditions are fulfilled.

A similar multidimensional problem was considered by Shilnikov [**15**]. From the modern point of view, one should immediately obtain a result similar to the two-dimensional one in the case where a smooth, normally-hyperbolic two-dimensional invariant manifold exists near the homoclinic loop (see Figure 2). However, the existence of such a manifold requires some extra conditions. First, the negative eigenvalue λ_1 nearest to the imaginary axis must be real and simple. The orbit Γ should not lie in the strong stable submanifold W^{ss} that corresponds to the eigenvalues $\lambda_n, \dots, \lambda_2$. Moreover, some transversality conditions must be satisfied by the flow map near Γ (see [**21, 23, 11, 12, 7**], [**10**] (this also includes the PDE case), [**13**]).

In fact, the existence of a two-dimensional invariant manifold is not so relevant to the dynamics near a homoclinic loop. It was a remarkable discovery of Shilnikov [**16, 19**] that if the characteristic exponents at the point O satisfy a condition which reads in our case as $\operatorname{Im}\lambda_1 \neq 0$, $-\operatorname{Re}\lambda_1 < \gamma$, then generically there exist nontrivial hyperbolic sets in a small neighborhood of the loop. In other words, the dynamics near a homoclinic loop to a saddle-focus with positive saddle value is quite opposite to that in dimension two. As of today, the Shilnikov homoclinic loop is a model of chaotic behavior, which is very simple to describe and which has a very complicated dynamics.

On the other hand, in the case of a negative saddle value, i.e., if

(C) $\sigma = \operatorname{Re}\lambda_1 + \gamma < 0$,

the bifurcation of the homoclinic loop leads to the appearance of only one stable periodic orbit, exactly as for the systems on the plane, no matter what the equilibrium state O is—a saddle or a saddle-focus [**15**].

In the present paper we give a proof of the corresponding result for C^1-smooth systems. In order to describe bifurcations of X_μ, we introduce the small parameter μ as described below. Namely, we suppose that

(D) *the separatrix* Γ *does not belong to* W^s *if* $\mu \neq 0$.

It follows from continuity with respect to μ that after leaving a small neighborhood of O, the separatrix Γ for $\mu \neq 0$ stays close to the locus of the homoclinic loop $\mathcal{L}$ until it enters the small neighborhood of O once again. Without loss of generality we assume that Γ enters U^+ at $\mu > 0$ and U^- at $\mu < 0$.

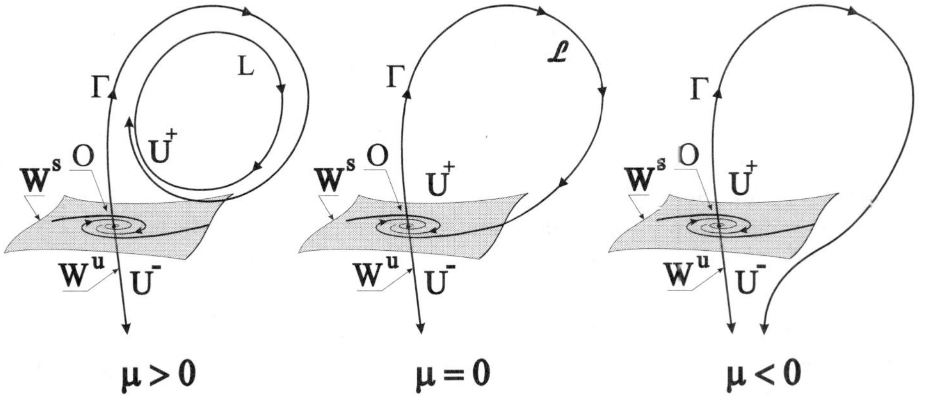

FIGURE 3. At $\mu > 0$, a stable periodic orbit L is born from the loop $\mathcal{L}$ ($\mu = 0$). All the orbits (except for those tending to O) leave at $\mu < 0$.

THEOREM 1.1 (see Figure 3). *If conditions* (A)–(D) *are fulfilled, then there exists a small neighborhood* U *of the homoclinic loop such that at all small* $\mu > 0$ *the* *system has a unique periodic orbit L, which is stable and, in particular, the separatrix Γ tends to L as $t \to +\infty$. The other orbits in U that do not lie in W^s either tend to L or leave U in finite time. For $\mu = 0$ the periodic orbit becomes a homoclinic loop (which may attract some orbits of $U \setminus W^s$, the other orbits leaving U). For all small $\mu < 0$ all orbits of $U \setminus W^s$ leave U in finite time.*

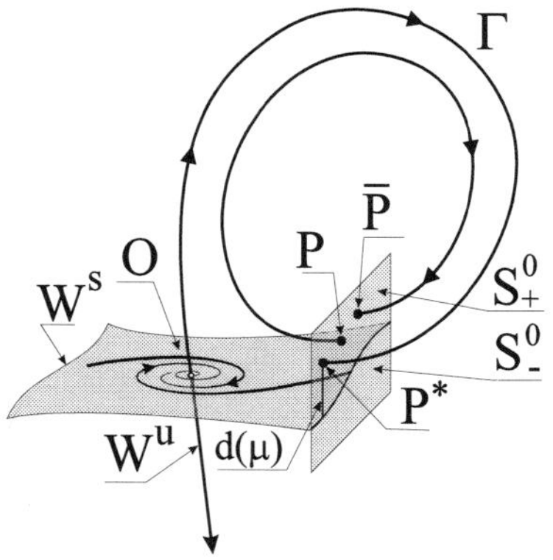

FIGURE 4. The Poincaré map $T\colon S^0_+ \to S^0$ is defined near $S^0 \cap W^s$. The image $T(W^s \cap S^0) = P^*_\mu$ is defined by continuity. The point $P^*_\mu = \Gamma \cap S^0$ lies at the distance $|d(\mu)|$ from $S^0 \cap W^s$.

PROOF. We follow the lines of the original proof in [**15**]. Take a small cross-section S^0 of the stable manifold W^s so as to intersect the homoclinic loop at $\mu = 0$. The stable manifold of O divides S^0 into two regions: $S^0_+ = S^0 \cap U_+$ and $S^0_- = S^0 \cap U_-$ (i.e., S^0_+ lies above W^s; see Figure 4). Let P^*_μ be the intersection point $\Gamma \cap S^0$. For $\mu = 0$ the separatrix Γ forms a homoclinic loop, so $P^*_0 \in \{S^0 \cap W^s\}$. Thus, the intersection point exists for all small μ. Let $d(\mu)$ be the distance from P^*_μ to $W^s \cap S^0$, taken with the positive sign when $P^*_\mu \in S^0_+$ and the negative one when $P^*_\mu \in S^0_-$. By assumption (D), the sign of $d(\mu)$ coincides with the sign of μ (Figure 4).

An orbit which starts at a point $P \in S^0_+$ goes near the stable manifold in a small neighborhood of O and then leaves the neighborhood staying close to the separatrix Γ. If μ is sufficiently small, then moving along Γ, such an orbit intersects S^0 again at some point $\bar{P}$ near the point P^*_μ. Thus, the Poincaré map $T\colon P \mapsto \bar{P}$ is defined on S^0_+ in a neighborhood of W^s. On $W^s \cap S^0$ the map T is defined by continuity: $T(W^s \cap S^0) = P^*_\mu$. The orbits which start on S^0_- leave a small neighborhood of O close to the other separatrix and, therefore, they leave the neighborhood U of the homoclinic loop under consideration. Thus, the Poincaré map T is not defined on S^0_-.

Shilnikov proved in [**15**] that if the saddle value σ (see (C)) is negative, then the map T is *strongly contracting* for small μ (i.e., $\operatorname{dist}(TP_1, TP_2) \leqslant K \operatorname{dist}(P_1, P_2)$, where the contraction factor K tends uniformly to zero as both P_1, P_2 tend to $W^s \cap S^0$). Then, he artificially defines the map T on S^0_-. We shall do the same, assuming, say, that $TP \equiv P^*_\mu$ at $P \in S^0_-$. This extended map is also contracting (with the same factor K). In particular, at $\mu = 0$, this map takes a small neighborhood of the point P^*_0 into itself. The same, obviously, holds for all small μ. Thus, the Banach principle gives the existence of a unique fixed point; moreover, this point attracts iterations (by the map T extended onto all S^0) of every initial point on S^0.

For $\mu \leqslant 0$ the fixed point is, by definition, the point P^*_μ. Since it lies in the region $S^0_- \cup (W^s \cap S^0)$ where the Poincaré map is not defined, no periodic orbit corresponds to this point; it is a homoclinic loop at $\mu = 0$ or just a fake at $\mu < 0$.

For $\mu > 0$, the fixed point is the limit of the iterations of the point P^*_μ. This point is the image of the line $W^s \cap S^0$ and it lies at the distance $d(\mu)$ from this line. Therefore, due to the contraction, all the iterations of this point (and their limit, the fixed point) lie in the ball of radius $K(1-K)^{-1}d(\mu)$ with center at P^*_μ. If μ is sufficiently small, one can assume that $K < 1/2$ and in this case the radius is less than $d(\mu)$. Thus, for $\mu > 0$ the fixed point of the extended map belongs to the region S^0_+. Hence, it is a fixed point of the true Poincaré map and there exists the corresponding periodic orbit of the system.

All this is in a complete correspondence with the statement of the theorem. The key point in the proof is to show that the Poincaré map is strongly contracting. For this, computations explicitly involving second derivatives of the right-hand sides of the system were used in [**15**]. Below (Sections 2 and 3) we prove the contraction in the case of minimal smoothness (C^1), by using Shilnikov's boundary value problem method discussed in [**17**]. □

At first glance, the passage from, say, C^2 to C^1 is an insignificant step. However, dynamical systems of low smoothness appear naturally when studying high-dimensional systems reduced to a normally hyperbolic invariant manifold (say, to the inertial manifold, or to a nonlocal center manifold as in the example below). The smoothness of such a manifold, and therefore the smoothness of the reduced system, does not correlate with the smoothness of the original system. In particular, the conditions for the existence of a C^2-smooth invariant manifold are much more restrictive than for a C^1 one. Thus, the study of the bifurcational problems in the case of the smallest possible smoothness may be crucial for a rigorous description of the high-dimensional dynamics.

As an example, consider a C^1-version of the result of Shilnikov [**18**]: a generalization of Theorem 1.1 to the case where the dimension of the unstable manifold of O is larger than one. Namely, let X_μ be a continuous family of C^1-smooth dynamical systems on an $(n+m)$-dimensional manifold. Let us modify conditions (A), (B) in the following way.

(A′) *The system X_μ has a hyperbolic equilibrium state O, and the characteristic exponents $\lambda_n, \dots, \lambda_1, \gamma, \gamma_2, \dots, \gamma_m$ at the point O for $\mu = 0$ satisfy the following condition*: $\operatorname{Re}\lambda_n \leqslant \cdots \leqslant \operatorname{Re}\lambda_1 < 0 < \gamma < \operatorname{Re}\gamma_2 \leqslant \cdots \leqslant \operatorname{Re}\gamma_m$.

(B′) *For $\mu = 0$ there exists a homoclinic orbit Γ, i.e., $\Gamma \subset (W^s \cap W^u)$.*

The conditions (C), (D) remain unchanged.

In this case the dimension of the unstable manifold W^u is equal to m and, moreover, there exists an $(m-1)$-dimensional strong unstable invariant submanifold $W^{uu} \subset W^u$. The characterizing feature of W^{uu} is that all orbits in it are tangent to the linear subspace that corresponds to the eigenvalues $\gamma_2, \dots, \gamma_m$, while all orbits of $W^u \setminus W^{uu}$ are tangent to the eigendirection corresponding to the *leading* eigenvalue γ. Let us assume the following condition.

(E) *The homoclinic orbit Γ does not belong to W^{uu}* (see Figure 5).

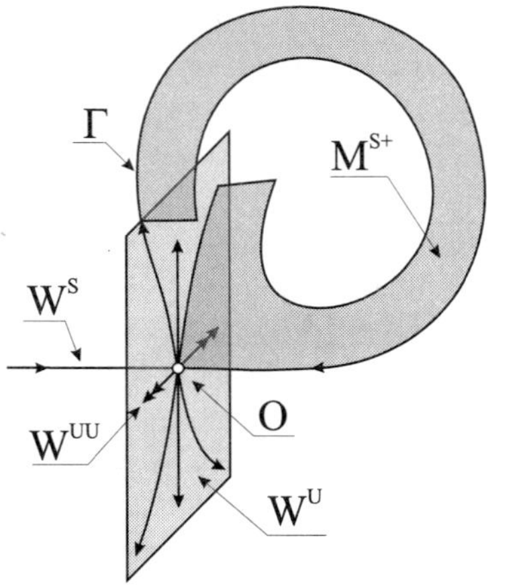

FIGURE 5. The orbit Γ does not lie in the strong unstable submanifold W^{uu}. The extended stable manifold M^{s+} is transverse to the unstable manifold W^u.

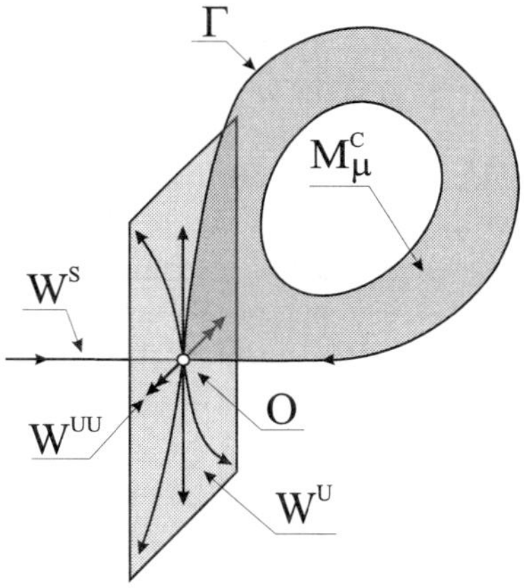

FIGURE 6. There exists an $(n+1)$-dimensional C^1-smooth center invariant manifold M^c_μ if conditions (A′), (B′), (E), and (F) are fulfilled.

The next assumption is necessary [**23**] to ensure the presence of an $(n+1)$-dimensional global invariant manifold (as well as condition (E)). Denote by $E^{s+} \subset R^{n+1}$ the invariant subspace of the system X_0 linearized at the point O, corresponding to the eigenvalues $\lambda_n, \ldots, \lambda_1, \gamma$. It is well known (see for instance [**5**]) that there exists an invariant C^1-smooth manifold M^{s+} tangent to E^{s+} at O (see Figure 5). The manifold M^{s+} contains W^s. It is not uniquely defined, but any two such submanifolds have the same tangent at each point of W^s. We require the following condition to be fulfilled.

(F) *The manifold M^{s+} is transverse to the manifold W^u at each point of* Γ (see Figure 5).

Note that the transversality must be verified only at one point on Γ because the manifolds M^{s+} and W^u are invariant with respect to the flow defined by the system X_0. One can check that condition (F) is equivalent to the requirement in [**18**] about the nonvanishing of a specific determinant. Note also that conditions (E) and (F) are not so restrictive, because they are fulfilled in a general position.

It is shown in [**13**] (in the case of higher smoothness in [**22, 12, 6, 7**]) that when conditions (A′), (B′), (E), and (F) are fulfilled, then *there exists a small neighborhood U of the homoclinic orbit Γ such that, for all μ small enough, the system X_μ has an $(n+1)$-dimensional repelling invariant C^1-manifold $\mathcal{M}_\mu$ depending continuously on μ and such that any orbit not lying in $\mathcal{M}_\mu$ leaves U as $t \to +\infty$. The manifold $\mathcal{M}_\mu$ is tangent at the point O to the linear subspace corresponding to the eigenvalues $(\lambda_n, \dots, \lambda_1, \gamma)$.*

Due to this result, the study of the $(n+m)$-dimensional system is reduced to the study of the $(n+1)$-dimensional system on the invariant manifold $\mathcal{M}_\mu$. Evidently, for the reduced system conditions (A)–(D) hold, and therefore, Theorem 1.1 immediately carries over to the multidimensional case. Note that the periodic orbit L born from the loop is now stable only on the invariant manifold $\mathcal{M}_\mu$ and since the manifold is repelling, the orbit L is unstable in the normal directions. Thus, in this case, L is a saddle periodic orbit with m-dimensional unstable and $(n+1)$-dimensional stable manifolds.

2. The Shilnikov boundary value problem

In order to prove Theorem 1.1, we need appropriate estimates (strong contraction) for the Poincaré map near the homoclinic loop $\mathcal{L}$. The study of the solutions near the equilibrium state is the most complicated point here because the flight time near O is unbounded and, therefore, we need the estimates which hold for arbitrarily large times. The question on the local estimates does not appear if the system can be linearized in a neighborhood of the equilibrium point. However, the smooth linearization requires a lot of resonance restrictions plus extra smoothness. Therefore, to find suitable estimates near the equilibrium, we use a method based on the consideration of a certain boundary value problem (see [**17, 19, 8, 9, 20**]). In this section we investigate solutions of the Shilnikov boundary value problem for the case in which the smoothness of the system is C^1 only.

Let us introduce local coordinates (x, y) ($x \in \mathbb{R}^n$, $y \in \mathbb{R}^1$) in a neighborhood of the saddle O so that the system X_μ takes the form

$$\begin{cases} \dot{x} = \Lambda x + f(x, y, \mu), \\ \dot{y} = \gamma y + g(x, y, \mu). \end{cases} \tag{2.1}$$

Here Λ is a matrix $(n \times n)$ such that $\operatorname{Spectr}(\Lambda) = \{\lambda_1 \dots \lambda_n\}$. The functions f and g are smooth with respect to (x, y) and depend continuously on μ along with the derivatives. Moreover,

$$f(0,0,\mu) = 0, \quad g(0,0,\mu) = 0, \quad \left.\frac{\partial(f,g)}{\partial(x,y)}\right|_{(x,y,u)=0} = 0. \tag{2.2}$$

According to [**17**], for any $\tau > 0$ and x^0 and y^1 small enough, in a small neighborhood of O there exists a unique orbit $\{x(t), y(t)\}_{t \in [0,\tau]}$ of system (2.1) which satisfies the following boundary conditions:

$$x(0) = x^0, \qquad y(\tau) = y^1. \tag{2.3}$$

Let us denote the solution of this boundary value problem by

$$x(t) = x(t; x^0, y^1, \tau, \mu), \qquad y(t) = y(t; x^0, y^1, \tau, \mu). \tag{2.4}$$

The theorem below follows from [**17**] (for completeness, we present a proof; $\|(x,y)\|$ denotes $\max(\|x\|, \|y\|)$).

THEOREM 2.1. *For any small $\varepsilon > 0$, if $\|(x^0, y^1)\| \leqslant \varepsilon$, then $\|(x(t), y(t))\| \leqslant \varepsilon$ in (2.4) for all $t \in [0, \tau]$ and all small μ. The solution (2.4) depends smoothly on $(t; x^0, y^1, \tau)$ and, along with the derivatives, depends continuously on μ. The following estimates hold for the derivatives:*

$$\left\| \frac{\partial(x,y)}{\partial x^0} \right\| \leqslant Ce^{-\alpha t}, \qquad \left\| \frac{\partial(x,y)}{\partial y^1} \right\| \leqslant Ce^{-\beta(\tau - t)}, \tag{2.5}$$

where C, α, and β are some constants such that

$$C > 0, \quad \operatorname{Re}\lambda_n \leqslant \cdots \leqslant \operatorname{Re}\lambda_1 < -\alpha < 0 < \beta < \gamma. \tag{2.6}$$

Moreover, as ε decreases, the constants α and β can be made arbitrarily close to $|\operatorname{Re}\lambda_1|$ and γ, respectively.

PROOF. It follows from (2.2), that for any small $\xi > 0$ there exists a small $\varepsilon > 0$ such that for $\|(x,y)\| \leqslant \varepsilon$ and for small μ

$$\|(f,g)\| \leqslant \xi\varepsilon, \qquad \left\| \frac{\partial(f,g)}{\partial(x,y)} \right\| < \xi. \tag{2.7}$$

Note also that for any λ so that

$$\max_{i=1,\dots,n} \operatorname{Re}\lambda_i < -\lambda, \tag{2.8}$$

the norm of $x \in \mathbb{R}^n$ may be defined so as to have

$$\|e^{\Lambda s}\| \leqslant e^{-\lambda s} \quad \text{for } s \geqslant 0. \tag{2.9}$$

Consider the Banach space H of continuous functions $(x(t), y(t))$ defined for $t \in [0, \tau]$, with the uniform norm

$$\|(x(t), y(t))\|_H = \sup_{t \in [0,\tau]} \|(x(t), y(t))\|. \tag{2.10}$$

Let H_ε be the ε-neighborhood of zero in H (i.e., H_ε is the set of continuous functions with norm not greater than ε). Let us take a small $\varepsilon > 0$ and introduce the integral operator $T\colon H_\varepsilon \to H$ that maps a function $(x(t), y(t))$ to the function $(\bar{x}(t), \bar{y}(t))$ defined by the following rule:

$$\begin{aligned} \bar{x}(t) &= e^{\Lambda t} x^0 + \int_0^t e^{\Lambda(t-s)} f(x(s), y(s), \mu)\, ds, \\ \bar{y}(t) &= e^{\gamma(t-\tau)} y^1 + \int_\tau^t e^{\gamma(t-s)} g(x(s), y(s), \mu)\, ds. \end{aligned} \tag{2.11}$$

It is easy to see that any solution of the boundary value problem (2.3) is a fixed point of the integral operator T and any fixed point of the operator (2.11) is a solution of the boundary value problem. Therefore, the existence and uniqueness of the solution of the boundary value problem follows from the fact that T is a *contraction operator that maps H_ε into itself.* Indeed, take any function (x,y) from H_ε. Using (2.7)–(2.9), its image $(\bar{x},\bar{y})$ by T is estimated as follows:

$$
\begin{aligned}
\|\bar{x}(t)\| &\leqslant e^{-\lambda t}\|x^0\| + \int_0^t e^{-\alpha(t-s)}\xi\,\|(x(s),y(s))\|\,ds \\
&\leqslant e^{-\lambda t}\|x^0\| + \frac{\xi}{\lambda}(1-e^{-\lambda t})\,\|(x,y)\|_H, \\
\|\bar{y}(t)\| &\leqslant e^{-\gamma(\tau-t)}\|y^1\| + \int_t^\tau e^{-\gamma(s-t)}\xi\,\|(x(s),y(s))\|\,ds \\
&\leqslant e^{-\gamma(\tau-t)}\|y^1\| + \frac{\xi}{\gamma}(1-e^{-\gamma(\tau-t)})\|(x,y)\|_H.
\end{aligned}
\tag{2.12}
$$

On the interval $0 \leqslant t \leqslant \tau$ the factors $e^{-\lambda t}$ and $e^{-\gamma(\tau-t)}$ are bounded in $[0,1]$. Therefore, if $\|(x,y)\|_H \leqslant \varepsilon$ and $\|(x^0,y^1)\| \leqslant \varepsilon$, then assuming that

$$
\xi\max(\lambda^{-1},\gamma^{-1}) < 1 \tag{2.13}
$$

we get $\|(\bar{x},\bar{y})\|_H \leqslant \varepsilon$; i.e., the ε-neighborhood of zero in H is indeed T-invariant.

To show contraction, take any functions (x_1,y_1) and (x_2,y_2) from H_ε. As above, we have the following estimates:

$$
\begin{aligned}
\|\bar{x}_1(t)-\bar{x}_2(t)\| &\leqslant \frac{\xi}{\lambda}(1-e^{-\lambda t})\|(x_1-x_2,y_1-y_2)\|_H, \\
\|\bar{y}_1(t)-\bar{y}_2(t)\| &\leqslant \frac{\xi}{\gamma}(1-e^{-\gamma(\tau-t)})\|(x_1-x_2,y_1-y_2)\|_H.
\end{aligned}
\tag{2.14}
$$

Thus, if ε is so small that (2.13) holds, the contraction follows, i.e.,

$$
\|(\bar{x}_1-\bar{x}_2,\bar{y}_1-\bar{y}_2)\|_H < q\|(x_1-x_2,y_1-y_2)\|_H, \quad \text{with } q<1.
$$

According to the Banach principle for contraction mappings, the operator T has a unique fixed point in H_ε for all (x^0,y^1,τ,μ), i.e., the boundary value problem (2.3) has a unique solution. It depends smoothly on the boundary data (x^0,y^1) because the integral operator T is smooth on H_ε (i.e., its Frechet derivative is uniformly continuous) and it depends smoothly on (x^0,y^1) so the latter is true for its fixed point as well.

Since T is a smooth contracting operator smoothly depending on the parameters (x^0,y^1), the iterations of any initial function in H_ε converge to the fixed point, along with the derivatives with respect to (x^0,y^1). Thus the sequence of functions $(x_0,y_0),(x_1,y_1),(x_2,y_2),\ldots$, obtained by the iterations

$$
(x_{n+1}(t),y_{n+1}(t)) = T(x_n(t),y_n(t)) \tag{2.15}
$$

with $(x_0(t),y_0(t)) = 0$, converges to the solution of the boundary value problem and the derivatives $\partial(x_n,y_n)/\partial(x^0,y^1)$ converge to the corresponding derivative of the solution.

Thus, to prove estimates (2.5), it is sufficient to check that for appropriately chosen constants C, α, and β (see (2.6)), if some function (x,y) satisfies (2.5), then its image by T satisfies (2.5) as well with the same constants.

By differentiating (2.11), we obtain

$$\frac{\partial \bar{x}(t)}{\partial x^0} = e^{\Lambda t} + \int_0^t e^{\Lambda(t-s)} \frac{\partial f}{\partial(x,y)} \frac{\partial(x(s),y(s))}{\partial x^0}\, ds, \tag{2.16}$$

$$\frac{\partial \bar{y}(t)}{\partial x^0} = \int_\tau^t e^{\gamma(t-s)} \frac{\partial g}{\partial(x,y)} \frac{\partial(x(s),y(s))}{\partial x^0}\, ds. \tag{2.17}$$

If the first inequality of (2.5) holds for $(x(s), y(s))$, then the equations above give (using (2.7))–(2.9)):

$$\left\| \frac{\partial(\bar{x}(t), \bar{y}(t))}{\partial x^0} \right\| \leqslant e^{-\lambda t} + C\xi \max\left\{ \int_0^t e^{-\lambda(t-s)} e^{-\alpha s}\, ds, \int_t^\tau e^{-\gamma(s-t)} e^{-\alpha s}\, ds \right\},$$

or if α is close to and less than λ, then

$$\left\| \frac{\partial(\bar{x}(t), \bar{y}(t))}{\partial x^0} \right\| \leqslant e^{-\alpha t} \left(1 + \xi \frac{C}{\lambda - \alpha} \right). \tag{2.18}$$

Similarly, if $(x(s), y(s))$ satisfies the second inequality of (2.5), then

$$\left\| \frac{\partial(\bar{x}(t), \bar{y}(t))}{\partial y^1} \right\| \leqslant e^{-\beta(\tau - t)} \left(1 + \xi \frac{C}{\gamma - \beta} \right) \tag{2.19}$$

for β close to and less than γ.

Thus, the image $(\bar{x}, \bar{y})$ satisfies estimates (2.5) with the new constant factor

$$C_{\text{new}} = 1 + Cq,$$

where $q = \xi \max((\lambda - \alpha)^{-1}, (\gamma - \beta)^{-1})$. Given α and β, assume ξ is so small that $q < 1$. In this case, if $C \geqslant (1 - q)^{-1}$, then $C_{\text{new}} \leqslant C$, which completes the proof of estimates (2.5).

It remains to prove the smoothness of the solution of the boundary value (2.1), (2.3) with respect to t and τ. Since $(x(t; x^0, y^1, \tau, \mu), y(t; x^0, y^1, \tau, \mu))$ is an orbit of the system X_μ, the smoothness with respect to t follows immediately. Let us now fix any initial point (x^0, y^0) and let $y^1(\tau)$ be the y-coordinate of its time τ shift by the flow X_μ. From the definition of $(x(t; x^0, y^1, \tau, \mu), y(t; x^0, y^1, \tau, \mu))$ as the unique solution of the boundary value problem (2.1), (2.3), we see that $(x(t; x^0, y^1(\tau), \tau, \mu), y(t; x^0, y^1(\tau), \tau, \mu))$ is the time t shift of (x^0, y^0), independently of the value of τ. Thus,

$$\frac{d}{d\tau}(x(t; x^0, y^1(\tau), \tau, \mu), y(t; x^0, y^1(\tau), \tau, \mu)) \equiv 0. \tag{2.20}$$

Now, since $y^1(\tau)$ depends smoothly on τ and

$$(x(t; x^0, y^1, \tau, \mu), y(t; x^0, y^1(\tau), \tau, \mu))$$

depends smoothly on y^1 (as we just have proved), the smoothness of

$$(x(t; x^0, y^1, \tau, \mu), y(t; x^0, y^1(\tau), \tau, \mu))$$

with respect to τ follows from (2.20) immediately (useful expressions for the derivatives are given by (3.13), (3.14) in the next section). □

3. The Poincaré map

Now let us prove that the Poincaré map near the homoclinic loop $\mathcal{L}$ is strongly contracting. This map is represented as a superposition of two maps: T_{loc} and T_{glo}, where T_{loc} is defined by the flow near the equilibrium point and T_{glo} is defined by the flow near the global piece of the homoclinic orbit Γ. These maps are defined on small cross-sections S^0 and S^1 (which we construct below): $T_{\mathrm{loc}}\colon S^0 \mapsto S^1$ and $T_{\mathrm{glo}}\colon S^1 \mapsto S^0$.

The n-dimensional stable manifold W^s of the point O is tangent to the plane $y = 0$ at the point $O = (0,0)$ at $\mu = 0$. Thus, W^s is locally the graph of a smooth function

$$y = y^s(x,\mu), \quad y^s(0,\mu) = 0, \quad \left.\frac{\partial y^s(x,\mu)}{\partial x}\right|_{(x,\mu)=0} = 0. \tag{3.1}$$

The unstable manifold W^u of O is locally the graph of a smooth function

$$x = x^u(y,\mu), \quad x^u(0,\mu) = 0, \quad \left.\frac{\partial x^u(y,\mu)}{\partial y}\right|_{(y,\mu)=0} = 0. \tag{3.2}$$

For $\mu = 0$, the orbit Γ tends to O as $t \to +\infty$. Therefore, the surface

$$S^0 = \{(x,y) : \|x\| = \xi,\ \|x - x^+, y - y^+\| \leqslant \delta\} \tag{3.3}$$

is a cross-section for the orbits close to Γ if μ is small enough. Here (x^+, y^+) are the coordinates of the first intersection of Γ with the surface $\|x\| = \xi$ at $\mu = 0$ (see Figure 7), and ξ and δ are small positive constants.

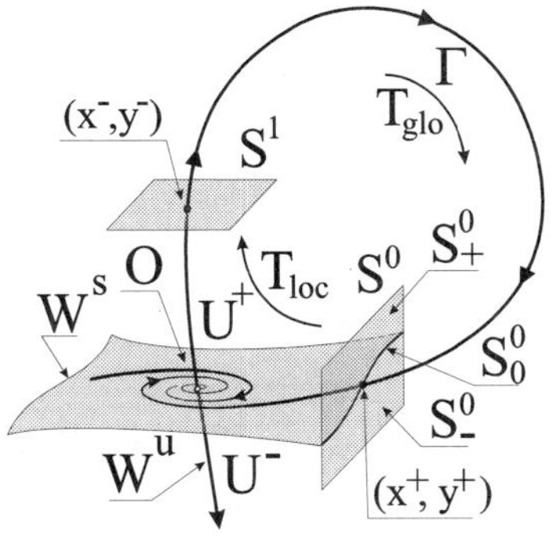

FIGURE 7. Two cross-sections can be constructed in the neighborhood of O: S^0 near the point (x^+, y^+) and S^1 near (x^-, y^-). The flow defines the maps $T_{\mathrm{loc}}\colon S^0_+ \to S^1$ and $T_{\mathrm{glo}}\colon S^1 \to S^0$.

The manifold W^u consists of three orbits: the equilibrium point O and two separatrices, one of which is the orbit Γ forming the homoclinic loop for $\mu = 0$. Without loss of generality we assume that the orbit Γ leaves O towards the positive y. So, for small positive δ and y^- and for small μ, the surface

$$S^1 = \{(x,y) : y = y^-,\ \|(x - x^-)\| \leqslant \delta\} \tag{3.4}$$

is a cross-section for the orbits close to Γ. Here (x^-, y^-) are the coordinates of the first intersection of Γ with the plane $y = y^-$ at $\mu = 0$.

Both cross-sections S^0 and S^1 are n-dimensional. Without loss of generality, we take $(x_1, \ldots, x_n)$ as the coordinates on the cross-section S^1 and $(x_1, \ldots, x_{n-1}, y)$

as the coordinates on S^0. Below we use the following notation (see Figure 7): S^0_0 for the intersection of W^s with S^0; S^0_+ for $S^0 \cap U^+$ and S^0_- for $S^0 \cap U^-$; also (x^0, y^0) for the coordinates on the cross-section S^0 and x^1 for the coordinates on S^1.

As we mentioned, the Poincaré map T near the homoclinic loop is the superposition of the two maps T_{loc} and T_{glo}. The local map is defined on S^0_+, it takes the region that corresponds to small y^0 greater than $y^s(x^0, \mu)$ to a small neighborhood of the point $x^1 = x^-$ on S^1 (the orbits starting with $y^0 < y^s(x^0, \mu)$, i.e., below the stable manifold, go close to the other separatrix and do not reach S^1). By continuity, the map T_{loc} may be defined at $y^0 = y^s(x^0, \mu)$:

$$T_{\text{loc}} S^0_0 = x^-.$$

The global map takes a small neighborhood of the point $x^1 = x^-$ on S^1 into S^0 (see Figure 7). The flight time from S^1 to S^0 is bounded, therefore, the map T_{glo} is a diffeomorphism. In particular, its derivatives are bounded. Thus, to show the required contraction of the Poincaré map $T = T_{\text{glo}} \circ T_{\text{loc}}$, it is sufficient to prove the following lemma which, basically, shows that the local map is arbitrarily strong contracting in a sufficiently small neighborhood of S^0_0.

LEMMA 3.1. *The map T_{loc} can be written as $x^1 = \varphi(x^0, y^0; \mu)$, where φ is a C^1 function of (x^0, y^0) defined on $S^0_+ \cup S^0_0$ and its first derivatives vanish on S^0_0.*

PROOF. According to Section 2, given $\tau > 0$ and small x^0, y^1, there exists a unique orbit $(x(t), y(t)) = (x(t; x^0, y^1, \tau, \mu), y(t; x^0, y^1, \tau, \mu))$ which, at $t = 0$, starts with the point $(x^0, y(0))$ and reaches the point $(x(\tau), y^1)$ for $t = \tau$. Thus, fixing $y^1 = y^-$ and $\|x^0\| = \xi$, we see that the orbit of a point $(x^0, y^0) \in S^0_+$ reaches the cross-section S^1 at a point x^1 at the time $\tau(x^0, y^0, \mu)$ if and only if

$$y^0 = y(0; x^0, y^-, \tau(x^0, y^0, \mu), \mu) \tag{3.5}$$

and

$$x^1 = \varphi(x^0, y^0, \mu) \equiv x(\tau(x^0, y^0, \mu); x^0, y^-, \tau(x^0, y^0, \mu), \mu). \tag{3.6}$$

It follows from (3.5) that

$$\frac{\partial \tau}{\partial x^0} = -\left(\frac{\partial y}{\partial \tau}\bigg|_{t=0}\right)^{-1} \frac{\partial y}{\partial x^0}\bigg|_{t=0}, \qquad \frac{\partial \tau}{\partial y^0} = \left(\frac{\partial y}{\partial \tau}\bigg|_{t=0}\right)^{-1}. \tag{3.7}$$

By (3.6) and (3.7)

$$\begin{aligned}
\frac{\partial \varphi}{\partial x^0} &= \frac{\partial x}{\partial x^0}\bigg|_{t=\tau} + \left(\frac{\partial x}{\partial t}\bigg|_{t=\tau} + \frac{\partial x}{\partial \tau}\bigg|_{t=\tau}\right)\frac{\partial \tau}{\partial x^0} \\
&= \frac{\partial x}{\partial x^0}\bigg|_{t=\tau} - \left(\frac{\partial x}{\partial t}\bigg|_{t=\tau} + \frac{\partial x}{\partial \tau}\bigg|_{t=\tau}\right)\left(\frac{\partial y}{\partial \tau}\bigg|_{t=0}\right)^{-1} \frac{\partial y}{\partial x^0}\bigg|_{t=0}, \\
\frac{\partial \varphi}{\partial y^0} &= \left(\frac{\partial x}{\partial t}\bigg|_{t=\tau} + \frac{\partial x}{\partial \tau}\bigg|_{t=\tau}\right)\frac{\partial \tau}{\partial y^0} = \left(\frac{\partial x}{\partial t}\bigg|_{t=\tau} + \frac{\partial x}{\partial \tau}\bigg|_{t=\tau}\right)\left(\frac{\partial y}{\partial \tau}\bigg|_{t=0}\right)^{-1}.
\end{aligned} \tag{3.8}$$

To prove Lemma 3.1, we must show that

$$\lim_{\tau \to +\infty} \frac{\partial \varphi}{\partial (x^0, y^0)} = 0 \tag{3.9}$$

(because, the limit $\tau = +\infty$ corresponds to a starting point on the stable manifold W^s, or, what is the same, to $(x^0, y^0) \in S_0^0$). According to Theorem 2.1,

$$\text{(3.10)}\quad \begin{aligned} &\left\| \frac{\partial(x,y)}{\partial x^0}\bigg|_{t=0} \right\| \leqslant C, && \left\| \frac{\partial(x,y)}{\partial y^1}\bigg|_{t=0} \right\| \leqslant Ce^{-\beta\tau}, \\ &\left\| \frac{\partial(x,y)}{\partial x^0}\bigg|_{t=\tau} \right\| \leqslant Ce^{-\alpha\tau}, && \left\| \frac{\partial(x,y)}{\partial y^1}\bigg|_{t=\tau} \right\| \leqslant C. \end{aligned}$$

Thus, by virtue of (3.8), (3.10), it is sufficient to show that

$$\text{(3.11)}\quad \lim_{\tau\to+\infty} \left(\frac{\partial x}{\partial t}\bigg|_{t=\tau} + \frac{\partial x}{\partial \tau}\bigg|_{t=\tau} \right) \left(\frac{\partial y}{\partial \tau}\bigg|_{t=0} \right)^{-1} = 0.$$

To find estimates for the derivatives of x, y with respect to τ, note that by the definition of the function (x, y) as the unique solution of the boundary value problem (2.1), (2.3), we have the identities

$$\text{(3.12)}\quad \begin{aligned} y(t; x^0, y^1, \tau, \mu) &\equiv y(t; x^0, y(\tau + \Delta\tau; x^0, y^1, \tau, \mu), \tau + \Delta\tau, \mu), \\ x(t; x^0, y^1, \tau, \mu) &\equiv x(t + \Delta\tau; x(-\Delta\tau; x^0, y^1, \tau, \mu), y^1, \tau + \Delta\tau, \mu). \end{aligned}$$

Differentiation of (3.12) with respect to $\Delta\tau$ at $\Delta\tau = 0$ gives

$$\text{(3.13)}\quad \frac{\partial y}{\partial \tau} = -\frac{\partial y}{\partial y^1}\, \dot{y}|_{t=\tau}$$

and

$$\text{(3.14)}\quad \frac{\partial x}{\partial t} + \frac{\partial x}{\partial \tau} = \frac{\partial x}{\partial x^0}\, \dot{x}|_{t=0}.$$

Now, by (3.10) and (3.14) we have

$$\text{(3.15)}\quad \frac{\partial x}{\partial t}\bigg|_{t=\tau} + \frac{\partial x}{\partial \tau}\bigg|_{t=\tau} = O(e^{\alpha\tau}) \quad \text{as } \tau \to +\infty.$$

Since $\dot{y}|_{t=\tau}$ is bounded away from zero (this is the value of $\dot{y}$ on the cross-section S^1), it remains to estimate $\partial y/\partial y^1$ *from below*. For this, let us consider the orbit $(x^*(t; x^0, y^0, \mu), y^*(t; x^0, y^0, \mu))$ which starts at the point (x^0, y^0) for $t = 0$, i.e., the solution of the initial value problem.

All the time during which the orbit $(x^*(t), y^*(t))$ is in a small neighborhood of the equilibrium state O, the following estimate holds for any fixed $\gamma^* > \gamma$:

$$\text{(3.16)}\quad \frac{d}{dt}\left\| \frac{\partial(x^*(t), y^*(t))}{\partial y^0} \right\| \leqslant \gamma^* \left\| \frac{\partial(x^*(t), y^*(t))}{\partial y^0} \right\|$$

(this is true because the spectrum of the linearization matrix of the system (2.1) at the point O lies to the left of the straight line $\mathrm{Re}(\,\cdot\,) = \gamma$ on the complex plane). Since γ^* may be chosen arbitrary close to γ and α arbitrary close to $-\operatorname{Re}\lambda_1$ (see Theorem 2.1), we may assume by condition (C) that

$$\text{(3.17)}\quad \alpha + \gamma^* < 0.$$

Inequality (3.16) implies

$$\text{(3.18)}\quad \left\| \frac{\partial y^*(t)}{\partial y^0} \right\| \leqslant c e^{\gamma^* t}$$

for some positive constant c.

By definition,

$$y^1 \equiv y^*(\tau; x^0, y(0; x^0, y^1, \tau, \mu), \tau, \mu) \tag{3.19}$$

(recall that the star indicates the solution of the initial value problem, whereas y without the star corresponds to the solution of the boundary value problem). Identity (3.19) implies

$$\left.\frac{\partial y^*}{\partial y^0}\right|_{t=\tau} \left.\frac{\partial y}{\partial y^1}\right|_{t=0} \equiv 1. \tag{3.20}$$

By (3.18) and (3.20),

$$\left\| \left.\frac{\partial y}{\partial y^1}\right|_{t=0} \right\| \geqslant \frac{1}{c} e^{-\gamma^* \tau}. \tag{3.21}$$

Now, by (3.15),

$$\left(\left.\frac{\partial x}{\partial t}\right|_{t=\tau} + \left.\frac{\partial x}{\partial \tau}\right|_{t=\tau} \right) \left(\left.\frac{\partial y}{\partial \tau}\right|_{t=0} \right)^{-1} = O(e^{(\alpha+\gamma^*)\tau}), \tag{3.22}$$

which, along with (3.17), gives the lemma. □

Acknowledgments. The authors are grateful to L. P. Shilnikov who pointed out the necessity of solving this problem.

References

[1] A. A. Andronov and E. A. Leontovich, *To the theory of changes of the qualitative structure of trajectories on the plane*, Dokl. Acad. Nauk USSR **21** (1938), no. 9. (Russian)

[2] E. A. Leontovich, *On the appearance of limit cycles from a separatrix*, Dokl. Akad. Nauk SSSR (N.S.) **78** (1951), 641–644; English transl., Soviet Math. Dokl. **28** (1951), no. 4.

[3] A. A. Andronov and E. A. Leontovich, *On the appearance of limit cycles from a loop of separatrix to a saddle and from a loop of separatrix to a saddle-node*, Mat. Sb. **48 (90)** (1959), 335–376. (Russian)

[4] A. A. Andronov, E. A. Leontovich, I. I. Gordon, and A. G. Maier, *Theory of bifurcations of dynamical systems on the plane*, "Nauka", Moscow, 1967; English transl., Israel Program of Scientific Translations, Jerusalem and London, 1973.

[5] M. Hirsch, C. Pugh, and M. Shub, *Invariant manifolds*, Lecture Notes in Math., vol. 583, Springer-Verlag, Heidelberg, 1977.

[6] A.-J. Homburg, *Some global aspects of homoclinic bifurcations of vector fields*, Preprint, 1993.

[7] ———, *Global aspects of homoclinic bifurcations of vector fields*, Mem. Amer. Math. Soc. **578**, 1996.

[8] I. M. Ovsyannikov and L. P. Shilnikov, *On systems with a saddle-focus homoclinic curve*, Mat. Sb.; English transl., Math. USSR-Sb. **58** (1986), 557–574.

[9] ———, *Systems with a homoclinic curve of multidimensional saddle-focus type, and spiral chaos*, Mat. Sb.; English transl., Math. USSR-Sb. **73** [1991], 415–443.

[10] B. Sandstede, *Center manifolds for homoclinic solutions*, Preprint, 1994.

[11] M. V. Shashkov, *On the existence of a smooth invariant two-dimensional attractive manifold for systems with a separatrix contour*, Methods of Qualitative Theory and Theory of Bifurcations (Nizhny Novgorod State University), 1991, pp. 61–73.

[12] ———, *Bifurcations of separatrix contours*, PhD Thesis, Nizhny Novgorod University, 1994.

[13] M. V. Shashkov and D. V. Turaev, *An existence theorem of smooth nonlocal center invariant manifolds for systems close to a system with a homoclinic loop*, J. Nonlinear Sci. **9** (1998), no. 5, 525–573.

[14] L. P. Shilnikov, *Some instances of generation of periodic motions in n-space*, Dokl. Akad. Nauk SSSR **143** (1962), 289–292; English transl. in Soviet Math. Dokl. **3** (1962).

[15] ———, *Some cases of generation of periodic motion from singular trajectories*, Mat. Sb. **61** (1963), 443–466. (Russian)

[16] ———, *A case of the existence of a denumerable set of periodic motions*, Dokl. Akad. Nauk SSSR **160** (1965), 558–561; English transl., Soviet Math. Dokl. **6** (1965), 163–166.

[17] ———, *On a Poincaré–Birkhoff problem*, Mat. Sb. (N.S.) **74 (116)** (1967), 378–397; English transl., Math. USSR-Sb. **3** (1967), 415–443.

[18] ———, *On the generation of a periodic motion from a trajectory doubly asymptotic to an equilibrium state of saddle type*, Mat. Sb. (N.S.) **77 (119)** (1968), 461–472; English transl., Math. USSR-Sb. **6** (1968), 415–443.

[19] ———, *A contribution to the problem of the structure of an extended neighborhood of a rough equilibrium state of saddle-focus type*, Mat. Sb. (N.S.) **81 (123)** (1970), 92–103; English transl., Math. USSR-Sb. **10** (1970), 91–102.

[20] L. Shilnikov, A. Shilnikov, D. Turaev, and L. Chua, *Methods of qualitative theory in nonlinear dynamics. Part I*, World Scientific, Singapore, 1998.

[21] D. V. Turaev, *On a case of bifurcations of a contour composed by two homoclinic curves of a saddle*, Methods of Qualitative Theory of Differential Equations, Gor′kov. Gos. Univ., 1984, pp. 162–175.

[22] ———, *On bifurcations of dynamical systems with two homoclinic curves of a saddle*, PhD Thesis, Nizhny Novgorod University, 1991.

[23] ———, *On dimension of nonlocal bifurcational problems*, Internat. J. Bifur. Chaos Appl. Sci. Engrg. **6** (1996), 911–948.

Department of Differential Equations, Institute for Applied Mathematics and Cybernetics, 10 Ul′yanov st., Nizhny Novgorod 603005, Russia
E-mail address: shashkov@focus.nnov.su

The Weierstrass Institute for Applied Analysis and Stochastics, Mohrenstr. 39, Berlin 10117, Germany
E-mail address: turaev@wias-berlin.de

Translated by MIKHAIL V. SHASHKOV

Amer. Math. Soc. Transl.
(2) Vol. **200**, 2000

A New Simple Bifurcation of a Periodic Orbit of "Blue Sky Catastrophe" Type

Leonid P. Shilnikov and Dmitry V. Turaev

Abstract. In this paper, we study a global bifurcation of codimension one connected with the disappearance (for positive values of a parameter μ) of a saddle-node periodic orbit L_0 under the condition that all orbits from the locally unstable manifold W^u of L_0 tend to L_0 as $t \to +\infty$. Conditions are presented which guarantee the blue sky catastrophe: the appearance of a stable periodic orbit L_μ which exists for any small positive values of μ but its length and period unboundedly increase as $\mu \to +0$.

1. Introduction

One of the main questions of nonlinear dynamics concerns the structure of the boundaries of stability regions of periodic orbits. It was the question which gave an initial impulse to the development of bifurcation theory, when Andronov and Leontovich [**1**] discovered that for two-dimensional systems of ODE's there are exactly four principal boundaries of stability of periodic orbits: on the first boundary the stable periodic orbit bifurcates from a stable equilibrium (which, in turn, loses stability), on the second boundary the periodic orbit coalesces with an unstable one and then disappears, on the third boundary the periodic orbit disappears merging into a homoclinic loop of a simple saddle-node equilibrium state and on the fourth boundary the stable periodic orbit merges into a homoclinic loop of a saddle equilibrium state with negative saddle value.

For the multidimensional case this list is extended in the following way. Obviously, there may be two types of stability boundaries: at the moment of bifurcation the periodic orbit either does or does not exist. In the first case, the intersection of the periodic orbit with a local cross-section is the fixed point of the Poincaré map, so the loss of stability corresponds to a multiplier on the unit circle. We have exactly three principal (i.e., those which are typical for generic one-parameter families) stability boundaries here: one corresponds to one multiplier of the periodic orbit equal to $(+1)$ and the remaining multipliers lying strictly inside the unit circle; this is the

---

2000 *Mathematics Subject Classification.* Primary 37J20.

The first author acknowledges the support of the RFBR (grants 96-01135 and 99-01-00231), INTAS-93-0570-ext, and the program "Universities of Russia" (project No. 1905). The second author acknowledges the support of the DFG Schwerpunktprogramme "Ergodentheorie, Analysis und Efficiente Simulation Dynamischer Systeme".

©2000 American Mathematical Society

saddle-node bifurcation (Figure 1) analogous to the two-dimensional case; another boundary corresponds to one multiplier equal to (-1), when the periodic orbit does not disappear as it crosses this stability boundary and a period-doubling bifurcation takes place; the third stability boundary corresponds to a pair of complex-conjugate multipliers; this bifurcation gives rise to the formation of an invariant torus.

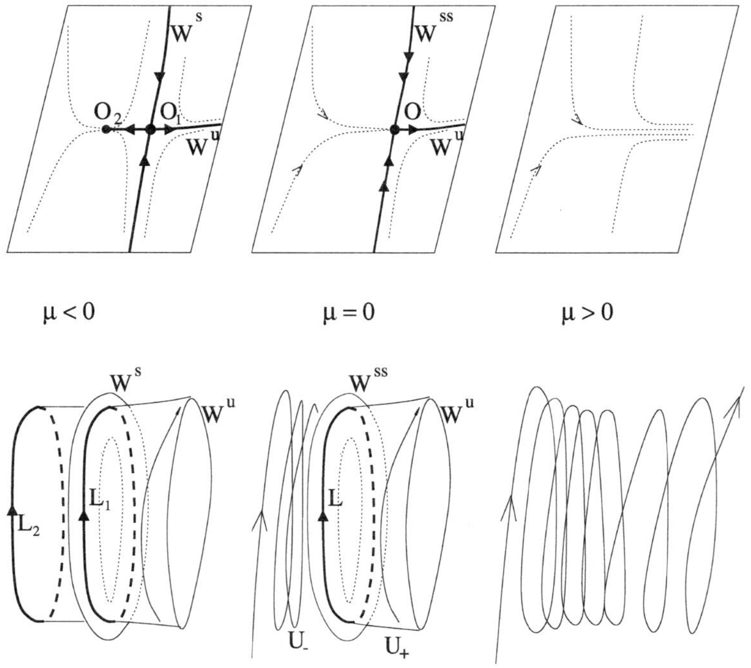

FIGURE 1. The bifurcation of a saddle-node periodic orbit. a) At $\mu < 0$ there exist stable and saddle periodic orbits. b) At $\mu = 0$ the periodic orbits unite into the saddle-node. Its strong stable manifold W^{ss} divides the neighborhood into the node region U_- and the saddle region U_+. The unstable manifold W^u lies in U_+. c) At $\mu > 0$ the saddle-node disappears and all the orbits leave the small neighborhood.

Stability boundaries of the other type correspond, as in the two-dimensional case, to the birth of a periodic orbit from the stable equilibrium state (the Andronov-Hopf bifurcation) or to the merging of the periodic orbit into a homoclinic loop [**2**] of either a simple saddle-node equilibrium state (Figure 2) or a hyperbolic equilibrium state with one-dimensional unstable manifold and with negative saddle value.

The following question immediately arises: Can there be other types of stability boundaries of codimension one? It can be shown that the list above gives *all* the principal stability boundaries for the case in which the *length* of the periodic orbit remains bounded when approaching the bifurcation moment (although the period may tend to infinity if the orbit merges into a homoclinic loop). Thus, the search for new stability boundaries must presuppose the unbounded growth of the orbit length.

In the present paper we do find such a stability boundary, which does not have two-dimensional analogs. The boundary is an open subset of a codimension

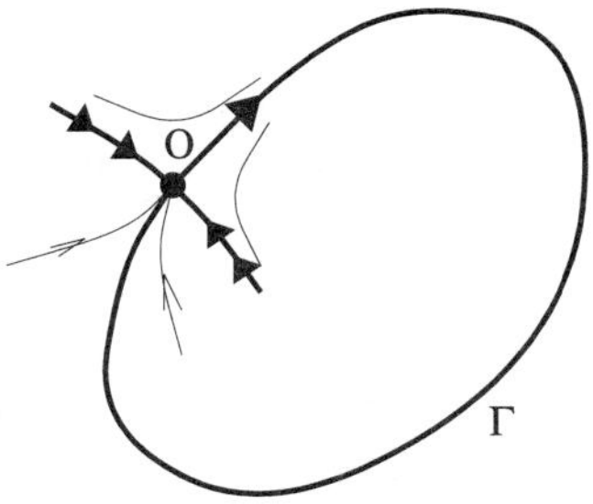

FIGURE 2. A homoclinic loop Γ to a saddle-node equilibrium O.

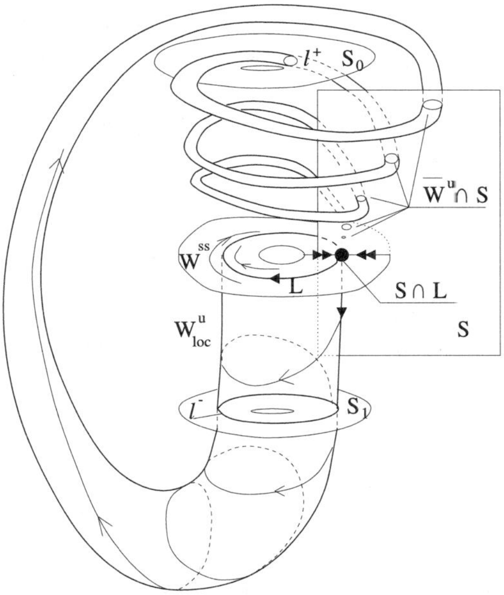

FIGURE 3. The global structure of the set $\overline{W}^u$ in the case under consideration. The image l^+ of the intersection l^- of the local unstable manifold W^u_{loc} of the saddle-node L with the cross-section S_1 is homotopic to zero on the cross-section S_1. The intersection of $\overline{W}^u$ with the local cross-section S in the node region is a countable set of circles which accumulate at $S \cap L$.

one bifurcational surface corresponding to the existence of a saddle-node periodic orbit. This open set is distinguished by some qualitative conditions determining the geometry of the unstable set of the saddle-node (see Figure 3) and also by some quantitative restrictions (a certain value should be less than 1, see below). We shall show under these conditions that when the saddle-node disappears, a new stable periodic orbit arises whose period and length both tend to infinity when approaching the bifurcation moment (Theorem 1).

This is one of the possible scenarios for the global bifurcation involving the disappearance of a saddle-node periodic orbit when all the orbits of its unstable set return to the saddle-node as $t \to +\infty$.

The study of this global bifurcation has a long history. Originally, this problem was raised in the twenties in connection with the study of the transition from synchronization to the amplitude modulation regime in the van der Pol equation

$$\ddot{x} + \mu(1 - x^2)\dot{x} + \omega_0^2 x = \mu A \cos \omega t.$$

Under the assumption that μ is a small parameter and that $1:1$ resonance occurs (i.e., $\omega - \omega_0 \sim \mu$), Andronov and Vitt showed that the transition from the synchronization to the amplitude modulation regime is connected with a bifurcation involving the birth of a stable limit cycle from a homoclinic loop to a saddle-node equilibrium state (as in Figure 2) in the time-averaged system. Returning to the initial equation, one can see that a similar picture occurs for the two-dimensional Poincaré map, where the saddle-node is now the fixed point of the map and the homoclinic loop is not a single orbit but forms a continuum of orbits that constitute the unstable set of the saddle-node. At that time, this kind of analysis was not carried out.

The study of this bifurcation started in [**3**] under the assumption that the dynamical system with the saddle-node is either nonautonomous and periodically depends on time, or autonomous but possesses a global cross-section (at least at the part of the phase space under consideration). Essentially, the problem was reduced to the study of a one-parameter family of C^r-diffeomorphisms ($r \geqslant 2$) which has, at $\mu = 0$, a saddle-node fixed point such that all orbits of the unstable set of the saddle-node return to it as the number of iterations tends to $+\infty$.

Recall that the saddle-node point has one multiplier equal to 1, the remaining multipliers lying inside the unit circle. Near the fixed point, the diffeomorphism (the Poincaré map) has the form

$$\begin{cases} \bar{y} = Ay + H(y,z), \\ \bar{z} = z + G(y,z) \end{cases} \tag{1}$$

where $z \in \mathbb{R}^1$, $y \in \mathbb{R}^n$, A is a matrix whose eigenvalues lie strictly inside the unit circle, $H(0,0) = 0$, $H'_{(y,z)}(0,0) = 0$, $G(0,0) = 0$, $G'_{(y,z)}(0,0) = 0$. Here, the fixed point O is at the origin. It is well known that there exists a C^r-smooth invariant center manifold of the form $y = \eta(z)$, where $\eta(0) = 0$, $\eta'(0) = 0$. The map on the center manifold takes the form

$$\bar{z} = z + g(z), \tag{2}$$

where $g(z) \equiv G(\eta(z), z) \in C^r$, $g(0) = 0$, $g'(0) = 0$.

The fixed point O (and the corresponding periodic orbit) is called a *saddle-node* if $g(z)$ has a strict extremum at $z = 0$ (a strict minimum, to be definite); i.e., $g(z) > 0$ at $z \neq 0$. The saddle-node is called *simple* if $g''(0) \neq 0$. In this case equation (2) takes the form

$$\bar{z} = z + l_2 z^2 + \dots, \tag{3}$$

where $l_2 = g''(0)/2 \neq 0$. Without loss of generality one can assume that $l_2 > 0$.

One can see (Figure 1b) that a small neighborhood of O is split by the strong-stable invariant manifold $\{z = \xi(y)\}$ ($\xi(0) = 0$, $\xi'(0) = 0$) into two regions: the node region $\{x < \xi(y)\}$ and the saddle region $\{z > \xi(y)\}$. All orbits from the node region tend to O along the z-axis. The one-dimensional local unstable manifold $\{y = \eta(z),\ z > 0\}$ lies in the saddle region, and all its orbits tend to O with the iterations of the inverse map. All the other orbits from the saddle region leave the neighborhood of O with the iterations of both map (1) itself and its inverse.

Let W^u denote the global unstable manifold of O (the union of all iterations of the local unstable manifold). As we mentioned, all orbits of W^u are supposed to return to the node region, i.e., $\partial W^u = O$. Thus, here the closure $\overline{W^u}$ is homeomorphic to a circle. It turns out that $\overline{W^u}$ can be a smooth circle (Figure 4) or it can be nonsmooth (Figure 5).

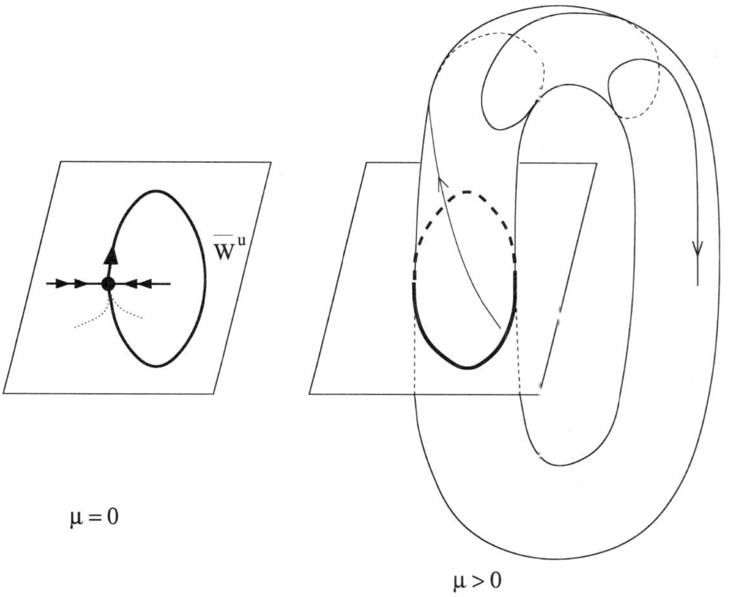

FIGURE 4. When (a) the set $\overline{W^u}$ is smooth, a smooth invariant curve is born (b) at $\mu > 0$.

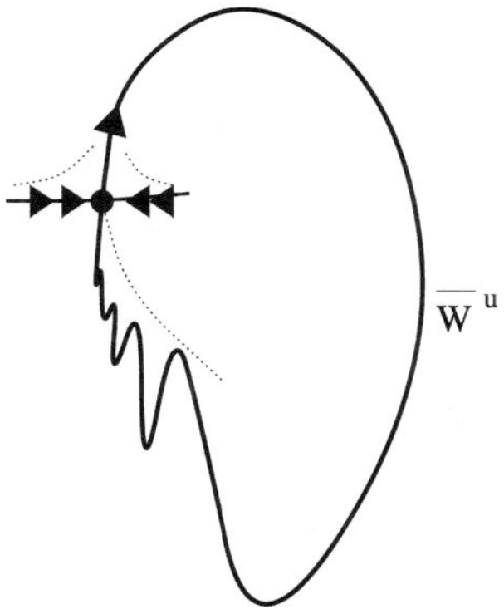

FIGURE 5. The set $\overline{W^u}$ may be nonsmooth; it may have "folds" which accumulate at the saddle-node.

To study bifurcations in a small neighborhood of $\overline{W^u}$ one must introduce a small parameter μ. The functions H and G in (1) should be viewed as depending on μ. We assume that μ is chosen so that the saddle-node disappears when $\mu > 0$

(Figure 1c). Generically, the saddle-node is simple and for $\mu < 0$, it degenerates into two fixed points, one being a saddle and another being stable (Figure 1a). The map on the center manifold is rewritten in this case as

$$\bar{z} = z + \mu + l_2 z^2 + \dots . \tag{4}$$

For the case where $\overline{W}^u$ is smooth, it was found in [**3**] that when the saddle-node disappears, an attractive smooth invariant curve inherits to $\overline{W}^u$. If the map under consideration is the Poincaré map of a global cross-section for some system of ODE's, then the invariant curve is the line of intersection of an invariant two-dimensional torus with the cross-section (Figure 4b). The Poincaré rotation number on the torus tends to zero as $\mu \to +0$.

This result gave a rigorous explanation of the transition from synchronization to amplitude modulation in periodically forced nonlinear systems: when $\mu < 0$, the only stable regime is the stable periodic orbit which corresponds to synchronization, and the invariant torus that exists at $\mu > 0$ corresponds to the modulation regime (see the discussion in [**4**]).

For the case where $\overline{W}^u$ is a nonsmooth manifold, it was established in [**3**] (under the so-called "big lobe" condition) that there exists a sequence of intervals (μ_i, μ_i') accumulating at $\mu = +0$ such that the system has nontrivial hyperbolic sets at $\mu \in (\mu_i, \mu_i')$. Without the big lobe condition (but for one-parameter families of a special kind), this result was proved in [**5**], on the basis of a theorem due to Block on the existence of periodic orbits for endomorphisms of a circle. In [**6**] the results of [**3, 5**] were extended to the general case; there it was also shown that for a "sufficiently small lobe" there exist both intervals of parameter values corresponding to complex dynamics (hyperbolic sets) and those corresponding to simple dynamics (a continuous invariant curve with rational Poincaré rotation number).

An important feature in the nonsmooth case is the existence [**5, 6, 7**] of parameter values arbitrarily close to $\mu = +0$, which correspond to the existence of saddle periodic orbits with *nontransverse homoclinic orbits*. According to our current knowledge (see [**8**]), this leads to extremely complicated dynamics: to the Newhouse phenomenon (persistence of homoclinic tangencies, coexistence of infinitely many sinks) [**9, 10, 11, 12, 13, 14**], to Hénon-like attractors [**15, 16, 7**] and to infinite degeneracies [**17, 14**], which make it impossible to give a complete description of the bifurcations that may occur in this case.

In the present paper we show that if an autonomous system with a saddle-node does *not* have a global cross-section, there may be considerably more different possible cases. Let

$$\dot{x} = X_\mu(x)$$

be a one-parameter family of n-dimensional C^r-smooth ($r \geqslant 2$) dynamical systems with a saddle-node periodic orbit L at $\mu = 0$. We assume that μ is the governing parameter for local bifurcations. Thus (Figure 1), at $\mu < 0$ there exist stable and saddle periodic orbits which combine in one orbit L at $\mu = 0$. The local unstable set $W^u_{L,\text{loc}}$ is homeomorphic to the half-cylinder $\mathbb{R}^+ \times S^1$. The orbit L also has a strong-stable manifold W^{ss}_L that divides the neighborhood of L into two regions: saddle and node. When $\mu < 0$, the saddle-node disappears and all orbits leave its small neighborhood. Thus, for the Poincaré map on the center manifold

$$\bar{z} = z + g(z; \mu), \tag{5}$$

the nonlinear part $g(z; \mu)$ is strictly positive at $\mu > 0$.

Note that if the starting point of some orbit lies in the node region, then the time which the orbit spends in a small fixed neighborhood of the saddle-node tends to infinity as $\mu \to +0$, as well as the length of the corresponding piece of the orbit.

Suppose that, at $\mu = 0$, all the orbits of W^u_L return to the node region and tend to L as $t \to +\infty$, not lying in W^{ss}_L. The union $\overline{W}^u$ of these orbits may, for instance, be a smooth two-dimensional surface: a torus, or a Klein bottle (the latter can happen if the phase space is nonorientable or if the dimension n of the phase space is not less than four). As in [**3**], the smooth invariant two-dimensional surface is preserved for $\mu > 0$. As above, if the set $\overline{W}^u$ is a nonsmooth torus, then saddle periodic orbits with homoclinic curves may appear at $\mu > 0$; the same can happen if $\overline{W}^u$ is a nonsmooth Klein bottle under some additional conditions [**6**].

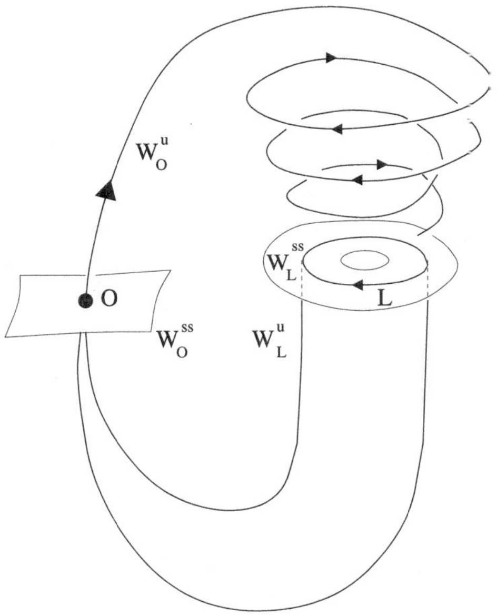

FIGURE 6. A codimension two bifurcation; the unstable manifold of the saddle-node periodic orbit L tends to the saddle-node equilibrium O whose unstable manifold, in turn, tends to L. When O disappears, the configuration shown in Figure 3 emerges.

Previously we did not know of any essentially different situation in which the set $\overline{W}^u$ is not a manifold. First, consider the following example. Let a *two*-parameter family of three-dimensional vector fields have, at some value of the parameters, a saddle-node periodic orbit L and a saddle-node equilibrium state O (Figure 6). Suppose that all orbits of W^u_L tend to O as $t \to +\infty$ and that the one-dimensional separatrix of O tends to L. If one of the parameters is varied so that O disappears and L does not, then the set $\overline{W}^u$ will have the form shown in Figure 3. The intersection of $\overline{W}^u$ with a local cross-section S to L will be a union of a countable set of circles accumulating at the point $S \cap L$ (Figure 3). Evidently, any neighborhood of this point in the set $\overline{W}^u$ is not homeomorphic to a disk. Therefore, in this case $\overline{W}^u$ is not a manifold.

Systems having a simple saddle-node periodic orbit with the set $\overline{W}^u$ as shown in Figure 3 constitute codimension one surfaces in the space of smooth flows in $\mathbb{R}^n$

($n \geqslant 3$). Below, we shall show (see Theorem 1) how open subsets are distinguished on these surfaces so that *for any one-parameter family* X_μ *that intersects such a subset transversely at* $\mu = 0$, *the system* X_μ *has (at all small* $\mu > 0$*) an attractive periodic orbit whose period and length tend to infinity as* $\mu \to +0$.

Note the relation of this result to the problem of the "blue sky catastrophe" [**18**]. The original formulation was as follows: *Does there exist a continuous one-parameter family of smooth vector fields on a compact manifold that have a closed orbit* L_μ *at all* $\mu > 0$ *and, as* $\mu \to +0$, *the period of* L_μ *tends to infinity so that* L_μ *disappears at* $\mu = +0$ *at a finite distance from the equilibrium states*[1]?

Virtual bifurcations of such kind were called *blue sky catastrophes* by Abraham. The first example of such a catastrophe was constructed by Medvedev [**19**] for a one-parameter family of vector fields on a Klein bottle with a saddle-node periodic orbit at $\mu = 0$. Medvedev's family was of a rather special kind: the system that corresponds to $\mu = 0$ is also embedded in a one-parameter family of conservative vector fields all orbits of which are closed. The Poincaré map for this conservative family has the form

$$\bar{\varphi} = -\varphi + \omega(\mu) \mod 1,$$

where $\omega \to \infty$ as $\mu \to +0$. This map has two fixed points, all other points are of period two. Basically, Medvedev used the fact that this family can be perturbed so as to have only two periodic orbits: one stable and one unstable fixed point; the stable fixed point corresponds to a stable periodic orbit whose period and length tend to infinity as $\mu \to +0$.

In a generic perturbation of Medvedev's one-parameter family both fixed points bifurcate infinitely many times as $\mu \to +0$, changing their stability (this was noticed in [**20**] and studied in more detail in [**21**]). Formally, the blue sky catastrophe also takes place here because the structural stability of the periodic orbit under consideration was not required in the original formulation.

The construction proposed in the present paper gives another solution to the blue sky problem. At the same time, it seems to be more adequate because the periodic orbit in Theorem 1 is stable (attractive) for all $\mu > 0$ and this property holds for an open set of one-parameter families.

The precise formulation of the result is given in the next section. The proof is based mostly on the evaluation of the local (Lemma 1) and global (Theorem 2) "through" map at small $\mu > 0$; the proofs are presented in Section 3. The computations are quite straightforward if the smoothness of the system with respect to the phase variables and the parameter is sufficiently high (in this case reduction to an almost autonomous normal form is possible) [**22, 23, 7, 26**]; the general case of low smoothness which we treat in the present paper requires more delicate calculations.

In fact, the results of Section 3 are applied to an arbitrary configuration of the set $\overline{W^u}$, allowing one to establish a correspondence between the dynamics near $\overline{W^u}$ and the dynamics of the one-dimensional *essential* map introduced in Section 2. The essential map is defined on a circle and its degree m defines the topological type of $\overline{W^u}$. Thus, at $m = 1$, the set $\overline{W^u}$ is homeomorphic to a torus, at $m = -1$ it is a Klein bottle. The present paper deals with $m = 0$, which corresponds to $\overline{W^u}$ as shown in Figure 3. In this case, under the conditions of Theorem 1, the dynamics of the essential map is trivial, so the main result follows immediately from

[1] The latter implies that the length of L_μ also tends to infinity.

the reduction principle of Theorem 2. More examples of the use of this principle can be found in [**6**] for the case in which $\overline{W}^u$ is a torus or a Klein bottle and in [**22, 23**] for other cases.

In particular, these papers treat the case $|m| \geqslant 2$, which is possible in $\mathbb{R}^n$ for $n \geqslant 4$. In this case hyperbolic attractors (the Smale–Williams solenoids) may appear via an analog of the blue sky catastrophe. More examples based on the disappearance of saddle-node invariant tori are given in [**23**].

2. Main results

Let U be a small neighborhood of $\overline{W}^u$ and U_0 a small neighborhood of L, $U_0 \in U$. Let us cut U_0 by a local cross-section S and consider the coordinates (y, z, φ), where $\varphi \in [0, 1]$ is the angular variable and (y, z) are the normal coordinates, $z \in \mathbb{R}^1$ is a coordinate on the center manifold, $y \in \mathbb{R}^{n-2}$ is a vector of coordinates corresponding to the multipliers less than 1 in absolute value; the values $\varphi = 0$ and $\varphi = 1$ correspond to the points lying on S.

The surfaces $\varphi = 0$ and $\varphi = 1$ are assumed to be glued by some involution, namely, by changing the sign of a number of components of the vector y. If this number is even (this is the orientable case, where the product of the multipliers of L is positive), then U_0 is a solid torus. If this number is odd (the nonorientable case, where the product of the multipliers of L is negative), then U_0 is the product of a Möbius strip and a disc D^{n-2}.

As it is well-known (see, for example, [**24**]), under an appropriate choice of the involution that glues the surfaces $\{\varphi = 0\}$ and $\{\varphi = 1\}$ one can without loss of smoothness introduce coordinates such that the linear part of the system near L is independent of φ. Thus, the system near L takes the form

$$\begin{cases} \dot{y} = \mathcal{A}(\mu)y + \widetilde{H}(z, y, \varphi; \mu), \\ \dot{z} = \widetilde{G}(z, y, \varphi; \mu), \\ \dot{\varphi} = 1, \end{cases} \tag{6}$$

where $\widetilde{H}$ and $\widetilde{G}$ vanish at $(y = 0,\ z = 0,\ \mu = 0)$ along with the first derivatives with respect to (z, y); the eigenvalues of the matrix $\mathcal{A}$ lie strictly to the left of the imaginary axis. We also assume that the center manifold W^c is locally straightened, so that it takes the form $\{y = 0\}$. Correspondingly,

$$\widetilde{H}(z, 0, \varphi; \mu) \equiv 0. \tag{7}$$

Further, let us straighten the strong stable invariant foliation [**27, 28**] transverse to the center manifold. The leaves of the foliation are given by

$$\{z = \mathcal{Q}(y; \varphi, z', \mu),\ \varphi = \text{const}\},$$

where z' is the coordinate of intersection of a leaf with the center manifold; $\mathcal{Q}$ is a C^{r-1}-function (it is C^r with respect to y). The straightening is the coordinate transformation $z \mapsto z'$. It brings the invariant foliation to the form $\{z = \text{const},\ \varphi = \text{const}\}$. Thus, the last two equations in (6) now become independent of y and the system can be rewritten as follows:

$$\begin{cases} \dot{y} = \mathcal{A}(\mu)y + \widetilde{H}(z, y, \varphi; \mu), \\ \dot{z} = \widetilde{G}(z, \varphi; \mu), \\ \dot{\varphi} = 1. \end{cases} \tag{8}$$

By construction, the new function $\widetilde{G}$ is the former one taken at $\{y = 0\}$; so it is still a C^r-function.

In the new coordinates, the strong-stable invariant manifold W_L^{ss} is the surface $\{z = 0\}$; the node region U_- now corresponds to small negative z and the saddle region U_+ corresponds to small positive z.

As we mentioned, the invariant foliation is *a priori* C^{r-1}. However, but it can be shown (Lemma 4) that it is, in fact, C^r everywhere except for W_L^{ss} at $\mu = 0$. The coordinate transformation which reduces (6) to (8) has the same smoothness.

According to the theorem "on the embedding into an autonomous flow" (see [**5**]; cf. Lemma 5 of the present paper), there exists a transformation of the coordinate $z\colon z \mapsto \mathcal{Z}(z, \varphi)$ which brings the second equation of (8) to an autonomous form for $\mu = 0$ (note that this transformation is identical at $\varphi = 0$ and is *uniquely* defined by the nonlinear part g of the Poincaré map on the center manifold (see (2))). Therefore, we assume

$$\left.\frac{\partial \widetilde{G}}{\partial \varphi}\right|_{\mu=0} \equiv 0. \tag{9}$$

In other words, at $\mu = 0$ the last two equations of (8) take the form

$$\dot{z} = \tilde{g}(z), \qquad \dot{\varphi} = 1, \tag{10}$$

where $\tilde{g}(0) = 0$, $\tilde{g}'(0) = 0$. If $z \neq 0$, then $\tilde{g}(z) > 0$. It is shown in Section 3 (see the comments after Lemma 5) that the "embedding" transformation $z \mapsto \mathcal{Z}$ is C^r at $z \neq 0$. Thus, system (8) after the transformation is C^{r-1} at $z \neq 0$, whereas the flow map between any two cross-sections that do not intersect $\{z = 0\}$ remains C^r.

Take small positive ε^+ and ε^-. Consider two cross-sections $S_0\colon \{z = -\varepsilon^-\}$ and $S_1\colon \{z = \varepsilon^+\}$ to the flow X_μ. In the orientable case, S_0 and S_1 are solid tori $S^1 \times D^{n-2}$; in the nonorientable case, they are homeomorphic to the product of the Möbius strip and the disc D^{n-3}.

At $\mu = 0$ (hence, at all small μ), all orbits of W_L^u return to the node region $U^- = \{z < 0\}$ in finite time. Therefore, the flow X_μ defines a diffeomorphism T_1 by which a small neighborhood of the intersection line $l^-\colon \{y = 0\} = W^u \cap S_1$ is mapped into S_0. This map has the form

$$y_0 = p(\varphi_1, y_1; \mu), \qquad \varphi_0 = q(\varphi_1, y_1; \mu) \mod 1, \tag{11}$$

where the coordinates on S_0 and S_1 are denoted by (φ_0, y_0) and (φ_1, y_1) respectively; C^r-smooth functions p and $q \mod 1$ are 1-periodic in φ.

The curve $l^+ = T_1 l^-\colon \{y_0 = p(\varphi_1, 0; 0),\ \varphi_0 = q(\varphi_1, 0; 0) \mod 1\}$ is the intersection of $\overline{W}^u$ and S_0. Note that the function q can be written in the form

$$q(\varphi, y; \mu) = m\varphi + q_0(\varphi, y; \mu), \tag{12}$$

where q_0 is periodic in φ. The integer m defines the homotopy class of l^+ in S_0 (the sign of m defines the orientation of l^+ with respect to l^-).

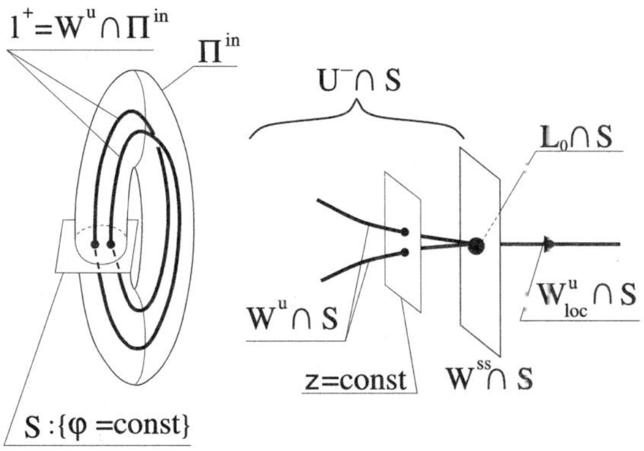

FIGURE 7. If the phase space is four-dimensional (or higher), the cross-sections S_0 and S_1 are at least three-dimensional. Therefore, the image l^+ of the curve l^- may be a multi-round closed curve. The figure gives an example with two rounds ($m = 2$).

If the dimension n of the phase space is greater than three, then S_0 is at least three-dimensional and the integer m may be arbitrary (see Figure 7 for the case $m = 2$). At $n = 3$, the cross-section S_0 is a two-dimensional annulus in the orientable case. Therefore, in this case m can only take the values 1 (Figures 4, 5), 0 (Figure 3) and -1 (the latter only on nonorientable manifolds). In the nonorientable case at $n = 3$, the cross-section S_0 is a Möbius strip, therefore, we can only have $m = 0, \pm 1, \pm 2$. At $n = 2$, S_0 coincides with l^+ and we can only have $m = \pm 1$.

Note that the structure of the set $\overline{W^u}$ is completely determined by the way $\overline{W^u}$ adjoins to L from the side of the node region. It is not hard to see that the intersection of $\overline{W^u} \cap U_-$ with any cross-section of the kind $\{\varphi = \text{const}\}$ consists, at $m \neq 0$, of $|m|$ pieces glued at the point $\{z = 0,\ y = 0\} = L_0 \cap \{\varphi = \text{const}\}$. It is clear that pieces of $\overline{W^u}$ corresponding to different values of m are mutually nonhomeomorphic. It is also clear that $\overline{W^u}$ is a manifold if and only if $m = \pm 1$ (a torus or a Klein bottle respectively).

We define the *essential map*

$$\varphi \mapsto f(\varphi) \equiv m\varphi + q_0(\varphi, 0; 0). \tag{13}$$

It is defined at $\mu = 0$. By construction, it is obtained as follows: apply the map T_1 to the intersection line of the unstable manifold $W^u = W^c \cap U_+$ with the cross-section S_1 and then project the image on the center manifold along the leaves of the strong-stable foliation. The projection is done in S_0, which lies in the node region where the foliation is defined uniquely [**27, 28**]. Thus, once the cross-sections S_0 and S_1 are fixed, the essential map is defined uniquely modulo coordinate transformations on the center manifold (the center manifold in the node region is not unique, but systems on different center manifolds are conjugate by the projection along the strong stable invariant foliation, therefore the choice of another center manifold is equivalent to a coordinate transformation on the initial one).

In fact, the set of coordinate transformations which keep the system at $\mu = 0$ in the form (10) is rather small. Indeed, a new coordinate φ must satisfy

$$\frac{d}{dt}(\varphi_{\text{new}} - \varphi) = 0,$$

hence the difference $\varphi_{\rm new} - \varphi$ must be constant along the orbits of the system. In particular, it is constant on L. Now, since any orbit on the center manifold tends to L either as $t \to +\infty$ or as $t \to -\infty$, it follows that $\varphi_{\rm new} - \varphi = \text{const}$ everywhere on W^c. Further, since the equation for z in (10) must remain autonomous, one can show that only autonomous (independent on φ) transformations of the variable z are allowed. Indeed, consider first a transformation which is identical at $\varphi = 0$. By definition, it does not change the Poincaré map of the cross-section $S\colon \{\varphi = 0\}$, therefore, by the uniqueness of the embedding into the flow, if such a transformation keeps the system autonomous, it cannot change the right-hand side. We see that if $z_{\rm new} \equiv z$ at $\varphi = 0$, then the time evolution of $z_{\rm new}$ and the time evolution of z is governed by the same equation, which immediately implies that $z_{\rm new} \equiv z$ for all φ in this case. Since an arbitrary transformation is a superposition of an autonomous transformation and a transformation of the kind we have just considered, this proves the claim.

Thus, the only possible coordinate transformation is

$$\varphi \mapsto \varphi + \text{const}, \qquad z \mapsto Z(z). \tag{14}$$

For the essential map, the effect of such a transformation of z is equivalent to the shift of the cross-sections S_0 and S_1 to $z = Z^{-1}(-\varepsilon^-)$ and $z = Z^{-1}(\varepsilon^+)$ respectively. Since the evolution of z is autonomous, the flight time from a cross-section $\{z = \text{const}\}$ to any other cross-section of this form depends only on the position of the cross-sections, but does not depend on the initial point on the cross-section. Thus, any shift of S_0 or S_1 is equivalent to a rigid rotation of φ_0 or φ_1 respectively. We see finally that

the essential map is uniquely defined by the system at $\mu = 0$, modulo an arbitrary additive constant and a shift of the origin:

$$f(\varphi) \to c_0 + f(\varphi + c_1).$$

The essential map carries most of the information on the global saddle-node bifurcations. As we mentioned, its degree m defines the topological type of $\overline{W^u}$. If $|m| = 1$, then $\overline{W^u}$ is smooth if and only if $f(\varphi)$ does not have critical points (cf. with [**5, 6**]). Below (Theorem 2), we give a precise formulation to the following *reduction principle*:

the bifurcations in $U(\overline{W^u})$ at $\mu > 0$ follow the bifurcations in the family of one-dimensional maps

$$\bar{\varphi} = \omega(\mu) + f(\varphi) \mod 1, \tag{15}$$

where ω increases to infinity as $\mu \to +0$.

This was used explicitly in [**6**] for the case $|m| = 1$; the study in [**5**] was essentially based on the same idea. In terms of [**25**], for the orbit of $\overline{W^u}$ intersecting S_1 at $\varphi = \varphi_1$, the derivative $f'(\varphi_1)$ equals the conventional multiplier of the orbit, i.e., the quantity which determines the value of the multiplier of a periodic orbit that may be born nearby when L disappears.

The theorem below is another consequence of the reduction principle.

THEOREM 1. *Let $m = 0$ and $|f'(\varphi)| < 1$ for all φ. Then, for all small $\mu > 0$, the system X_μ has a stable periodic orbit L_μ (nonhomotopic to L in U) which attracts all orbits of U.*

Notice that we do not require here that the saddle-node L be simple. If it is, however, a simple saddle-node, then systems close to X_0 and having a saddle-node periodic orbit close to L constitute a codimension one bifurcational surface in the space of dynamical systems. By construction, the function f depends continuously on the system on the bifurcational surface. Thus, if the conditions of Theorem 1 are fulfilled for some system X_0, they are also fulfilled for all close systems on the bifurcational surface. This implies that Theorem 1 is valid for any one-parameter family that intersects the surface transversely near X_0. In other words, our blue sky catastrophe occurs generically in one-parameter families.

The proof of Theorem 1 is based on the calculation of the Poincaré map $T = T_0 \circ T_1$ of the cross-section S_1, which is defined by the orbits of X_μ for all small $\mu > 0$. Here, T_1 is a global map defined by (11) and $T_0 \colon S_0 \to S_1$ is the through map defined locally near L at $\mu > 0$.

Since the last two equations in (8) are independent of y, the map $T_0 \colon (y_0, \varphi_0) \mapsto (y_1, \varphi_1)$ is written in the form (for some C^r-function Y)

$$(16) \qquad \begin{cases} y_1 = Y(\varphi_0, y_0, \mu), \\ \varphi_1 = \varphi_0 + \tau(\varphi_0, \mu) \mod 1. \end{cases}$$

The function τ is the *flight time* from S_0 to S_1. It is a smooth function periodic in φ_0. Clearly, we have $\tau(\varphi_0, \mu) \to \infty$ as $\mu \to +0$. In the next section we prove the following

LEMMA 1. *If* (9) *is satisfied at* $\mu = 0$, *then* $\partial\tau/\partial\varphi$ *uniformly tends to zero as* $\mu \to +0$ *in the* C^{r-1}*-topology.*

Denoting $\omega(\mu) = \tau(0, \mu)$, from this lemma we obtain

$$(17) \qquad \tau(\varphi_0, \mu) = \omega(\mu) + o(1).$$

Besides, since the spectrum of the matrix $\mathcal{A}(\mu)$ from (6) lies strictly to the left of the imaginary axis and since, by (7), $\dot{y} = (\mathcal{A} + \tilde{h})y$ for some C^{r-1}-function $\tilde{h}$, it is routinely shown that

$$(18) \qquad \|Y\|_{C^{r-1}} \leqslant O(e^{-\alpha\omega})$$

for some positive α. In fact, we shall note in the next section that

$$(19) \qquad \|Y\|_{C^r} \to 0 \quad \text{as } \mu \to 0.$$

Collecting formulas (11), (12), (13), (16), (17), and (19) we obtain the following result.

THEOREM 2 (reduction principle). *As* $\mu \to +0$, *the Poincaré map* $T = T_0 \circ T_1$ *approaches* (*along with all derivatives*) *the map*

$$(20) \qquad \bar{y} = 0, \qquad \bar{\varphi} = \omega + f(\varphi) \mod 1.$$

In the case of Theorem 1 ($|f'(\varphi)| < 1$), the map (20) is contracting. Hence, for each ω it has a unique attractor, namely, the unique stable fixed point. The same is clearly valid for all close maps, in particular, for the map T at small $\mu > 0$. Since the map T is defined by the orbits of the flow X_μ, the fixed point corresponds to the attractive periodic orbit L_μ of X_μ. This proves the theorem: the period of L_μ grows in proportion to $\omega(\mu)$ and it tends to infinity as $\mu \to +\infty$; since the vector field of X_μ does not vanish in U, it follows that the length of L_μ also tends to infinity.

The rest of the paper is devoted to the proof of Theorem 2. For greater generality, we consider also the case where the system depends on μ smoothly: we assume that

> *the first derivatives of the right-hand sides of* (6) *with respect to the phase variables* (y, z, φ) *are* C^{r-1} *with respect to all variables and* μ.

Assume also that the local Poincaré map near L depends monotonically on μ, i.e.,

$$g'_\mu(0; 0) > 0 \tag{21}$$

in (5). In this case we prove that $\partial\omega/\partial\mu \neq 0$, i.e., ω can be taken as a new parameter and μ may then be considered as a function of ω which tends to zero as $\omega \to \infty$. We prove (Lemma 9) that all derivatives of μ with respect to ω tend to zero also. Lemma 1 remains valid, concerning now the derivatives of τ'_φ with respect to both ω and φ_0. Theorem 2 reads exactly as before, "all the derivatives" now include the derivatives with respect to ω.

3. Calculation of the through map

In this section we give the proof of Lemma 1 and other facts related to the proof of Theorem 2. Let $\{z_0, z_1, \dots\}$ be an orbit of the local Poincaré map (5) of the system X_μ on the center manifold at $\mu \geqslant 0$:

$$z_{j+1} = z_j + g(z_j; \mu), \tag{22}$$

where the variable z is assumed to belong to some small interval $[-\varepsilon^-, \varepsilon^+]$.

Since $g > 0$ if $\mu > 0$ or if $\mu = 0$ and $z \neq 0$, the sequence $\{z_0, z_1, \dots\}$ is monotonically increasing: $z_0 < z_1 < \dots < z_j < \dots$ in this case. We shall use the following simple estimate:

$$\sum_{i=0}^{j-1} g(z_i; \mu) = \sum_{i=0}^{j-1} (z_{i+1} - z_i) = z_j - z_0 \leqslant \varepsilon^+ + \varepsilon^-. \tag{23}$$

LEMMA 2. *For some smooth function* ψ

$$\frac{\partial z_j}{\partial z_0} = \frac{g(z_j; \mu)}{g(z_0; \mu)} \exp\left\{\sum_{i=0}^{j-1} \psi(z_i; \mu)\right\}, \tag{24}$$

where $\Psi_j = \sum_{i=0}^{j-1} \psi(z_i; \mu)$ *is uniformly bounded for all* $\mu \geqslant 0$ *and for all* z_0 *and* $j \geqslant 0$ *such that* $-\varepsilon^- \leqslant z_0 \leqslant z_j \leqslant \varepsilon^+$.

PROOF. By differentiating (22), we obtain

$$\frac{\partial z_{j+1}}{\partial z_0} = (1 + g'_z(z_j; \mu)) \frac{\partial z_j}{\partial z_0} \quad \text{and} \quad \frac{\partial z_j}{\partial z_0} = \prod_{i=0}^{j-1} (1 + g'_z(z_i; \mu)). \tag{25}$$

Thus relation (24) is fulfilled for

$$\psi = \ln \frac{(1 + g'_z(z; \mu))g(z; \mu)}{g(\bar{z}; \mu)}. \tag{26}$$

Since

$$g(\bar{z}; \mu) = g(z + g(z; \mu); \mu) = g(z; \mu)\left(1 + \int_0^1 g'_z(z + sg(z; \mu); \mu)\, ds\right),$$

it follows that

$$|\psi(z_i;\mu)| \leqslant \frac{1}{1+\min_{z\in[-\varepsilon^-,\varepsilon^+]} g'_z(z,\mu)} \max_{s\in[0,1]} |g'_z(z_i;\mu) - g'_z(z_i + sg(z_i;\mu);\mu)| \leqslant Cg(z_i;\mu), \tag{27}$$

where

$$C = \frac{\max_{z\in[-\varepsilon^-,\varepsilon^+]} |g''_{z^2}(z;\mu)|}{1-\max_{z\in[-\varepsilon^-,\varepsilon^+]} |g'_z(z;\mu)|}.$$

Since g'_z is small by assumption, the factor C is finite. Hence, the uniform boundedness of $\sum_{i=0}^{j-1}\psi(z_i,\mu)$ follows from (27) and (23), which yields the lemma. □

LEMMA 3. *At small $\mu \geqslant 0$, for any $k = 1,\dots,r$ the following estimate holds:*

$$\left|\frac{\partial^k z_j}{\partial z_0^k}\right| \leqslant \text{const}\cdot g(z_j;\mu) \tag{28}$$

uniformly for all z_0 and z_j from $[-\varepsilon^-,\varepsilon^+]$, provided z_0 is bounded away from zero.

PROOF. By Lemma 2, the estimate (28) holds at $k=0$. Suppose it is fulfilled for all $k \leqslant k_0$ for some k_0. In this case similar estimates holds for the same k for any smooth function of z_j. Indeed, if β is some C^k-function of z, then the kth derivative of $\beta(z_j)$ with respect to z_0 is represented as the sum of terms of the form

$$\text{const}\cdot\frac{\partial^s\beta}{\partial z_j^s}\left(\frac{\partial^{k_1}z_j}{\partial z_0^{k_1}}\right)\cdots\left(\frac{\partial^{k_s}z_j}{\partial z_0^{k_s}}\right),$$

where $1\leqslant s\leqslant k$, $k_1\geqslant 1$, ..., $k_s \geqslant 1$, and $k_1+\cdots+k_s = k$. By (28), the absolute value of such term is estimated from above by $O(g(z_j;\mu)^s)$. Indeed, since g is small, it follows that

$$\left|\frac{\partial^k\beta}{\partial z_0^k}\right| \leqslant \text{const}\cdot g(z_j;\mu).$$

In particular, the validity of (28) at all $k\leqslant k_0\leqslant r-1$ implies that

$$\left|\frac{\partial^k g(z_j;\mu)}{\partial z_0^k}\right| \leqslant \text{const}\cdot g(z_j;\mu). \tag{29}$$

$$\left|\frac{\partial^k \psi(z_j;\mu|}{\partial z_0^k}\right| \leqslant \text{const}\cdot g(z_j;\mu) \tag{30}$$

where ψ is given by (26). By Lemma 2,

$$\frac{\partial^{k_0+1}z_j}{\partial z_0^{k_0+1}} = \frac{\partial^{k_0}}{\partial z_0^{k_0}}\left(\frac{g(z_j;\mu)}{g(z_0;\mu)}e^{\Psi_j(z_0;\mu)}\right).$$

Thus, the (k_0+1)th derivative of z_j is represented as the sum of terms

$$\left(\frac{\partial^{k_1}}{\partial z_0^{k_1}}g(z_j;\mu)\right)\left(\frac{\partial^{k_2}}{\partial z_0^{k_2}}\frac{1}{g(z_0;\mu)}\right)\left(\frac{\partial^{k_3}}{\partial z_0^{k_3}}e^{\Psi_j(z_0;\mu)}\right) \tag{31}$$

taken with some constant coefficients; here, $k_1+k_2+k_3=k_0$.

The first factor in this product is estimated by (29), the second factor is bounded if z_0 is bounded away from zero. The derivative $\partial^k\Psi_j/\partial z_0^k$ is uniformly bounded at any $k\leqslant k_0$ because

$$\left|\frac{\partial^s}{\partial z_0^s}\Psi_j\right| = \left|\sum_{i=0}^{j-1}\frac{\partial^k\psi_i}{\partial z_0^k}\right| \leqslant O\left(\sum_{i=0}^{j-1}g(z_i;\mu)\right) \leqslant \text{const}\cdot(\varepsilon^+ + \varepsilon^-) \tag{32}$$

(see (23), (30)). Therefore, the third factor in (31) is uniformly bounded. Thus, estimate (28) remains fulfilled at $k = k_0 + 1$. By induction, this proves the lemma. □

Let us now prove that the strong stable invariant foliation of system (6) is C^r-smooth outside the strong stable manifold. This is a necessary step to ensure the transition from (6) to (8) without an essential loss of smoothness.

LEMMA 4. *Let the leaves of the strong stable invariant foliation of system* (6) *be given by* $\{z = Q(y, z', \varphi, \mu),\ \varphi = \text{const}\}$, *where* z' *is the intersection of the leaf with* $\{y = 0\}$. *Then* Q *is* C^r *with respect to* (y, z', φ) *in the region* $\{\mu > 0\} \cup \{\mu = 0, z \neq 0\}$.

PROOF. It suffices to prove the smoothness of the strong stable foliation $z = Q(y, z', \mu)$ of the local Poincaré map $\Pi\colon S \to S$ given by (1), where the center manifold is straightened to $\{y = 0\}$ so that

$$H(0, z, \mu) \equiv 0, \quad \text{and} \quad G(0, z, \mu) \equiv g(z; \mu). \tag{33}$$

Thus, $\bar{y} = (A(\mu) + h(y, z, \mu))y$ for some C^{r-1}-function h. Since z grows sub-exponentially, this equation gives

$$\|y_j\|_{C^{r-1}} \leqslant O(e^{-\alpha j}), \tag{34}$$

where y_j, as a function of a point (y, z), is the y-coordinate of the jth iteration of this point by Π; here $\alpha > 0$ is such that the spectrum of A lies strictly inside the circle $|\cdot| = e^{-\alpha}$ in the complex plane.

By invariance of the strong stable foliation, we must have

(35)
$$Q(y, z', \mu) + G(y, Q(y, z', \mu), \mu) = Q(A(\mu)y + H(y, Q(y, z', \mu), \mu), z' + g(z'; \mu)).$$

Thus, at $\mu > 0$, once Q is defined on the fundamental domain $\varepsilon^+ < z' \leqslant \varepsilon^+ + g(\varepsilon^+, \mu)$ in an arbitrary (smooth) way, by applying (35) repeatedly we can define it on a whole neighborhood of the origin (because at $\mu > 0$, for any $z' \in [-\varepsilon^-, \varepsilon^+]$ some iteration of it by (5) enters the fundamental domain). At $\mu = 0$ the function Q is defined by continuity. It is well-known [**27, 28**], that this procedure indeed defines a C^{r-1}-function Q. Moreover, Q'_y is also C^{r-1}. Thus, we have only to prove the existence and continuity of $Q^{(r)}_{(z')^r}$. By (35)

$$\frac{\partial Q}{\partial z'}(y, z', \mu) = \frac{\partial Q}{\partial z'}(y_J, z_J, \mu) \cdot \prod_{j=0}^{J-1} \Phi(y_j, z_j, \mu), \tag{36}$$

where

$$\Phi = \frac{1 + g'_z(z; \mu)}{1 + G'_z(y, Q(y, z, \mu), \mu) - Q'_y(y, z, \mu) H'_z(y, Q(y, z, \mu), \mu)};$$

here, $z_0 = z', z_1, \ldots, z_J$ is the orbit of z' by the map (5) with the last point z_J in the fundamental domain, and y_j is the y-coordinate of the jth iteration of the point $(y, Q(y, z', \mu))$ by the map (1): $(y_j, Q(y_j, z_j, \mu)) = \Pi^j(y, Q(y, z', \mu))$.

This formula defines $Q'_{z'}$ as a C^{r-1}-function at $\mu > 0$ and at $\mu = 0$, $z' > 0$. At $z' < 0$, the limit $\mu \to +0$ gives

$$\frac{\partial Q}{\partial z'}(y, z', 0) = \prod_{j=0}^{+\infty} \Phi(y_j, z_j, 0) \tag{37}$$

(since y_j tends to zero exponentially as $j \to +\infty$, uniformly in μ, it follows from (33) that Φ_j tends exponentially to 1, i.e., the product converges uniformly; moreover, $Q'_{z'}(y_J, z_J, \mu) \to 1$, because $Q(0, z, \mu) \equiv z$ by definition).

It remains to prove that (37) defines a C^{r-1}-function at $z' < 0$. For this, one must show that the series

$$\sum_{j=0}^{J} \frac{\partial^k}{\partial z_0^k} \Phi(y_j, z_j, \mu)$$

converges absolutely and uniformly in μ for all $k \leqslant r - 1$. Since Φ is smooth, it suffices to prove that the series

$$\sum_{j=0}^{J} \frac{\partial^k (y_j, z_j)}{\partial z_0^k}$$

converges absolutely and uniformly for $k \leqslant r - 1$. The latter immediately follows from (34), (23), and (28). The lemma is proved. □

Another result necessary to properly define the essential map (13) is given by the following lemma.

LEMMA 5. *There exists a unique smooth function $\tilde{g}(z)$ such that at $\mu = 0$ the local Poincaré map on the center manifold*

$$\bar{z} = z + g(z; 0) \tag{38}$$

coincides with the time-1 map of the flow

$$\dot{z} = \tilde{g}(z). \tag{39}$$

The function $\tilde{g}$ is given by

$$\tilde{g}(z) = \begin{cases} g(z;0)\exp\Big\{-\sum\limits_{i=0}^{\infty} \psi(z_i;0)\Big\} & \text{at } z < 0, \\ g(z;0)\exp\Big\{-\sum\limits_{i=-1}^{-\infty} \psi(z_i;0)\Big\} & \text{at } z > 0, \end{cases} \tag{40}$$

where $\{z_j\}$ is the orbit of $z = z_0$ by the map (38); ψ is the function defined by (26).

PROOF. The map (38) coincides with the time-1 map of flow (39) if and only if

$$\int_z^{\bar{z}} \frac{ds}{\tilde{g}(s)} = 1. \tag{41}$$

Since $z = 0$ is the fixed point of (38) with multiplier equal to 1, it follows that $\tilde{g}$ must vanish at zero along with $\tilde{g}'$. Taking the limit $z \to 0$ in (41), we get

$$\lim_{z\to 0} \frac{\tilde{g}(z)}{g(z;0)} = 1 \tag{42}$$

because

$$\int_z^{\bar{z}} \frac{ds}{\tilde{g}(s)} = \frac{\bar{z} - z}{\tilde{g}(z) + \frac{1}{2}\tilde{g}'(z)(\bar{z} - z) + o(\bar{z} - z)} \sim \frac{g(z;0)}{\tilde{g}(z)};$$

here we use the smoothness of g at zero.

Differentiating (41), we obtain

$$\frac{1 + g'_z(z;0)}{\tilde{g}(\bar{z})} - \frac{1}{\tilde{g}(z)} = 0. \tag{43}$$

Therefore, for $z < 0$,

$$\tilde{g}(z_j) = \tilde{g}(z) \prod_{i=0}^{j-1} (1 + g'_z(z_i; 0))$$

which gives (40) by virtue of (42), (24), and (25). The series $\sum_{i=0}^{+\infty} \psi(z_i; 0)$ is convergent due to (27), (23). For $z > 0$ formula (40) is obtained in the same way.

Thus, the smooth function $\tilde{g}$ satisfying (41) is indeed defined uniquely and it must be given by (40). It remains to prove that the function $\tilde{g}$ defined by (40) is smooth (obviously, it satisfies (43) and (42), which implies (41) for *smooth* functions). The first derivative is given, for $z < 0$, by the formula (we use (24)):

$$\tilde{g}'(z) = \left[g'(z) - \sum_{j=0}^{+\infty} \psi'(z_j; 0) g(z_j; 0) \exp\left\{ \sum_{i=0}^{j-1} \psi(z_i; 0) \right\} \right] \exp\left\{ - \sum_{i=0}^{+\infty} \psi(z_j; 0) \right\}.$$

The series here are uniformly convergent and bounded by virtue of (27) and (23). Moreover, since $g'_z(0;0) = 0$ and $\psi'_z(0;0) = 0$, it follows that $\tilde{g}'(-0) = 0$. Analogously, $\tilde{g}'(+0) = 0$; i.e., the function $\tilde{g}$ defined by (40) is smooth at zero. The lemma is proved. □

Note that the series $\sum_{i=0}^{\infty} \psi(z_i; 0)$ for $z_0 < 0$ and $\sum_{i=1}^{+\infty} \psi(z_{-i}; 0)$ for $z_0 > 0$ absolutely converge together with the derivatives with respect to z_0 up to the order $r-1$ because all partial sums are uniformly bounded according to (30), (23). Thus, the function $\tilde{g}$ is C^{r-1} at $z \neq 0$. Hence, the time t shift $z \mapsto \pi_t^*(z)$ of the system (39) is C^r with respect to z at $z \neq 0$.

If $z \mapsto \pi_t(z, \varphi)$ denotes the time t shift of the point (z, φ) on the center manifold of system (8), then $\pi_{-\varphi}(z, \varphi)$ is the projection of the point (z, φ) onto the local cross-section S: $\{\varphi = 0\}$ by the backward orbit. By uniqueness of $\tilde{g}$, the transformation $z \mapsto \mathcal{Z}(z, \varphi) \equiv \pi_\varphi^* \circ \pi_{-\varphi}$ is the unique 1-periodic transformation in φ which is identical at $\varphi = 0$ and which brings the last two equations of (8) to the autonomous form (10). As we just showed, the transformation is C^r at $z \neq 0$, hence, the essential map and the map T of Theorem 2 are C^r indeed.

Now let us proceed directly to the proof of Theorem 2 (first, for the case where the right-hand sides depend only continuously on μ). At $\mu > 0$, for any z_0 there exists an integer $N(z_0)$ such that the Nth iteration z_N of z_0 belongs to the interval $I_\mu = [0, g(0;\mu)[$ ($N > 0$ for $z_0 < 0$ and $N \leqslant 0$ for $z_0 \geqslant 0$). Let

$$\zeta = \frac{z_N}{g(0;\mu)} \,.$$

Since $|z_N| \leqslant g(0;\mu)$ and $g(z_N, \mu) = g(0;\mu) + g'_z(0;\mu) z_N + o(z_N)$, it follows that $g(z_N;\mu)/g(0;\mu) \to 1$ as $\mu \to +0$. Thus, by Lemma 3, we get

LEMMA 6. *All derivatives of $\zeta(z;\mu)$ are uniformly bounded.*

Let us define

$$\sigma(z;\mu) = \theta(\zeta(z);\mu) - N(z), \tag{44}$$

where

$$\theta(\zeta;\mu) = \int_0^\zeta \frac{ds}{\mathcal{G}(s;\mu)} + \chi(\zeta)\left[\int_\zeta^{\zeta + \mathcal{G}(\zeta;\mu)} \frac{ds}{\mathcal{G}(s;\mu)} - 1 \right], \tag{45}$$

$$\mathcal{G}(\zeta;\mu) = \frac{g(\zeta g(0;\mu);\mu)}{g(0;\mu)}, \tag{46}$$

and $\chi(\zeta)$ is a C^r-function equal to 1 at $|\zeta| \leqslant \delta$ and equal to zero at $|\zeta - 1| \leqslant \delta$ for some small $\delta > 0$. Note that

$$\lim_{\mu \to +0} \mathcal{G}(\zeta; \mu) = 1, \tag{47}$$

$$\lim_{\mu \to +0} (\theta(\zeta; \mu) - \zeta) = 0, \tag{48}$$

where the limits are taken in C^r-topology.

By definition, $\sigma(z; \mu)$ satisfies

$$\sigma(\bar{z}; \mu) = \sigma(z; \mu) + 1. \tag{49}$$

LEMMA 7. *The function $\sigma(z; \mu)$ is C^r. Moreover, for small $z \neq 0$ we have*

$$\lim_{\mu \to +0} \frac{d\sigma}{dz} = \frac{1}{\tilde{g}(z)}, \tag{50}$$

where the limit is taken in the C^{r-1}-topology; $\tilde{g}$ is the right-hand side of system (10), defined by (40).

PROOF. By definition, σ is C^r on I_μ. Thus, by virtue of (49) it is sufficient to check the continuity of the derivatives of σ at $z = 0$. By (45) and (46), for small $z > 0$ we have

$$\sigma(z; \mu) = \int_0^z \frac{ds}{g(s; \mu)} + \chi\left(\frac{z}{g(0; \mu)}\right)\left[\int_z^{\bar{z}} \frac{ds}{g(s; \mu)} - 1\right]$$

and for small $z < 0$

$$\sigma(z; \mu) = \int_0^{\bar{z}} \frac{ds}{g(s; \mu)} - 1 + \chi\left(\frac{\bar{z}}{g(0; \mu)}\right)\left[\int_{\bar{z}}^{\bar{z}+g(\bar{z};\mu)} \frac{ds}{g(s; \mu)} - 1\right].$$

By the definition of χ, in both cases we see that if, say, $|z| < \delta g(0; \mu)/2$, then

$$\sigma(z; \mu) = \int_0^{\bar{z}} \frac{ds}{g(s; \mu)} - 1$$

which is indeed a C^r-function for all small z.

Now, by (44)–(46) and (24),

$$\sigma'_z(z; \mu) = \theta'_\zeta(\zeta; \mu)\mathcal{G}(\zeta; \mu) \frac{1}{g(z; \mu)} \exp\left\{\sum_{i=0}^{N-1} \psi(z_i; \mu)\right\} \tag{51}$$

for $z < 0$. Thus, (50) follows immediately from (47), (48), Lemma 6 and (40). The case $z > 0$ is analysed in the same way. The lemma is proved. □

Now note that Lemma 1 follows from Lemma 7 almost immediately. Indeed, let the orbit by system (8) of the point $(z = -\varepsilon^-, \varphi_0, y_0)$ of S_0 intersect S_1 at the point $(z = \varepsilon^+, \varphi_1, y_1)$. Then, by construction

$$\sigma(z^0, \mu) + N(z^0) = \sigma(z^1, \mu) - N(z^1), \tag{52}$$

where

$$z^0 = \pi_{1-\varphi_0}(-\varepsilon^-, \varphi_0), \qquad z^1 = \pi_{-\varphi_1}(\varepsilon^+, \varphi_1); \tag{53}$$

here, as above, N is the number of iterations necessary to get into I_μ, and $\pi_t(z, \varphi)$ is the time t shift of z by the flow (8).

Differentiating (52), we get

$$\sigma_z'(z^0,\mu)\,\frac{\partial z^0}{\partial\varphi_0}=\sigma_z'(z^1,\mu)\,\frac{\partial z^1}{\partial\varphi_1}\,\frac{\partial\varphi_1}{\partial\varphi_0}\,. \tag{54}$$

As $\mu\to+0$,

$$\pi_t(z,\varphi)\,\underset{C^r}{\longrightarrow}\,\pi_t^*(z) \tag{55}$$

where π_t^* is the time t shift by the autonomous flow (10). Thus, by (50) and (53)–(55)

$$\frac{\partial\varphi_1}{\partial\varphi_0}\,\underset{C^{r-1}}{\longrightarrow}\,1 \tag{56}$$

as $\mu\to+0$, which is exactly the statement of Lemma 1.

To complete the proof of Theorem 2, it remains to prove (19). It is sufficient to show that as $j\to+\infty$,

$$\|y_j\|_{C^r}\to 0 \tag{57}$$

uniformly for all μ; here y_j, as a function of (y_0,z_0), is the y-coordinate of the jth iteration (y_j,z_j) of the point (y_0,z_0) by the local Poincaré map $\Pi\colon S\to S$ given by (1). We assume that the center manifold is straightened, i.e., (33) holds. By differentiating of (1) r times, we obtain

(58)

$$\frac{\partial^r y_{j+1}}{\partial(y_0,z_0)^r}=(A+H_y'(y_j,z_j))\,\frac{\partial^r y_j}{\partial(y_0,z_0)^r}+H_{z^r}^{(r)}(y_j,z_j)\left(\frac{\partial z_j}{\partial(y_0,z_0)}\right)^r+O(e^{-\alpha j})$$

(we have used that by (34), (33) all the derivatives of y_j and of H up to order $r-1$ tend to zero exponentially as $j\to+\infty$). Now note that $\partial z_j/\partial(y_0,z_0)$ is uniformly bounded (z_j is independent of y_0 when the strong stable foliation is straightened, and $\partial z_j/\partial z_0$ is bounded according to Lemma 2), whereas $H_{z^r}^{(r)}(y,z)$ tends to zero as $y\to 0$ (by (33)). Thus, since H_y' is small, (58) can be rewritten as

$$\left\|\frac{\partial^r y_{j+1}}{\partial(y_0,z_0)^r}\right\|\leqslant e^{-\alpha}\left\|\frac{\partial^r y_j}{\partial(y_0,z_0)^r}\right\|+o_{j\to+\infty}(1),$$

which gives (57).

This proves Theorem 2 for the case of continuous dependence on μ. Let us now consider the case where the first derivatives of the right-hand sides of the system (6) with respect to (y,z,φ) are C^{r-1} with respect to all the variables and μ. By (21), one may assume

$$\mu=g(0;\mu). \tag{59}$$

Let

$$\xi(z^1,z^2)=\int_{z^1}^{z^2}\frac{ds}{g^2(s;\mu)}; \tag{60}$$

this is a well-defined expression for $\mu>0$. Since g_z' is small, it follows that

$$g(z;\mu)\sim g(z';\mu)$$

for all $z\in[-\varepsilon^-,\varepsilon^+]$ and $z'\in[z,\bar z=g(z;\mu)]$ (here "$\sim$" means that the ratio of the left-hand and right-hand sides is bounded away from zero and infinity).

This implies that

$$\sum_{i=0}^{j} \frac{1}{g(z_i;\mu)} \sim \xi(z_0, z_j), \tag{61}$$

where z_j is the jth iteration of z_0, given by (22). It is immediately seen that as $j \to +\infty$, while z_0 and z_j remain in the small interval $[-\varepsilon^-, \varepsilon^+]$, we have

$$\frac{\xi(z_0, z_j)}{j} \to +\infty \tag{62}$$

(because in the left-hand side of (61) $g_i \equiv z_{i+1} - z_i$ must be unboundedly small for an unboundedly large number of i's in order to have an unboundedly large number of iterations on the bounded interval) and

$$\frac{\xi(z_0, z_j)}{\max_{z\in[z_0,z_j]}(1/g(z;\mu))} \to +\infty \tag{63}$$

(because g'_z is small provided g is small, so the number of points z_i for which $g(z_i;\mu) \sim \min_{z\in[z_0,z_j]} g$ grows unboundedly as $\min_{z\in[z_0,z_j]} g \to +0$).

Now note that by (22), we have

$$\frac{\partial z_{j+1}}{\partial \mu} = (1 + g'_z(z_j;\mu))\frac{\partial z_j}{\partial \mu} + g'_\mu(z_j;\mu).$$

Comparing this formula with (25) we obtain (see (24), (21), and (61))

$$\begin{aligned} \frac{\partial z_j}{\partial \mu} &= \frac{\partial z_j}{\partial z_0}\sum_{i=0}^{j} g'_\mu(z_i;\mu)\Big/\frac{\partial z_{i+1}}{\partial z_0} \\ &= g(z_j,\mu)\exp\left\{\sum_{i=0}^{j-1}\psi(z_i;\mu)\right\}\sum_{i=0}^{j}\frac{g'_\mu(z_i;\mu)}{g(z_{i+1};\mu)}\exp\left\{-\sum_{l=0}^{i}\psi(z_l;\mu)\right\} \\ &\sim g(z_j;\mu)\xi(z_0,z_j). \end{aligned} \tag{64}$$

LEMMA 8. *For any $k_1 = 0, \dots, r-1$, $k_2 = 0, \dots, r$, $k_1 + k_2 \leqslant r$ the estimate*

$$\left|\frac{\partial^{k_1+k_2} z_j}{\partial \mu^{k_1}\partial z_0^{k_2}}\right| \leqslant \text{const}\cdot g(z_j;\mu)\xi^{k_1}(z_0, z_j) \tag{65}$$

holds uniformly for all $\mu > 0$ and for all z_0 and z_j from $[-\varepsilon^-, \varepsilon^+]$, provided z_0 is bounded away from zero.

PROOF. The case $k_1 = 0$ is given by Lemma 3; ($k_1 = 1$, $k_2 = 0$) is given by (64). The lemma is proved by induction, in the same way as Lemma 3. Assume that it is proved for all $k_1 \leqslant k_1^0$. Then for any smooth function β of z and μ

$$\left|\frac{\partial^{k_1+k_2}\beta(z_j;\mu)}{\partial \mu^{k_1}\partial z_0^{k_2}}\right| \leqslant \text{const}\cdot g(z_j;\mu)\xi^{k_1}(z_0, z_j) \tag{66}$$

for these values of k_1; for a proof note that this derivative is the sum of terms of the kind

$$\text{const}\cdot\frac{\partial^{s_1+s_2}\beta}{\partial\mu^{s_1}\partial z_j^{s_2}}\left(\frac{\partial^{k_{11}+k_{12}}z_j}{\partial\mu^{k_{11}}\partial z_0^{k_{12}}}\right)\cdots\left(\frac{\partial^{k_{s_21}+k_{s_22}}z_j}{\partial\mu^{k_{s_21}}\partial z_0^{k_{s_22}}}\right),$$

where $0 \leqslant s_1 \leqslant k_1$, $1 \leqslant s_1 + s_2 \leqslant k_1 + k_2$, $k_{11} + \dots + k_{s_21} = k_1 - s_1$, $k_{12} + \dots + k_{s_22} = k_2$. By (66), the absolute value of such a term is estimated from above by

$O(g(z_j;\mu)^{s_2}\xi^{k_1-s_1}(z_0,z_j))$. Since $g\xi$ is bounded away from zero by (63) and since g is small, (66) follows.

By (66) and (23)

$$\sum_{i=0}^{j}\left|\frac{\partial^{k_1+k_2}\beta(z_i;\mu)}{\partial\mu^{k_1}\partial z_0^{k_2}}\right| \leqslant \text{const}\cdot\xi^{k_1}(z_0,z_j). \tag{67}$$

Applying (66) and (67) with $\beta=\psi$, $\beta=g$ and $\beta=g'_\mu$ to formula (64), we get for $k_1=k_1^0+1$

$$\left|\frac{\partial^{k_1+k_2}z_{j+1}}{\partial\mu^{k_1}\partial z_0^{k_2}}\right| \leqslant \text{const}\cdot g(z_j;\mu)\xi^{k_1-1}(z_0,z_j)\sum_{i=0}^{j-1}\frac{1}{g(z_i;\mu)}$$

which, by (61), coincides with (65). By induction, this gives the lemma. □

Let $\omega(\mu)$ denote the flight time of the point $(z_0=-\varepsilon^-,\varphi_0=0)$ on S_0 to the point $(\varepsilon^+,\varphi_1=\omega-M)$ on the cross-section S^1, where M denotes the integral part of ω. In our notation

$$z_M(-\varepsilon^-)=\pi_{-(\omega-M)}(\varepsilon^+,\omega-M),$$

where z_M is the Mth iteration of z_0 by the map (5) and $\pi_{-\varphi}(z,\varphi)$ is the projection on the local cross-section $S:\{\varphi=0\}$ by the flow of (8). Thus,

$$\omega'_\mu \sim \frac{\partial z_M}{\partial\mu} \tag{68}$$

or, by (64),

$$\omega'_\mu \sim \xi^* \equiv \xi(-\varepsilon^-,\varepsilon^+) \tag{69}$$

(since z_M is close to ε^+, it follows that z_M and $g(z_M;\mu)$ are bounded away from zero).

By (69), ω'_μ never vanishes, so ω can be regarded as a new parameter and μ becomes a function of ω. By (68), (65), and (69),

$$\omega^{(k)}_{\mu^k} \leqslant \text{const}\cdot(\omega'_\mu)^k,$$

and it is easy to see (just differentiate the identity $\mu(\omega(\mu))=\mu$ k times) that the following estimate holds.

LEMMA 9. *For all $k\leqslant r-1$, we have*

$$\mu^{(k)}_{\omega^k} \leqslant \text{const}\,\frac{1}{\xi^*}\,. \tag{70}$$

In particular, this implies that for any function smooth with respect to μ, its derivatives with respect to ω tend to zero as $\omega\to+\infty$. Moreover, plugging (70) into (66), (67) yields

$$\left|\frac{\partial^{k_1+k_2}\beta(z_j;\mu)}{\partial\omega^{k_1}\partial z_0^{k_2}}\right| \leqslant \text{const}\cdot g(z_j;\mu), \tag{71}$$

$$\sum_{i=0}^{j}\left|\frac{\partial^{k_1+k_2}\beta(z_i;\mu)}{\partial\omega^{k_1}\partial z_0^{k_2}}\right| \leqslant \text{const} \tag{72}$$

for any smooth function β of (z,μ). Revisiting the proof above for the case of continuous dependence on μ, we now immediately reprove Lemmas 4 and 6, in the sense that $\mathcal{Q}'_{(y,z,\varphi)}$ is C^{r-1} with respect to all variables and ω (Lemma 4) and all derivatives of ζ'_z with respect to z and ω are uniformly bounded (Lemma 6). Formula (19) (where C^r now refers to the space of functions whose first derivatives with respect to (y_0, φ_0) have $r-1$ continuous derivatives with respect to (y_0, φ_0, ω)) is proved in the same way as before.

To get Lemma 7, where the "C^{r-1}-topology" now refers to the derivatives both with respect to z and ω, note that the transition from (51) to (50) is justified by (72) applied with $\beta = \psi$ and by (47) (the relation (48) follows from (47), (45)). To prove the ω-dependent version of the limit (47), note that $\mathcal{G}'_\zeta(\zeta; \mu) = g'_z(\zeta\mu; \mu)$ and $\mathcal{G}(0; \mu) = 1$; hence, the validity of (47) follows from (71) applied with $\beta = g'_z$.

Lemma 7 implies Lemma 1 which (together with (19)) gives the theorem.

References

[1] A. A. Andronov, E. A. Leontovich, I. I. Gordon, and A. G. Maier, *Theory of bifurcations of dynamical systems on the plane*, "Nauka", Moscow, 1967; English transl., Israel Program of Scientific Translations, Jerusalem and London, 1973.

[2] L. P. Shilnikov, *Some cases of generation of periodic motions from singular trajectories*, Mat. Sb. (N.S.) **61 (103)** (1963), 443–466; English transl. in Math. USSR-Sb.

[3] V. S. Afraimovich and L. P. Shilnikov, *On some global bifurcations connected with the disappearance of a fixed point of saddle-node type*, Dokl. Akad. Nauk SSSR **219** (1974), 1281–1284; English transl., Soviet Math. Dokl. **15** (1974), 1761–1765.

[4] ———, *Invariant two-dimensional tori, their breakdown and stochasticity*, Amer. Math. Soc. Transl. Ser. 2, vol 149, Amer. Math. Soc., Providence, RI, 1991, pp. 201–211.

[5] S. Newhouse, J. Palis, and F. Takens, *Bifurcations and stability of families of diffeomorphisms*, Inst. Hautes Études Sci. Publ. Math. **57** (1983), 5–71.

[6] D. V. Turaev and L. P. Shilnikov, *Bifurcations of torus-chaos quasiattractors*, in: Mathematical Mechanisms of Turbulence (modern nonlinear dynamics in application to turbulence simulation), Kiev, 1986, pp. 113–121. (Russian)

[7] L. J. Diaz, J. Rocha, and M. Viana, *Strange attractors in saddle-node cycles: Prevalence and globality*, Invent. Math. **125** (1996), no. 1, 37–74.

[8] S. V. Gonchenko, L. P. Shilnikov, and D. V. Turaev, *Dynamical phenomena in systems with structurally unstable Poincaré homoclinic orbits*, Chaos **6** (1996), no. 1, 23–52.

[9] S. E. Newhouse, *The abundance of wild hyperbolic sets and nonsmooth stable sets for diffeomorphisms*, Inst. Hautes Études Sci. Publ. Math. **50** (1979), 101–151.

[10] S. E. Newhouse, *Diffeomorphisms with infinitely many sinks*, Topology **13** (1974), 9–18.

[11] J. Palis and M. Viana, *High dimension diffeomorphisms displaying infinitely many periodic attractors*, Ann. of Math. (2) **140** (1994), no. 1, 207–250.

[12] S. V. Gonchenko, D. V. Turaev, and L. P. Shilnikov, *On the existence of Newhouse domains in a neighborhood of systems with a structurally unstable Poincaré homoclinic curve* (*the higher-dimensional case*), Dokl. Akad. Nauk **329** (1993), no. 4, 404–407; English transl., Russian Acad. Sci. Dokl. Math. **47** (1993), no. 2, 268–273.

[13] S. V. Gonchenko and L. P. Shilnikov, *On dynamical systems with structurally unstable homoclinic curves*, Dokl. Akad. Nauk SSSR **286** (1986), no. 5, 1049–1053; English transl., Soviet Math. Dokl. **33 (1)** (1986), 234–238.

[14] S. V. Gonchenko, D. V. Turaev, and L. P. Shilnikov, *Dynamical phenomena in multidimensional systems with a structurally unstable homoclinic Poincaré curve*, Dokl. Akad. Nauk **330** (1993), no. 2, 144–147; English transl., Russian Acad. Sci. Dokl. Math. **47** (1993), no. 3, 410–415.

[15] M. Benedics and L. Carleson, *The dynamics of the Hénon map*, Ann. Of Math. **133** (1991), 73–169.

[16] L. Mora and M. Viana, *Abundance of strange attractors*, Acta Math. **171** (1993), no. 1, 1–71.

[17] S. V. Gonchenko, L. P. Shilnikov, and D. V. Turaev, *On models with nonrough Poincaré homoclinic curves*, Physica D **62** (1993), 1–14.

[18] J. Palis and C. Pugh, *Fifty problems in dynamical systems*, Lecture Notes in Math. vol. 468, Springer-Verlag, Heidelberg, 1975, pp. 345–353.

[19] V. S. Medvedev, *On a new type of bifurcations on manifolds*, Mat. Sb. (N.S.) **113 (155)** (1980), no. 3 (11), 487–492; English transl., Math. USSR-Sb. **41** (1982), 403–407.

[20] V. S. Afraimovich and L. P. Shilnikov, *On a bifurcation of codimension* 1 *leading to the appearance of a countable set of tori*, Dokl. Akad. Nauk SSSR **262** (1982), no. 4, 777–780; English transl., Soviet Math. Dokl. **25** (1982), 101–105.

[21] Wei Gu Li and Zhi Fen Zhang, *The blue sky catastrophe on closed surfaces*, Res. Rep., vol. 19, Inst. of Math., Peking University, 1990.

[22] D. V. Turaev and L. P. Shilnikov, *Blue sky catastrophes*, Dokl. Akad. Nauk **342** (1995), no. 5, 596–599; English transl., Russian Acad. Sci. Dokl. Math. **51** (1995), no. 3, 404–407.

[23] L. P. Shilnikov and D. V. Turaev, *On simple bifurcations leading to hyperbolic attractors*, Comput. Math. Appl. **34** (1997), no. 2–4, 441–457.

[24] L. Shilnikov, A. Shilnikov, D. Turaev, and L. Chua, *Methods of qualitative theory in nonlinear dynamics. Part I*, World Scientific, Singapore, 1998.

[25] V. S. Afraimovich, S. N. Chow, and W. Liu, *Lorenz type attractors from codimension one bifurcation*, J. Dynam. Differential Equations **7** (1995), no. 2, 375–407.

[26] Yu. S. Il'yashenko and S. Yu. Yakovenko, *Nonlinear Stokes phenomena in smooth classification problems*, Nonlinear Stokes Phenomena, Adv. Soviet Math., vol. 14, Amer. Math. Soc., Providence, RI, 1993, pp. 235–287.

[27] V. I. Luk′yanov, *Bifurcations of dynamical systems with a saddle-node-separatrix loop*, Differentsial′nye Uravneniya **18** (1982), no. 9, 1493–1506; English transl., Differential Equations **18** (1983), 1049–1059.

[28] M. Hirsh, C. Pugh, and M. Shub, *Invariant manifolds*, Lecture Notes in Math., vol. 583, Springer-Verlag, Heidelberg, 1991.

Department of Differential Equations, Institute for Applied Mathematics and Cybernetics, 10 Ul′yanova st., Nizhny Novgorod 603005, Russia
E-mail address: shilnikov@focus.nnov.ru

The Weierstrass Institute for Applied Analysis and Stochastics, Mohrenstr. 39, Berlin 10117, Germany
E-mail address: turaev@wias-berlin.de

Translated by DMITRY V. TURAEV

Amer. Math. Soc. Transl.
(2) Vol. **200**, 2000

On the Splitting of the Complex Loop of a Separatrix

V. P. Tareev

1. The paper deals with the problem of splitting the loop of a separatrix for the complex system

$$(A_0) \qquad \frac{dw}{dt} = Q(z,w), \qquad \frac{dz}{dt} = P(z,w)$$

of holomorphic class as it passes from (A_0) to the modified system

$$(A_\mu) \qquad \frac{dw}{dt} = Q(z,w) + \mu q(z,w), \qquad \frac{dz}{dt} = P(z,w) + \mu p(z,w)$$

of the same class ($|\mu|$ is arbitrarily small). It is known that for real systems of differential equations in the plane, each separatrix passing through two saddle points or forming a loop can be split by rotation of the vector field [**1**]. However, for complex systems, the notion of separatrix itself has a different meaning, and therefore the problem of splitting for complex separatrices is solved differently from the real case.

Let us clarify the meaning of the notion of separatrix for a complex system. We say that a relation

$$(1) \qquad \Phi(z,w) = 0$$

is a holomorphic integral (partial integral) on a domain $G \subset \mathbb{C}^2$ for the system (A_0), and the set of all points (z,w) satisfying (1) is a holomorphic integral curve in G if

a) the function $\Phi(z,w)$ is holomorphic on G and irreducible (i.e., cannot be represented as $\Phi(z,w) = \Phi_1(z,w)\Phi_2(z,w)$, where $\Phi_1(z,w)$ and $\Phi_2(z,w)$ are nonconstant holomorphic functions on G);
b) relation (1) satisfies the system (A_0), i.e., $\Phi_z P + \Phi_w Q \equiv 0$ for (z,w) satisfying (1).

In a neighborhood of a singular point $O_0(z_0, w_0)$ for which the ratio of eigenvalues λ (which will be called the characteristic root) is not a positive real number(A_0) can always be reduced, by a holomorphic transformation of variables, to the canonical form

$$\dot z = z, \qquad \dot w = -w(\lambda + H(z,w)),$$

where $H(z,w)$ is a holomorphic function in a neighborhood G of the point $O(0,0)$, and $H(0,0) = 0$. Clearly, $z = 0$ and $w = 0$ are holomorphic integral curves in G. They are called separatrices passing through the singular point O of the system.

2000 *Mathematics Subject Classification.* Primary 37F20.

©2000 American Mathematical Society

Turning back to the original system (A_0), which is not in canonical form, we conclude that, according to [**9**], there are exactly two separatrices passing through the singular point O_0 of system (A_0), provided that its eigenvalues are nonresonant. By a linear transformation of variables one can always guarantee that both separatrices take the form $w = \varphi_k(z)$, $k = 1, 2$, where $\varphi_k(z)$ are holomorphic functions in a neighborhood of the point z_0. From now on we assume that this condition always holds.

Let us extend the definition of a separatrix to the case in which it passes through two singular points of the system. Consider the system (A_0) in a bounded part of $\mathbb{C}^2$ in which it has only finitely many singular points. Suppose that $O_0(z_0, w_0)$ and $O_1(z_1, w_1)$ are regular singular points of (A_0) with nonresonant eigenvalues.

DEFINITION 1. A holomorphic integral (1) in a certain domain $G \subset \mathbb{C}^2$ containing the points O_0 and O_1 for the system of equations (A_0) is called a *separatrix passing through two singular points* O_0 *and* O_1 *of* (A_0) if equation (1) can be solved with respect to one of the variables, say, $w = \varphi(z)$, where $\varphi(z)$ is a holomorphic function on a simply connected domain U in the plane $\mathbb{C}(z)$, $z_k \in U$, and $\varphi(z_k) = w_k$, $k = 0, 1$.

We say that a separatrix passing through two singular points of (A_0) *splits* when passing to system (A_μ) if (A_μ) has no separatrix passing through the corresponding two singular points.

The problem of splitting a complex separatrix passing through two singular points of (A_0) was formulated in [**2**], and in the case of rotation of the field, i.e., for the system of equations (A_μ) with

$$(*) \qquad p(z, w) = -Q(z, w), \qquad q(z, w) = P(z, w),$$

it was considered in [**3**]. Specific forms of the right-hand sides of (A_0) and conditions for splitting the separatrix by rotating the field were established in [**4**]. The problem of splitting the separatrix passing through two singular points of (A_0) without the rotation condition $(*)$ was treated in [**5**].

If in the previous definition of a separatrix passing through two singular points of the system (A_0) not only a solution $w = \varphi(z)$ holomorphic in a simply connected domain $U \subset \mathbb{C}(z)$ is considered, but also its analytic continuation, then we obtain the definition of a separatrix in a broader sense. A generalized separatrix can have branch points. The first object in studying generalized separatrices is the class of separatrices forming a loop. If a separatrix of (A_0) forms a loop, then it can sometimes be split by rotating the field. This follows from the necessary condition in [**6**] for the preservation of the loop of a separatrix under rotation of the field and a specific example in which this condition does not hold. In [**6**] no sufficient condition in the form of a nontrivial integral inequality was given for the splitting of the loop of a separatrix.

In this paper we present a detailed deduction of a condition for the preservation of a separatrix loop of the system (A_0) without the condition $(*)$ of rotating the field. This allows us to derive a sufficient condition in the form of an integral inequality for splitting the loop of a separatrix by holomorphic additional terms.

DEFINITION 2. A complex separatrix $w = \varphi(z)$, $\varphi(z_0) = w_0$ of the singular point $O_0(z_0, w_0)$ of (A_0) *forms a loop* if a holomorphic element $\varphi_1(z)$ of the function $\varphi(z)$ is defined in a neighborhood $U_1(z_0) \subset \mathbb{C}(z)$ and there is a closed curve $l \subset \mathbb{C}(z)$ starting and ending at the point z_0 such that the analytic continuation of $\varphi_1(z)$

along this curve results in another holomorphic element $w = \varphi_2(z)$ of $\varphi(z)$ in a neighborhood $U_2(z_0) \subset \mathbb{C}(z)$.

It follows from the definition of a separatrix loop that together with the curve l there is a domain $V \subset \mathbb{C}(z)$ containing it such that the analytic continuation of the element $\varphi_1(z)$ holomorphic at the point z_0 along any curve that starts and ends at z_0 and homotopic to the curve l leads to another holomorphic element $\varphi_2(z)$. We shall suppose that in such a domain $V \subset \mathbb{C}(z)$ entirely containing l, the solution $w = \varphi(z)$ defining the loop of the separatrix of O_0 has a single critical point z_* of second order. It can be assumed without loss of generality that O_0 coincides with the origin $O(0,0)$ and V contains no points $z' \neq 0$, $z' \neq z_*$ such that $\widetilde{\varphi}_k(z') = w'_k$, where $\widetilde{\varphi}_k(z)$ are analytic continuations of $\varphi_k(z)$, $k = 1, 2$, and $O'_k(z', w'_k)$ are singular points of (A_0). We assume that $O(0,0)$ is a regular singular point of (A_0) with the characteristic root λ corresponding to nonresonant eigenvalues (e.g., complex ones). Assume that z_* is a branch point of second order, $z_* \neq 0$, and $\varphi(z_*) = w_0$. Then the analytic continuations $\widetilde{\varphi}_k(z)$, $k = 1, 2$, of the corresponding holomorphic elements $\varphi_k(z)$ to the domain V can be represented as

$$
\begin{aligned}
w &= \widetilde{\varphi}_1(z) = \frac{w_0}{z_*} z + z\sqrt{z - z_*}\, \widetilde{\Phi}_1(z), \\
w &= \widetilde{\varphi}_2(z) = \frac{w_0}{z_*} z + z\sqrt{z - z_*}\, \widetilde{\Phi}_2(z),
\end{aligned} \tag{2}
$$

where $\widetilde{\Phi}_k(z)$ are analytic functions on V with the point z_* deleted. It is clear that $\widetilde{\varphi}_k(z)$ can be expanded in generalized Laurent series in fractional powers of $z - z_*$ **[7]**[1].

Let us consider the system (A_μ) for $|\mu| < \mu^*$, where $\mu_* > 0$ is sufficiently small. We assume without loss of generality that the singular point $O(\mu)$ for the system (A_μ) does not depend on μ, the point $(0,0)$ remains singular with nonresonant eigenvalues and, consequently, there are two local separatrices **[9]**. Let $w = \varphi_k(z, \mu)$ be the separatrix of the singular point O of (A_μ) that becomes the separatrix $w = \varphi_k(z)$ of the singular point O of (A_0) for $\mu = 0$, $k = 1, 2$.

We say that the separatrix $w = \varphi(z)$ forming a loop at the singular point O of (A_0) is *preserved* as we pass to the system (A_μ) if, for sufficiently small $|\mu|$, the analytic continuation of one of the two elements $\varphi_k(z, \mu)$, $k = 1, 2$, of $\varphi(z, \mu)$ holomorphic at the point $z = 0$ along any curve $\tilde{l} \subset V$ that starts and ends at $z = 0$ and homotopic to l results in another of two above elements of $\varphi(z, \mu)$.

If there exists a path l' in V that starts and ends at $z = 0$ such that l' is homotopic to l, and the analytic continuation of $\varphi_k(z, \mu)$ along l' does not give $\varphi_j(z, \mu)$, $j \neq k$, then we say that the separatrix is *split* in passing from (A_0) to (A_μ).

The main goal of the present paper is to prove the following theorem.

THEOREM 1. *Assume that the system* (A_0) *of holomorphic class has a solution* $w = \varphi(z)$ *determining a separatrix loop for the singular point* $O(0,0)$ *and having only one critical point* z_* *of second order in a domain* $V \subset \mathbb{C}(z)$. *Then there exists a modified system* (A_0) *close to* (A_μ) *of the same class such that as* (A_0) *passes to* (A_μ) *the loop of the separatrix is split provided that*

$$\int_0^{z_*} e^{-\widehat{\mathcal{L}}_1(z)}\, dz \neq 0,$$

[1]These series contain no principal part because of the nature of the point z_*.

where $\widehat{\mathcal{L}}_1(z) = \int_\varepsilon^z \mathcal{L}_1(z)\,dz$, *$\varepsilon$ is close to 0, and $\mathcal{L}_1(z)$ is a known coefficient determined from the equation of the first variation on the loop of the separatrix.*

2. Let us find a necessary condition for the preservation of the loop of the separatrix $w = \varphi(z)$ of the singular point O for system (A_0). To do this we use the equation of the first variation for the branches (2) of the separatrix $w = \varphi(z)$ in passing from (A_0) to an arbitrary nearby system (A_μ). We obtain (see [**5, 6**])

$$\frac{df_k}{dz} = \mathcal{L}_k(z)f_k + \mathcal{M}_k(z), \qquad k = 1, 2, \tag{3}$$

where

$$\mathcal{L}_k(z) = \frac{Q'_w(z, \widetilde{\varphi}_k(z)) - P'_w(z, \widetilde{\varphi}_k(z))\widetilde{\varphi}_k(z)}{P(z, \widetilde{\varphi}_k(z))}, \tag{4}$$

$$\mathcal{M}_k(z) = \frac{P(z, \widetilde{\varphi}_k(z))q(z, \widetilde{\varphi}_k(z)) - Q(z, \widetilde{\varphi}_k(z))p(z, \widetilde{\varphi}_k(z))}{P^2(z, \widetilde{\varphi}_k(z))}. \tag{5}$$

If we take $p(z, w)$ and $q(z, w)$ so that $\mathcal{M}_k(z) \not\equiv 0$, then under the additional condition $\operatorname{Re}\lambda < 1$ we can find, as in [**6**], a necessary condition for the preservation of the loop of the separatrix. Equations (3) for the first variation differ from the corresponding equations in the case of "rotation of the field" only in the constant terms $\mathcal{M}_k(z)$ (see (2) in [**6**]). From the general integrals of (3), let us find particular solutions $f_k = f_k(z)$, $k = 1, 2$, holomorphic in some neighborhoods $U'_k(0) \subset V$. Continue the solution $f_1(z)$ along a curve $L_1 \subset V$ starting at the point $\varepsilon \in U'_1 \cap U'_2$ and ending at the point $z^* \in V$. Continue the other holomorphic solution $f_2(z)$ along a curve $L_2 \subset V$ also starting at ε and ending at the same point z^*. Denote the values of the resulting analytic continuations at z^* by $\delta_{L_1} f_1(z^*)$ and $\delta_{L_2} f_2(z^*)$ respectively. The expressions (6) for these values differ from those in [**6**] only in that $R_k(z)$ is replaced by $\mathcal{M}_k(z)$, namely,

(6)

$$\delta_{L_n} f_k(z^*) = \exp\left(\int_{L_k} \mathcal{L}_k(z)dz\right)\left(C_k(\varepsilon) + \int_{L_k} \mathcal{M}_k(z)\exp\left(-\int_\varepsilon^z \mathcal{L}_k(u)\,du\right)dz\right),$$

where $(n, k) \in \{(1,1), (2,2), (3,1)\}$. Here $L_1 = j_0 \cup l_0 \cup j_1 \cup l_1$, $L_2 = j_0 \cup l_2$, and $L_3 = j_0 \cup l_0 \cup l_1$, where j_0 and j_1 are closed curves going around the points $z = 0$ and z_* respectively. The curve l_0 joins the points ε and δ, while the curve l_1 joins the points δ and z^*. The parts of the curve L_1 join at the points ε and $\delta = z_* + \varepsilon$, and the parts of the curve L_2 join at the point ε.[2]

If the loop of the separatrix $w = \varphi(z)$ does not split, then the following relations must hold:

$$\delta_{L_1} f_1(z^*) = \delta_{L_2} f_2(z^*), \qquad \delta_{L_3} f_1(z^*) = \delta_{L_2} f_2(z^*).$$

It follows that

$$\delta_{L_1} f_1(z^*) = \delta_{L_3} f_1(z^*) \tag{7}$$

is a necessary condition for nonsplitting of the loop of the separatrix. The condition (7) holds for any paths of the above form, in particular, if j_0 and j_1 are circles contracting to the points $z = 0$ and z^* respectively (i.e., as $\varepsilon \to 0$ and $\delta = z_* + \varepsilon \to z_*$).

Furthermore, we use (7) to derive a condition for nonsplitting of the loop of the separatrix $w = \varphi(z)$ for the singular point O of (A_0). To this end we rewrite

---

[2]We can always assume that $|\varepsilon| < |z_*|$.

(7) in terms of (6). We first use the condition that $f_1(z)$ is holomorphic in the neighborhood $U'(0)$ to find the value of $C_1(\varepsilon)$,

$$\exp\left(\oint_{j_0} \mathcal{L}_1(z)\,dz\right)\left(C_1(\varepsilon)+\oint_{j_0} \mathcal{M}_1(z)\exp\left(-\int_\varepsilon^z \mathcal{L}_1(u)\,du\right)dz\right)=C_1(\varepsilon).$$

Therefore, taking into account the fact that $\operatorname{res}\mathcal{L}_1(0)=\lambda$, we get

$$C_1(\varepsilon)=\frac{e^{2\pi\lambda i+\widehat{\mathcal{L}}_1(\varepsilon)}}{1-e^{2\pi\lambda i}}I_0, \tag{8}$$

where $\widehat{\mathcal{L}}_1(\varepsilon)$ is found from

$$\widehat{\mathcal{L}}_1(z)=\int_\varepsilon^z \mathcal{L}_1(z)\,dz. \tag{9}$$

Denoting $\mathcal{M}_1(z)e^{-\widehat{\mathcal{L}}_1(z)}=\Psi_1(z)$, we write

$$I_0=\oint_{j_0}\mathcal{M}_1(z)e^{-\widehat{\mathcal{L}}_1(z)}\,dz=\oint_{j_0}\Psi_1(z)\,dz. \tag{10}$$

Denoting $\int_a^b \mathcal{L}_1(u)\,du=\int_a^b$, we obtain[3]

$$\begin{aligned}
&\int_{L_1}\mathcal{M}_1(z)\exp\left(-\int_\varepsilon^z \mathcal{L}_1(u)\,du\right)dz\\
&=\oint_{j_0}\mathcal{M}_1(z)\exp\left(-\int_\varepsilon^z\right)dz+\int_\varepsilon^\delta \mathcal{M}_1(z)\exp\left(-\oint_{j_0}-\int_\varepsilon^z\right)dz\\
&\qquad+\oint_{j_1}\mathcal{M}_1(z)\exp\left(-\left(\oint_{j_0}+\int_\varepsilon^\delta+\int_\delta^z\right)\right)dz\\
&\qquad+\int_\delta^{z^*}\mathcal{M}_1(z)\exp\left(-\left(\oint_{j_0}+\oint_{j_1}+\int_\varepsilon^\delta+\int_\delta^z\right)\right)dz\\
&=J_0+J_1+J_2+J_3.
\end{aligned} \tag{11}$$

Taking into account the form of $\Psi_1(z)$ and formulas (9) and (10), we find the following expressions for J_i in (11):

$$\begin{aligned}
J_0&=I_0e^{\widehat{\mathcal{L}}_1(\varepsilon)},\\
J_1&=\int_\varepsilon^\delta \Psi_1(z)\,dz\,e^{-2\pi\lambda i+\widehat{\mathcal{L}}_1(\varepsilon)}=J_1^*e^{-2\pi\lambda i+\widehat{\mathcal{L}}_1(\varepsilon)},\\
J_2&=\oint_{j_1}\Psi_1(z)\,dz\,e^{-2\pi\lambda i+\widehat{\mathcal{L}}_1(\varepsilon)}=I_1e^{-2\pi\lambda i+\widehat{\mathcal{L}}_1(\varepsilon)},\\
J_3&=\int_\delta^{z^*}\Psi_1(z)\,dz\,e^{-(2\pi\lambda i+A)+\widehat{\mathcal{L}}_1(\varepsilon)}=J_3^*e^{-(2\pi\lambda i+A)+\widehat{\mathcal{L}}_1(\varepsilon)},
\end{aligned} \tag{12}$$

where

$$A=\oint_{j_1}\mathcal{L}_1(z)\,dz. \tag{13}$$

---

[3]Our notation does not indicate the branch that we choose as we integrate branching functions since this choice is always clear from the context.

By the same arguments we obtain

$$(14)\qquad \int_{L_3} \mathcal{M}_1(z)\exp\Big(-\int_\varepsilon^z \mathcal{L}_1(u)\,du\Big)dz = J_0 + J_1 + J_3^* e^{-2\pi\lambda i+\widehat{\mathcal{L}}_1(\varepsilon)}.$$

Further, we have

$$(15)\qquad \int_{L_1} \mathcal{L}_1(z)\,dz = 2\pi\lambda i + A + B(z^*),$$

where

$$(16)\qquad B(z^*) = \int_\varepsilon^{z^*} \mathcal{L}_1(z)\,dz.$$

Similarly,

$$(17)\qquad \int_{L_3} \mathcal{L}_1(z)\,dz = 2\pi\lambda i + B(z^*).$$

REMARK 1. The terms $B(z^*)$ in (15) and (17) have two components each. The first components, which are determined by the curve l_0, coincide, whereas the second components, which are determined by the curve l_1, can be different if the other branch of the function f is taken in moving along the portion l_1 of the curve L_1. To be definite, in what follows we consider the case in which $B(z^*)$ in (15) and (17) coincide. Otherwise, the first factor in the first term of (19) changes, but we again get the required formula (24) after passing to the limit as $\varepsilon \to 0$.

Let us rewrite (7) using (6) and taking into account the fact that the values of $C_1(\varepsilon)$ determined by (8) are the same for the curves L_1 and L_3, since we take the same branch $\widetilde{\varphi}_1(z)$ of the solution and the same path j_0. In view of (11)–(17), on canceling the factor $e^{\widehat{\mathcal{L}}_1(\varepsilon)}$, we obtain

$$(18)\qquad \begin{aligned} &e^{2\pi\lambda i+A+B(z^*)}\Big(\frac{e^{2\pi\lambda i}}{1-e^{2\pi\lambda i}}\,I_0 + I_0 + J_1^* e^{-2\pi\lambda i} + I_1 e^{-2\pi\lambda i} + J_3^* e^{-(2\pi\lambda i+A)}\Big)\\ &\quad = e^{2\pi\lambda i+B(z^*)}\Big(\frac{e^{2\pi\lambda i}}{1-e^{2\pi\lambda i}}\,I_0 + I_0 + J_1^* e^{-2\pi\lambda i} + J_3^* e^{-2\pi\lambda i}\Big).\end{aligned}$$

Canceling the factor $e^{2\pi\lambda i+B(z^*)}$ in (18) and transferring all the terms to the left-hand side, we get

$$(19)\qquad (e^A-1)\Big(\frac{1}{1-e^{2\pi\lambda i}}\,I_0 + J_1^* e^{-2\pi\lambda i}\Big) + I_1 e^{A-2\pi\lambda i} = 0$$

after straightforward simplifications.

Let us pass to the limit as $\varepsilon \to 0$ in (19). First we show that $\lim_{\varepsilon\to 0} I_0 = 0$. To accomplish this, taking into account (4), we write $\mathcal{L}_1(z)$ in the form

$$\mathcal{L}_1(z) = \frac{\lambda}{z} + \mathcal{L}_1^*(z),$$

where $\mathcal{L}_1^*(z)$ is a holomorphic function on a small neighborhood of $z = 0$. Then, by (10),

$$(20)\qquad I_0 = \oint_{j_0} \widehat{\mathcal{M}}_0(z) z^{-\lambda}\,dz,$$

where

$$\widehat{\mathcal{M}}_0(z) = \mathcal{M}_1(z) e^{-\widetilde{\mathcal{L}}_0(z)}, \qquad \widetilde{\mathcal{L}}_0(z) = \int_\varepsilon^z \mathcal{L}_1^*(z)\,dz.$$

Combining (20) with (5) and taking for $p(z,w)$ and $q(z,w)$ holomorphic functions on a disk enclosed by the curve j_0 such that $|\widehat{\mathcal{M}}_0(z)| < M$ (where M is a finite number), we get the inequality

$$|I_0| \leqslant M' \oint_{j_0} \frac{|dz|}{|z|^{\lambda_1}}, \tag{21}$$

where $\lambda_1 = \operatorname{Re}\lambda$ and the number M' does not depend on z. Substituting $z = r_0 e^{i\varphi}$, $\varphi \in [0, 2\pi]$, where r_0 is the radius of the circle j_0, into (21), and taking into account the fact that $r_0 \to 0$ as j_0 contracts to the point $z = 0$, we see that the integral on the right-hand side of (21) tends to zero if $\operatorname{Re}\lambda < 1$. Consequently, $I_0 \to 0$ as $\varepsilon \to 0$.

Now we use (13) to show that $\lim_{\varepsilon\to 0}(e^A - 1) \neq 0$. To do this we find $\lim_{\varepsilon \to 0} A$. In view of (4), we have the following expression for $\mathcal{L}_1(z)$ on a punctured neighborhood of the point $z = z_*$:

$$\mathcal{L}_1(z) = \frac{\sigma_{-1}}{z - z_*} + \frac{\sigma_{-1/2}}{\sqrt{z - z_*}} + \mathcal{L}_1^{**}(z), \tag{22}$$

where $\mathcal{L}_1^{**}(z)$ is an analytic function on the punctured neighborhood of the point z_*, in which it has a branching of the second order. Expanding $\mathcal{L}_1(z)$ as a generalized Laurent series in powers of $(z - z_*)^{1/2}$ [**7**], we easily find $\sigma_{-1} = -1/2$, whence

$$A = \oint_{j_1} \mathcal{L}_1(z)\, dz = -\frac{1}{2} \oint_{j_1} \frac{dz}{z - z_*}\, dz + \oint_{j_1} \left(\frac{\sigma_{-1/2}}{\sqrt{z - z_*}} + \mathcal{L}_1^{**}(z) \right) dz.$$

The contour j_1 contracts to the point z_* as $\varepsilon \to 0$; therefore the second integral on the right-hand side tends to 0, and so $\lim_{\varepsilon\to 0} A = -\pi i$. Hence, $\lim_{\varepsilon \to 0}(e^A - 1) = -2 \neq 0$.

Finally, we find $\lim_{\varepsilon \to 0} I_0$ in the same way as $\lim_{\varepsilon\to 0} I_1$. It suffices to use (12) and (22) as we used (10) and (20) to represent I_1 in the form

$$I_1 = \oint_{j_1} \widehat{\mathcal{M}}_1(z) \sqrt{z - z_*}\, dz, \tag{23}$$

where $\widehat{\mathcal{M}}_1(z) = \mathcal{M}_1(z) e^{-\widetilde{\mathcal{L}}_1(z)}$. Here

$$\widetilde{\mathcal{L}}_1(z) = \int_\varepsilon^z \left(\frac{\sigma_{-1/2}}{\sqrt{z - z_*}} + \mathcal{L}_1^{**}(z) \right) dz.$$

If we take analytic additional terms $p(z,w)$ and $q(z,w)$ such that $\widehat{\mathcal{M}}_1(z)$ is bounded in the disk enclosed by the circle j_1, then (23) implies that $\lim_{\varepsilon\to 0} I_1 = 0$.

Taking into account the above limits of $e^A - 1$, I_0, and I_1 as $\varepsilon \to 0$, we immediately derive from (19) and (12) the condition

$$Y = \int_0^{z_*} \mathcal{M}_1(z) e^{-\widehat{\mathcal{L}}_1(z)}\, dz = 0 \tag{24}$$

for the nonsplitting of the loop of the separatrix, where the function $\widehat{\mathcal{L}}_1(z)$ is given by (9).

3. The form (5) of the function $\mathcal{M}_1(z)$ shows that there exist $p(z,w)$ and $q(z,w)$ such that all the required conditions for deriving (24) are fulfilled. As an obvious example of such additional terms we can take the holomorphic functions

$$p(z,w) = 0, \qquad q(z,w) = cP(z,w) \quad (c = \text{const}).$$

on $\mathbb{C}^2$. In this case it follows from (5) that $\mathcal{M}_1(z) = c$. Combining this with the deduction of condition (24), we arrive at Theorem 1.

Now consider additional terms of the form

$$p(z,w) = 0, \qquad q(z,w) = cP(z,w)e^{\widehat{\mathcal{L}}_1(z)}. \tag{25}$$

Here $c = \text{const} \neq 0$ and $q(z,w)$ is an analytic function in z in V except possibly at the points $z = 0$, $z = z_*$ ($z_* \neq 0$). It follows from (5) that $\mathcal{M}_1(z) = ce^{\widehat{\mathcal{L}}_1(z)}$, and (24) implies $Y = c\int_0^{z_*} dz \neq 0$, which means that the condition (24) for the preservation of the loop of the separatrix does not hold for (25) and any value of λ. Thus we have proved the following result.

THEOREM 2. *Suppose system* (A_0) *of holomorphic class has a solution* $w = \varphi(z)$ *that determines a loop of a separatrix for the singular point* $O(0,0)$ *and has a unique critical point* z_* *of second order in a domain* $V \subset \mathbb{C}(z)$. *Then the loop of the separatrix is always split by the additional terms* (25) *analytic in* V.

REMARK 2. In the case of rotation of the field (∗), the function $\mathcal{M}_1(z)$ in (24) is replaced by $R_1(z) = 1 + (\widetilde{\varphi}_1'(z))^2$. Then the condition for the preservation of the loop of the separatrix under the rotation of the field (see also [**6**]) is written as

$$\int_0^{z_*} R_1(z)e^{-\widehat{\mathcal{L}}_1(z)}\,dz = 0. \tag{26}$$

A specific example of the splitting of the loop of a separatrix by the holomorphic additional terms (∗), for which (26) does not hold, was considered in [**8**].

The author is grateful to L. M. Lerman for useful discussions and to G. M. Polotovskiĭ for comments that helped to improve the text.

References

[1] A. A. Andronov, E. A. Leontovich, I. I. Gordon, and A. G. Maier, *Theory of bifurcations of dynamical systems on the plane*, "Nauka", Moscow, 1967; English transl., Israel Program of Scientific Translations, Jerusalem and London, 1973.

[2] V. P. Tareev, *Splitting of the separatrices in two-dimensional complex systems*, Uspekhi Mat. Nauk **42** (1977), no. 4, 247–248. (Russian)

[3] E. A. Leontovich and V. P. Tareev, *Remarks on the problem of splitting the complex separatrices*, Methods of the Qualitative Theory of Differential Equations (E. A. Leontovich-Andronova, ed.), Gor′kov. Gos. Univ., Gor′kiĭ, 1983, pp. 59–70. (Russian)

[4] V. P. Tareev, *On same cases of splitting the complex separatrices*, Differentsial′nye Uravneniya **30** (1994), no. 6, 1086–1088; English transl. in Differential Equations **30** (1994).

[5] ———, *Splitting of complex separatrices by holomorphic components*, in: Methods of the Qualitative Theory of Differential Equations (E. A. Leontovich-Andronova, ed.), Gor′kov. Gos. Univ., Gor′kiĭ, 1984, pp. 158–162. (Russian)

[6] E. A. Leontovich and V. P. Tareev, *On the problem of splitting the complex loop of a separatrix by "rotating the field"*, in: Methods of Qualitative Theory and Bifurcation Theory (L. P. Shilnikov, ed.), Gor′kov. Gos. Univ., Gor′kiĭ, 1983, pp. 76–83. (Russian)

[7] B. V. Shabat, *Introduction to complex analysis*, "Nauka", Moscow, 1969. (Russian)

[8] V. P. Tareev, *On a case of splitting the complex loop of a separatrix by rotating the field*, Methods of Qualitative Theory and Bifurcation Theory (L. P. Shilnikov, ed.), Nizhegorod. Gos. Univ., Nizhny Novgorod, 1990, pp. 33–40.

[9] L. Bieberbach, *Theorie der gewöhnlichen Differentialgleichungen*, Springer-Verlag, Berlin, 1953.

NIZHNY NOVGOROD STATE TECHNICAL UNIVERSITY, MININ STR., 24, NIZHNY NOVGOROD 603155, RUSSIA

Translated by A. B. SOSSINSKI

Selected Titles in This Subseries

(Continued from the front of this publication)

22 **N. N. Uraltseva, Editor,** Nonlinear Evolution Equations (TRANS2/164)

Published Earlier as Advances in Soviet Mathematics

21 **V. I. Arnold, Editor,** Singularities and bifurcations, 1994

20 **R. L. Dobrushin, Editor,** Probability contributions to statistical mechanics, 1994

19 **V. A. Marchenko, Editor,** Spectral operator theory and related topics, 1994

18 **Oleg Viro, Editor,** Topology of manifolds and varieties, 1994

17 **Dmitry Fuchs, Editor,** Unconventional Lie algebras, 1993

16 **Sergei Gelfand and Simon Gindikin, Editors,** I. M. Gelfand seminar, Parts 1 and 2, 1993

15 **A. T. Fomenko, Editor,** Minimal surfaces, 1993

14 **Yu. S. Il′yashenko, Editor,** Nonlinear Stokes phenomena, 1992

13 **V. P. Maslov and S. N. Samborskiĭ, Editors,** Idempotent analysis, 1992

12 **R. Z. Khasminskiĭ, Editor,** Topics in nonparametric estimation, 1992

11 **B. Ya. Levin, Editor,** Entire and subharmonic functions, 1992

10 **A. V. Babin and M. I. Vishik, Editors,** Properties of global attractors of partial differential equations, 1992

9 **A. M. Vershik, Editor,** Representation theory and dynamical systems, 1992

8 **E. B. Vinberg, Editor,** Lie groups, their discrete subgroups, and invariant theory, 1992

7 **M. Sh. Birman, Editor,** Estimates and asymptotics for discrete spectra of integral and differential equations, 1991

6 **A. T. Fomenko, Editor,** Topological classification of integrable systems, 1991

5 **R. A. Minlos, Editor,** Many-particle Hamiltonians: spectra and scattering, 1991

4 **A. A. Suslin, Editor,** Algebraic K-theory, 1991

3 **Ya. G. Sinaĭ, Editor,** Dynamical systems and statistical mechanics, 1991

2 **A. A. Kirillov, Editor,** Topics in representation theory, 1991

1 **V. I. Arnold, Editor,** Theory of singularities and its applications, 1990